Selected Titles in This Series

174 **Ya-Zhe Chen and Lan-Cheng Wu,** Second order elliptic equations and elliptic systems, 1998
173 **Yu. A. Davydov, M. A. Lifshits, and N. V. Smorodina,** Local properties of distributions of stochastic functionals, 1998
172 **Ya. G. Berkovich and E. M. Zhmud′,** Characters of finite groups. Part 1, 1998
171 **E. M. Landis,** Second order equations of elliptic and parabolic type, 1998
170 **Viktor Prasolov and Yuri Solovyev,** Elliptic functions and elliptic integrals, 1997
169 **S. K. Godunov,** Ordinary differential equations with constant coefficient, 1997
168 **Junjiro Noguchi,** Introduction to complex analysis, 1998
167 **Masaya Yamaguti, Masayoshi Hata, and Jun Kigami,** Mathematics of fractals, 1997
166 **Kenji Ueno,** An introduction to algebraic geometry, 1997
165 **V. V. Ishkhanov, B. B. Lur′e, and D. K. Faddeev,** The embedding problem in Galois theory, 1997
164 **E. I. Gordon,** Nonstandard methods in commutative harmonic analysis, 1997
163 **A. Ya. Dorogovtsev, D. S. Silvestrov, A. V. Skorokhod, and M. I. Yadrenko,** Probability theory: Collection of problems, 1997
162 **M. V. Boldin, G. I. Simonova, and Yu. N. Tyurin,** Sign-based methods in linear statistical models, 1997
161 **Michael Blank,** Discreteness and continuity in problems of chaotic dynamics, 1997
160 **V. G. Osmolovskiĭ,** Linear and nonlinear perturbations of the operator div, 1997
159 **S. Ya. Khavinson,** Best approximation by linear superpositions (approximate nomography), 1997
158 **Hideki Omori,** Infinite-dimensional Lie groups, 1997
157 **V. B. Kolmanovskiĭ and L. E. Shaĭkhet,** Control of systems with aftereffect, 1996
156 **V. N. Shevchenko,** Qualitative topics in integer linear programming, 1997
155 **Yu. Safarov and D. Vassiliev,** The asymptotic distribution of eigenvalues of partial differential operators, 1997
154 **V. V. Prasolov and A. B. Sossinsky,** Knots, links, braids and 3-manifolds. An introduction to the new invariants in low-dimensional topology, 1997
153 **S. Kh. Aranson, G. R. Belitsky, and E. V. Zhuzhoma,** Introduction to the qualitative theory of dynamical systems on surfaces, 1996
152 **R. S. Ismagilov,** Representations of infinite-dimensional groups, 1996
151 **S. Yu. Slavyanov,** Asymptotic solutions of the one-dimensional Schrödinger equation, 1996
150 **B. Ya. Levin,** Lectures on entire functions, 1996
149 **Takashi Sakai,** Riemannian geometry, 1996
148 **Vladimir I. Piterbarg,** Asymptotic methods in the theory of Gaussian processes and fields, 1996
147 **S. G. Gindikin and L. R. Volevich,** Mixed problem for partial differential equations with quasihomogeneous principal part, 1996
146 **L. Ya. Adrianova,** Introduction to linear systems of differential equations, 1995
145 **A. N. Andrianov and V. G. Zhuravlev,** Modular forms and Hecke operators, 1995
144 **O. V. Troshkin,** Nontraditional methods in mathematical hydrodynamics, 1995
143 **V. A. Malyshev and R. A. Minlos,** Linear infinite-particle operators, 1995
142 **N. V. Krylov,** Introduction to the theory of diffusion processes, 1995
141 **A. A. Davydov,** Qualitative theory of control systems, 1994
140 **Aizik I. Volpert, Vitaly A. Volpert, and Vladimir A. Volpert,** Traveling wave solutions of parabolic systems, 1994

(*Continued in the back of this publication*)

Second Order Elliptic Equations and Elliptic Systems

Translations of
MATHEMATICAL
MONOGRAPHS

Volume 174

Second Order Elliptic Equations and Elliptic Systems

Ya-Zhe Chen
Lan-Cheng Wu

Translated by
Bei Hu

American Mathematical Society
Providence, Rhode Island

二阶椭圆型方程与椭圆型方程组

(Second Order Elliptic Equations and Elliptic Systems)

Ya-Zhe Chen (陈亚浙) and Lan-Cheng Wu (吴兰成)

Originally published in Chinese by Science Press, Beijing, China (科学出版社), 1991

Translated from the Chinese by Bei Hu (胡钡)

1991 *Mathematics Subject Classification.* Primary 35-01, 35-02; Secondary 35B45, 35B65, 35J15, 35J60, 35K10.

ABSTRACT. This book is based on the authors' lecture notes at the Institute of Mathematics at Nankai University during the Partial Differential Equations Year in 1985, absorbing also the most recent materials from the lectures of experts.

There are two parts of the book. For the Dirichlet problem of second order elliptic partial differential equations, various kinds of a priori estimate methods, including the most recent techniques, are rather completely introduced in the first part. Linear, quasilinear and fully nonlinear equations are studied. In the second part, the existence and regularity theories of the Dirichlet problem for linear and nonlinear second order elliptic partial differential systems are introduced. This book chooses appropriate materials; it is a very good textbook for graduate students.

This book can also be used as a reference book for undergraduate mathematics majors, graduate students, professors and scientists.

Library of Congress Cataloging-in-Publication Data

Chen, Yazhe, 1939–
[Erh chieh t'o yüan hsing fang ch'eng yü t'o yüan hsing fang ch'eng tsu. English]
Second order elliptic equations and elliptic systems / Ya-Zhe Chen, Lan-Cheng Wu ; translated by Bei Hu.
p. cm. — (Translations of mathematical monographs, ISSN 0065-9282 ; v. 174)
Includes bibliographical references and index.
ISBN 0-8218-0970-9 (alk. paper)
1. Differential equations, Elliptic. I. Wu, Lancheng, 1934– . II. Title. III. Series.
QA377.C4413 1998
515′.353—dc21

97-46794
CIP

English translation published by the AMS, with the consent of Science Press.
Printed in the United States of America.

∞ The paper used in this book is acid-free and falls within the guidelines
established to ensure permanence and durability.
Information on copying and reprinting can be found in the back of this volume.
Visit the AMS home page at URL: http://www.ams.org/

10 9 8 7 6 5 4 3 2 1   03 02 01 00 99 98

Contents

Preface to the English Translation

We are very pleased that our graduate textbook has been chosen by the American Mathematical Society for translation into English. In this English edition, some obvious typographic errors in the Chinese edition are corrected; some names are modified according to the current convention. We also adopted the translator's suggestion and added very brief bibliographic notes for each chapter, together with updated references.

We would like to express our wholehearted gratitude to Professors Alice Chang, Tsit-Yuen Lam and other AMS personnel for their vast amount of work on this translation project. We would also like to thank the translator, Dr. Bei Hu, for his valuable suggestions; he typed the LaTeX manuscript for this English edition.

Owing to the authors' limited knowledge, errors and inappropriateness are hard to avoid; we welcome corrections from readers.

Ya-Zhe Chen, Lan-Cheng Wu
July 9, 1997

Preface

The theory of second order elliptic equations and systems is fundamental for studying partial differential equations, and therefore it was listed as a basic course for graduate students at the Institute of Mathematics at Nankai University during the Partial Differential Equations Year in 1985. The authors were invited to give lectures to graduate students for this course. At that time, the Institute at Nankai also invited many well-known experts from around the world to give lectures, which provided the course with the most recent results in the area. This book is based on the authors' lecture notes, but we have absorbed also the most recent materials from the lectures of the other invited experts.

There are many excellent books on second order elliptic partial differential equations and systems, such as [GT], [LU] and [GQ1], listed in the bibliography of this book. These books give a thorough introduction in this area. However, some of these books are too big to be suitable as textbooks for beginners. The purpose of this book is to provide a textbook for graduate students. This book includes both basic materials and the most recent results and methods, so as to bring graduate students to the frontier of this area.

There are two parts of the book. For the Dirichlet problem for second order elliptic partial differential equations, various kinds of a priori estimate methods are rather completely introduced in the first part. The Krylov-Safonov estimate and fully nonlinear elliptic equations, which appeared in the 80's, are introduced in detail, but concisely. In the second part, the existence and regularity theories of linear and nonlinear second order elliptic partial differential systems are introduced. Basic facts about Sobolev spaces are given in Appendix 1. In order to emphasize the main theme, the proofs of some theorems, such as the Stampacchia interpolation theorem and the reverse Hölder inequality, are included in the Appendix.

Owing to the authors' limited knowledge, errors are hard to avoid; we welcome suggestions from readers.

Under the leadership of Professor Li-Shang Jiang (姜礼尚), the partial differential equation seminar at Peking University played an important role in this book. We would like to express our deep gratitude to Professor Li-Shang Jiang and the seminar participants for their contributions. We would also like to express our wholehearted gratitude to Professor Guang-Lie Wang (王光烈) from Jilin University, who read the draft of this book and made valuable suggestions.

Ya-Zhe Chen, Lan-Cheng Wu
May 20, 1990

Part I

Second Order Elliptic Equations

CHAPTER 1

L^2 Theory

The study of solvability of the Dirichlet problem for elliptic equations is one of the central topics of this book. The introduction of Sobolev spaces (cf. Appendix 1) provides an effective tool for the study. In Sobolev spaces, we can seek solutions in a more general class of functions; this makes solvability problems much easier. Solutions of this kind are often referred to as *weak solutions* or *generalized solutions.* Of course, in order to obtain the existence of a classical solution, we must study the smoothness of weak solutions. Such a problem is called the regularity problem. In §4 of this chapter and next chapter, we will explore the basic methods for studying the regularity problem.

1. Lax-Milgram theorem

Let H be a real Hilbert space and H' its dual space. Denote by $\langle\cdot,\cdot\rangle$ the dual product between H and H'.

Definition 1.1. Let $a(u,v)$ be a bilinear form on the Hilbert space H,

(i) $a(u,v)$ is said to be *bounded,* if there exists $M>0$ such that

$$|a(u,v)| \leqslant M\|u\|_H\|v\|_H, \qquad \forall u,v\in H. \tag{1.1}$$

(ii) $a(u,v)$ is said to be *coercive,* if there exists $\delta>0$ such that

$$a(u,u) \geqslant \delta\|u\|_H^2, \qquad \forall u\in H. \tag{1.2}$$

Theorem 1.1 (Lax-Milgram theorem)**.** *Let $a(u,v)$ be a bounded, coercive bilinear form on H. Then for any $f\in H'$, there exists a unique $u\in H$ such that*

$$a(u,v) = \langle f,v\rangle, \qquad \forall v\in H, \tag{1.3}$$

and

$$\|u\|_H \leqslant \frac{1}{\delta}\|f\|_{H'}. \tag{1.4}$$

Proof. For each fixed $u\in H$, $a(u,\cdot)$ is a bounded linear functional on H, and therefore there exists a unique $Au\in H'$ such that

$$a(u,v) = \langle Au,v\rangle, \qquad \forall v\in H, \tag{1.5}$$

and

$$\|Au\|_{H'} \leqslant M\|u\|_H.$$

It can easily be verified that A is linear. By the coerciveness, we have

$$\|Au\|_{H'}\|u\|_H \geqslant a(u,u) \geqslant \delta\|u\|_H^2,$$

and therefore

$$\|Au\|_{H'} \geqslant \delta\|u\|_H. \tag{1.6}$$

It follows that A is one to one. We will prove that the range of A is $R(A) = H'$. First, we show that $R(A)$ is closed. If $Au_n \to v$, then by (1.6)

$$\|u_n - u_m\|_H \leqslant \frac{1}{\delta}\|Au_n - Au_m\|_{H'};$$

thus $\{u_n\}$ is a Cauchy sequence. From the completeness we conclude that its limit $u \in H$ exists. By the continuity of A, we derive that $Au_n \to Au$, and therefore $v = Au \in R(A)$. This shows that $R(A)$ is a closed subspace of H'. If $R(A) \neq H'$, then it follows from the orthogonal decomposition theorem that there exists $v' \in H'$ such that $v' \neq 0$ and $v' \perp R(A)$. Since H is reflexive, there exists $v \in H$ such that $(v', \cdot)_{H'} = \langle \cdot, v\rangle$. Using the coerciveness, we get

$$0 = (v', Av)_{H'} = \langle Av, v\rangle \geqslant \delta\|v\|_H^2 > 0.$$

This contradiction implies that $R(A) = H'$, and therefore there is a unique u such that $Au = f$. From (1.5) and (1.6) we immediately deduce (1.3) and (1.4). □

2. Weak solutions of elliptic equations

Let Ω be an open domain in $\mathbb{R}^n$. For simplicity, we shall always assume that $n \geqslant 3$. In this chapter we shall consider elliptic equations of divergence form on Ω:

$$Lu = -D_j(a^{ij}D_iu + d^ju) + (b^iD_iu + cu) = f + D_if^i. \tag{2.1}$$

We use the summation convention throughout this book; repeated subscripts and superscripts are summed from 1 through n, and $D_i = \dfrac{\partial}{\partial x_i}$. As for the operator L, we shall assume throughout this chapter that

$$a^{ij} \in L^\infty(\Omega),$$

and that there exist positive constants λ, Λ such that

$$\lambda|\xi|^2 \leqslant a^{ij}(x)\xi_i\xi_j \leqslant \Lambda|\xi|^2, \qquad \forall \xi \in \mathbb{R}^n, x \in \Omega, \tag{2.2}$$

$$\sum_{i=1}^n \|b^i\|_{L^n(\Omega)} + \sum_{i=1}^n \|d^i\|_{L^n(\Omega)} + \|c\|_{L^{n/2}(\Omega)} \leqslant \Lambda. \tag{2.3}$$

We shall denote the Sobolev space $W^{k,2}(\Omega)$ by $H^k(\Omega)$. For $u, v \in H^1(\Omega)$, we set

$$a(u,v) = \int_\Omega \left\{(a^{ij}D_iu + d^ju)D_jv + (b^iD_iu + cu)v\right\}dx.$$

From the proof of Lemma 2.1 below we shall see that the integrals are well defined in the above equality.

Definition 2.1. For $T \in H^{-1}(\Omega)$ (the dual space of $H_0^1(\Omega)$) and $g \in H^1(\Omega)$, we say that $u \in H^1(\Omega)$ is a *weak solution* of the Dirichlet problem

$$\begin{cases} Lu = T & \text{in } \Omega, \\ u = g & \text{on } \partial\Omega, \end{cases} \tag{2.4}$$

if u satisfies

$$\begin{cases} a(u,v) = \langle T, v\rangle, & \forall v \in H_0^1(\Omega), \\ u - g \in H_0^1(\Omega). \end{cases} \tag{2.5}$$

Lemma 2.1. *Let the assumptions* (2.2) *and* (2.3) *be in force, and let Ω be an open bounded domain in $\mathbb{R}^n$. Then $a(u,v)$ is a bounded bilinear form on $H_0^1(\Omega)$.*

Proof. Using Hölder's inequality and (2.2), we obtain

$$\left|\int_\Omega a^{ij} D_i u D_j v dx\right| \leqslant \Lambda \|u\|_{H_0^1(\Omega)} \|v\|_{H_0^1(\Omega)}.$$

Applying Hölder's inequality, (2.3) and the embedding theorem, we derive

$$\begin{aligned} \left|\int_\Omega d^j u D_j v dx\right| &\leqslant \sum_j \|d^j\|_{L^n(\Omega)} \|u\|_{L^{2^*}(\Omega)} \|D_j v\|_{L^2(\Omega)} \\ &\leqslant C\Lambda \|u\|_{H_0^1(\Omega)} \|v\|_{H_0^1(\Omega)}, \\ \left|\int_\Omega cuv dx\right| &\leqslant \|c\|_{L^{n/2}(\Omega)} \|u\|_{L^{2^*}(\Omega)} \|v\|_{L^{2^*}(\Omega)} \\ &\leqslant C\Lambda \|u\|_{H_0^1(\Omega)} \|v\|_{H_0^1(\Omega)}, \end{aligned}$$

where $2^* = 2n/(n-2)$, and the constant C depends only on n. The remaining terms can be estimated in a similar way. Thus we obtain

$$|a(u,v)| \leqslant C\Lambda \|u\|_{H_0^1(\Omega)} \|v\|_{H_0^1(\Omega)}. \tag{2.6}$$ □

Remark. For each fixed $u \in H^1(\Omega)$, $a(u,\cdot)$ is a bounded linear functional on $H_0^1(\Omega)$. Analogously to the proof of Theorem 1.1, there exists a bounded linear operator $\widetilde{L} : H^1(\Omega) \to H^{-1}(\Omega)$ such that

$$a(u,v) = \langle \widetilde{L}u, v\rangle, \qquad \forall u \in H^1(\Omega),\ v \in H_0^1(\Omega). \tag{2.7}$$

From now on we shall not distinguish the operator $\widetilde{L}$ from the operator L given in (2.1).

Lemma 2.2. *Let the assumptions* (2.2) *and* (2.3) *be in force, and let Ω be an open bounded domain in $\mathbb{R}^n$. Then there exists $\overline{\mu} > 0$ such that $a(u,v) + \mu(u,v)_{L^2(\Omega)}$ is coercive on $H_0^1(\Omega)$ for $\mu \geqslant \overline{\mu}$, where $(\cdot,\cdot)_{L^2(\Omega)}$ denotes the inner product on $L^2(\Omega)$.*

In order to prove the lemma, we need the following fact: for $f \in L^p(\Omega)$ and arbitrary positive $\varepsilon > 0$, there exists a decomposition $f = f_1 + f_2$ such that

$$\|f_2\|_{L^p(\Omega)} < \varepsilon, \qquad \sup_{x \in \Omega} |f_1(x)| < K(\varepsilon). \tag{2.8}$$

One can simply choose

$$f_1(x) = \begin{cases} f(x) & \text{for } |f(x)| < K, \\ 0 & \text{for } |f(x)| \geqslant K, \end{cases}$$

where K is sufficiently large.

Proof of Lemma 2.2. For any $\varepsilon > 0$, there are decompositions

$$b^i = b^i_1 + b^i_2, \qquad d^i = d^i_1 + d^i_2, \qquad c = c_1 + c_2,$$

such that

$$\begin{aligned} &\sum \|b^i_2\|_{L^n(\Omega)} + \sum \|d^i_2\|_{L^n(\Omega)} + \|c_2\|_{L^{n/2}(\Omega)} \leqslant \varepsilon, \\ &\sum \|b^i_1\|_{L^\infty(\Omega)} + \sum \|d^i_1\|_{L^\infty(\Omega)} + \|c_1\|_{L^\infty(\Omega)} \leqslant K(\varepsilon). \end{aligned}$$

Set

$$\begin{aligned} a_2(u,v) &= \int_\Omega \Big\{(a^{ij}D_i u + d^j_2 u)D_j v + (b^i_2 D_i u + c_2 u)v\Big\}dx, \\ a_1(u,v) &= a(u,v) - a_2(u,v). \end{aligned}$$

Using the positive-definiteness condition (2.2) and a similar computation as in Lemma 2.1, we easily obtain

$$a_2(u,u) \geqslant (\lambda - C\varepsilon)\|u\|^2_{H^1_0(\Omega)}.$$

We fix ε such that $C\varepsilon = \lambda/4$. For $a_1(u,u)$, we have

$$\begin{aligned} |a_1(u,u)| &\leqslant CK(\varepsilon)\Big[\int_\Omega \sum_i |D_i u||u|dx + \int_\Omega |u|^2 dx\Big] \\ &\leqslant \frac{\lambda}{4}\int |Du|^2 dx + C\Big(\frac{K^2(\varepsilon)}{\lambda} + K(\varepsilon)\Big)\int_\Omega |u|^2 dx. \end{aligned}$$

Combining all the above estimates, we derive

$$a(u,u) \geqslant \frac{\lambda}{2}\|u\|^2_{H^1_0(\Omega)} - C\Big(\frac{K^2(\varepsilon)}{\lambda} + K(\varepsilon)\Big)\|u\|^2_{L^2(\Omega)}.$$

Thus we can finish the proof of the lemma by taking $\overline{\mu} = C\Big(\frac{K^2(\varepsilon)}{\lambda} + K(\varepsilon)\Big)$. □

Using the Lax-Milgram theorem, we can derive the existence theorem for a weak solution.

Theorem 2.3. *Let the assumptions* (2.2) *and* (2.3) *be in force, and let* $\Omega \subset \mathbb{R}^n$ *be an open bounded domain on which the Sobolev embedding theorem is valid. Let* $T \in H^{-1}(\Omega)$ *and* $g \in H^1(\Omega)$. *Then there exists* $\overline{\mu} > 0$ *such that for* $\mu > \overline{\mu}$, *the nonhomogeneous Dirichlet problem*

$$\begin{cases} Lu + \mu u = T, \\ u - g \in H^1_0(\Omega) \end{cases} \tag{2.9}$$

has a unique weak solution.

Proof. By the definition of a weak solution, the bilinear form corresponding to (2.9) is given by $a(u,v)+\mu(u,v)_{L^2(\Omega)}$. A weak solution satisfies

$$\begin{cases} a(u,v)+\mu(u,v)_0=\langle T,v\rangle, & \forall v\in H_0^1(\Omega),\\ u-g\in H_0^1(\Omega), \end{cases} \tag{2.10}$$

where $(u,v)_0 \triangleq (u,v)_{L^2(\Omega)}$. Under the change of variables $w=u-g$, (2.10) is equivalent to finding a $w\in H_0^1(\Omega)$ such that

$$a(w,v)+\mu(w,v)_0=\langle T,v\rangle-a(g,v)-\mu(g,v)_0,\quad \forall v\in H_0^1(\Omega). \tag{2.11}$$

From Lemmas 2.1 and 2.2, we derive that $a(w,v)+\mu(w,v)_0$ is a bounded coercive bilinear form on $H_0^1(\Omega)$ for $\mu\geqslant\overline{\mu}$. It is not difficult to verify that $\langle T,v\rangle-a(g,v)-\mu(g,v)_0$ is a continuous linear functional on $H_0^1(\Omega)$. It follows from the Lax-Milgram theorem that the equation (2.11) has a unique solution $w\in H_0^1(\Omega)$, and $u=w+g$ is a unique weak solution of (2.9). □

3. The Fredholm Alternative

The *Fredholm alternative* in a Banach space is stated as follows:

Theorem 3.1. *Let V be a normed linear vector space and $A:V\to V$ be a linear compact operator. Then there are only two possibilities:*

(1) there exists $x\in V$, $x\neq 0$, such that $x-Ax=0$; or

(2) for any $y\in V$, there exists a unique $x\in V$ such that

$$x-Ax=y.$$

In the case (2), $(I-A)^{-1}$ is a bounded linear operator, where I is the identity operator; furthermore, the spectrum of A is discrete, 0 is its only possible limit point, and every eigenvalue is of finite order.

The proof of this theorem can be found in textbooks on functional analysis. Next, we will apply it to the Dirichlet problem of elliptic equations.

Theorem 3.2. *Suppose that L and Ω satisfy the assumptions in Theorem 2.3. Then there are only two possibilities for the problem (2.9):*

(1) for any $T\in H^{-1}(\Omega)$, $g\in H^1(\Omega)$, there exists a unique weak solution to (2.9); or

(2) there exists a nonzero $u\in H_0^1(\Omega)$ such that $Lu+\mu u=0$ (i.e., $a(u,v)+\mu(u,v)_0=0$, $\forall v\in H_0^1(\Omega)$).

Furthermore, in the case (2), μ is discrete, ∞ is the only possible limit point, and for each eigenvalue μ, its corresponding eigenfunction space is finite dimensional.

Proof. Without loss of generality we may assume that $g\equiv 0$ (cf. the proof of Theorem 2.3). For each fixed $u\in L^2(\Omega)$, $(u,\cdot)_0$ is a bounded linear functional on

$H_0^1(\Omega)$. Therefore there exists a bounded operator $P : L^2(\Omega) \to H^{-1}(\Omega)$ such that

$$(u, v)_0 = \langle Pu, v\rangle, \qquad \forall u \in L^2(\Omega),\ v \in H_0^1(\Omega).$$

Let I be the embedding operator from $H_0^1(\Omega)$ to $L^2(\Omega)$. From the remark to Lemma 2.1 and the above facts, we derive that (2.9) is equivalent to finding a $u \in H_0^1(\Omega)$ such that

$$Lu + \mu PIu = T. \tag{3.1}$$

By Theorem 2.3, there exists $\overline{\mu} > 0$ such that $(L + \overline{\mu}PI)^{-1}$ exists, and it is a bounded operator $H^{-1}(\Omega) \to H_0^1(\Omega)$. Let $G = (L + \overline{\mu}PI)^{-1}$. Applying G to the equation (3.1), we get

$$u - (\overline{\mu} - \mu)GPIu = GT. \tag{3.2}$$

The equations (3.1) and (3.2) are equivalent. Since the embedding from $H_0^1(\Omega)$ to $L^2(\Omega)$ is compact, GPI is a compact operator from $H_0^1(\Omega)$ to itself. Now we can apply Theorem 3.1 to the equation (3.2) to obtain the theorem. □

4. A maximum principle for weak solutions

There are many methods for proving a maximum principle; the De Giorgi iteration and the Moser iteration are two commonly used methods. These are important tools in studying partial differential equations. In this book, we shall introduce them in different chapters. We shall use the De Giorgi iteration in this section. It is often reduced to the following lemma:

Lemma 4.1. *Suppose that $\varphi(t)$ is a nonnegative nondecreasing function defined on $[k_0, \infty)$ satisfying*

$$\varphi(h) \leqslant \frac{C}{(h-k)^\alpha}[\varphi(k)]^\beta \quad \text{for } h > k \geqslant k_0, \tag{4.1}$$

where $\alpha > 0$, $\beta > 1$. Then

$$\varphi(k_0 + d) = 0, \tag{4.2}$$

where

$$d = C^{1/\alpha}[\varphi(k_0)]^{(\beta-1)/\alpha}2^{\beta/(\beta-1)}. \tag{4.3}$$

Proof. Set

$$k_s = k_0 + d - \frac{d}{2^s} \qquad (s = 0, 1, 2, \cdots).$$

From (4.1) we deduce

$$\varphi(k_{s+1}) \leqslant \frac{C2^{(s+1)\alpha}}{d^\alpha}[\varphi(k_s)]^\beta \qquad (s = 0, 1, 2, \cdots). \tag{4.4}$$

We shall prove by induction that

$$\varphi(k_s) \leqslant \frac{\varphi(k_0)}{r^s} \qquad (s = 0, 1, 2, \cdots), \tag{4.5}$$

where $r > 1$ is to be determined. Suppose that (4.5) is valid for s. Then by (4.4) and the induction assumption,

$$\varphi(k_{s+1}) \leqslant \frac{C2^{(s+1)\alpha}}{d^\alpha}\left[\frac{\varphi(k_0)}{r^s}\right]^\beta = \frac{\varphi(k_0)}{r^{s+1}} \frac{C2^{(s+1)\alpha}}{d^\alpha r^{s(\beta-1)-1}}[\varphi(k_0)]^{\beta-1}.$$

To conclude the induction, we take $r = 2^{\alpha/(\beta-1)}$ and require that

$$\frac{C2^{\alpha\beta/(\beta-1)}}{d^\alpha}[\varphi(k_0)]^{\beta-1} \leqslant 1.$$

Notice that d defined in (4.3) satisfies the above inequality. Thus (4.5) is valid. We obtain the lemma by letting $s \to \infty$. □

In order to state the weak maximum principle more precisely, we introduce the concept of subsolutions and supersolutions.

Definition 4.1. $u \in H^1(\Omega)$ is said to be a *weak subsolution* (*weak supersolution, weak solution*), if

$$a(u,\varphi) \leqslant (\geqslant, =)\ (f,\varphi)_0 - (f^i, D_i\varphi)_0, \quad \forall \varphi \in C_0^\infty(\Omega), \varphi \geqslant 0, \tag{4.6}$$

where $a(u,v)$ is defined in §2.

In fact, for any $\varphi \in H_0^1(\Omega)$, (4.6) is also valid for $\varphi = \varphi^+ = \max\{\varphi, 0\}$.

Definition 4.2. Let $u \in H^1(\Omega)$. We define

$$\begin{aligned} \sup_{\partial\Omega} u &= \inf\{l \mid (u-l)^+ \in H_0^1(\Omega)\}. \\ \operatorname*{ess\,sup}_{\Omega} u &= \inf\{l \mid (u-l)^+ = 0, \ \text{a.e. } \Omega\}. \end{aligned} \tag{4.7}$$

Theorem 4.2 (Weak maximum principle for weak solutions). *Let the assumptions* (2.2), (2.3) *be in force, and*

$$c - D_i d^i \geqslant 0 \qquad (\textit{in the sense of } \mathscr{D}'(\Omega)). \tag{4.8}$$

If $u \in H^1(\Omega)$ *is a weak subsolution of* (2.1), *then for any* $p > n$, *we have*

$$\operatorname*{ess\,sup}_{\Omega} u \leqslant \sup_{\partial\Omega} u^+ + C(\|f\|_{L^{np/(n+p)}(\Omega)} + \sum_i \|f^i\|_{L^p(\Omega)})|\Omega|^{(1/n)-(1/p)}, \tag{4.9}$$

where C *depends only on* $n, p, \lambda, \Lambda, \Omega$ *and* b^i, d^i, c, *and it is independent of the lower bound of* $|\Omega|$ (*if a family of domains* Ω *is considered*).

Remark. (4.8) means that for any $\varphi \in C_0^\infty(\Omega)$, $\varphi \geqslant 0$,

$$\int_\Omega (c\varphi + d^i D_i\varphi)dx \geqslant 0.$$

Proof. Let $l = \sup_{\partial\Omega} u^+$. Suppose that $\sup_{\Omega} u^+ > l$. For any $k > l$, we choose the test function $\varphi = (u-k)^+$ in (4.6). Then

$$\begin{aligned} a(u,\varphi) - \int_\Omega \Big\{(a^{ij}D_i\varphi + d^i\varphi)D_j\varphi + (b^iD_i\varphi + c\varphi)\varphi\Big\}dx \\ +k\int_\Omega (d^jD_j\varphi + c\varphi)dx. \end{aligned} \tag{4.10}$$

The second term in the right-hand side of the above equality is nonnegative, by (4.8). For the first term, we use a similar computation as in Lemma 2.2 to conclude that

$$a(u,\varphi) \geqslant \frac{\lambda}{2}\|D\varphi\|^2_{L^2(\Omega)} - C\lambda\|\varphi\|^2_{L^2(\Omega)}, \tag{4.11}$$

where C depends only on $n, \lambda, \Lambda, \partial\Omega$ and b^i, d^i, c. From the definition of a weak subsolution and (4.11), we have

$$\begin{aligned} &\frac{\lambda}{2}\|D\varphi\|^2_{L^2(\Omega)} - C\lambda\|\varphi\|^2_{L^2(\Omega)} \\ \leqslant\ & (f,\varphi) - (f^i, D_i\varphi) \\ \leqslant\ & \sum_i \|f^i\|_{L^p}\|D\varphi\|_{L^2}|A(k)|^{(1/2)-(1/p)} \\ & + \|f\|_{L^{p_*}}\|\varphi\|_{L^{2^*}}|A(k)|^{(1/2)-(1/p)}, \end{aligned} \tag{4.12}$$

where $p_* = np/(n+p)$, $2^* = 2n/(n-2)$,

$$A(k) = \{x \in \Omega \mid u(x) > k\}, \tag{4.13}$$

and $|A(k)|$ is the measure of the set $A(k)$. Set

$$F_0 = \frac{1}{\lambda}\Big(\sum_i \|f^i\|_{L^p} + \|f\|_{L^{p_*}}\Big).$$

Applying the embedding theorem and Cauchy's inequality to the right-hand side of (4.12), we obtain

$$\|D\varphi\|^2_{L^2} \leqslant C\|\varphi\|^2_{L^2} + \varepsilon\|D\varphi\|^2_{L^2} + C_\varepsilon F_0^2|A(k)|^{1-2/p}.$$

We take $\varepsilon = 1/2$ and then apply Hölder's inequality and the Sobolev embedding theorem to the first term on the right-hand side of the above inequality to obtain

$$\begin{aligned} \|D\varphi\|^2_{L^2} &\leqslant C|A(k)|^{2/n}\|\varphi\|^2_{L^{2^*}} + CF_0^2|A(k)|^{1-2/p} \\ &\leqslant \widetilde{C}|A(k)|^{2/n}\|D\varphi\|^2_{L^2} + CF_0^2|A(k)|^{1-2/p}. \end{aligned} \tag{4.14}$$

If there exists $k_0 \geqslant l$ such that

$$\widetilde{C}|A(k_0)|^{2/n} \leqslant \frac{1}{2}, \tag{4.15}$$

then from (4.14),

$$\|D\varphi\|_{L^2} \leqslant CF_0|A(k)|^{(1/2)-(1/p)} \quad \text{for } k \geqslant k_0.$$

Applying the Sobolev embedding theorem, we obtain

$$\|\varphi\|_{L^{2^*}} \leqslant CF_0|A(k)|^{(1/2)-(1/p)} \quad \text{for } k \geqslant k_0. \tag{4.16}$$

Recall that $\varphi = (u-k)^+$,

$$\|\varphi\|_{L^{2^*}} \geqslant (h-k)|A(h)|^{1/2^*} \quad \text{for } h > k.$$

Therefore

$$|A(h)| \leqslant \frac{(CF_0)^{2^*}}{(h-k)^{2^*}}|A(k)|^{n(p-2)/[p(n-2)]} \quad \text{for } h > k \geqslant k_0.$$

By Lemma 4.1,

$$A(k_0 + d) = 0,$$

where

$$d = CF_0|A(k_0)|^{(1/n)-(1/p)}2^{n(p-2)/[p(n-2)]}.$$

Thus

$$\operatorname*{ess\,sup}_{\Omega} u \leqslant k_0 + d \leqslant k_0 + CF_0|\Omega|^{(1/n)-(1/p)}. \tag{4.17}$$

The estimation of k_0 is divided into two steps.

Step 1. Since

$$k^2|A(k)| \leqslant \int_\Omega u^2 dx = \|u\|^2_{L^2(\Omega)},$$

we can take

$$k_0 \geqslant (2\widetilde{C})^{n/4}\|u\|_{L^2(\Omega)}, \quad \text{and } k_0 \geqslant \sup_{\partial\Omega} u^+,$$

so that (4.15) is valid. From (4.17) we immediately obtain

$$\operatorname*{ess\,sup}_{\Omega} u \leqslant \sup_{\partial\Omega} u^+ + C\|u\|_{L^2(\Omega)} + CF_0|\Omega|^{(1/n)-(1/p)}. \tag{4.18}$$

Step 2. (4.18) gives us the essential upper bound for u, and we must further estimate the second term on the right-hand side of (4.18). Let $M = \operatorname{ess\,sup} u - l$, $v = (u-l)^+$. We shall consider the equation satisfied by the function

$$w = \log \frac{M+\varepsilon+\widetilde{F}_0}{M+\varepsilon+\widetilde{F}_0 - v}, \tag{4.19}$$

where $\widetilde{F}_0 = F_0|\Omega|^{(1/n)-(1/p)}$, and ε is an arbitrary small positive constant. We choose the test function

$$\varphi = \frac{v}{M+\varepsilon+\widetilde{F}_0 - v} \in H_0^1(\Omega). \tag{4.20}$$

Analogously to (4.10), we have

$$\begin{aligned} a(u,\varphi) &\geqslant \int_\Omega \Big\{(a^{ij}D_iv + d^jv)D_j\varphi + (b^iD_iv + cv)\varphi\Big\}dx \\ &= \int_\Omega \Big\{a^{ij}D_ivD_j\varphi + (b^i - d^i)D_iv\cdot\varphi\Big\}dx \\ &\quad + \int_\Omega \Big\{d^jD_j(v\varphi) + cv\varphi\Big\}dx. \end{aligned} \tag{4.21}$$

Notice that $\varphi v \in H_0^1(\Omega)$ and $\varphi v \geqslant 0$; it follows from (4.8) that the second term on the right-hand side of the above inequality is nonnegative. Substituting (4.20) into (4.21), we obtain

$$a(u,\varphi) \geqslant \int_\Omega \Big\{(M+\varepsilon+\widetilde{F}_0)a^{ij}D_iwD_jw - (b^i - d^i)vD_iw\Big\}dx. \tag{4.22}$$

Since u is a weak subsolution,

$$
\begin{aligned}
a(u,\varphi) &\leqslant (f,\varphi)_0 - (f^i, D_i\varphi)_0 \\
&\leqslant \int_\Omega \frac{fv}{M+\varepsilon+\widetilde{F}_0 - v}dx + \int_\Omega \frac{(M+\varepsilon+\widetilde{F}_0)f^i D_i w}{M+\varepsilon+\widetilde{F}_0 - v}dx \qquad (4.23)\\
&\leqslant \frac{M}{\widetilde{F}_0}\int_\Omega |f|dx + \frac{\lambda}{4}(M+\varepsilon+\widetilde{F}_0)\int_\Omega |Dw|^2 dx \\
&\quad + \frac{C(M+\varepsilon+\widetilde{F}_0)}{\lambda \widetilde{F}_0^2}\int_\Omega \sum |f^i|^2 dx.
\end{aligned}
$$

Combining (4.22) and (4.23), using also the positive definiteness condition (2.2) and Hölder's inequality, we conclude that

$$
\begin{aligned}
\|Dw\|^2 &\leqslant \frac{1}{\lambda \widetilde{F}_0}\int_\Omega |f|dx + \frac{C}{\lambda^2}\int \sum_i (|b^i|^2 + |d^i|^2)dx \\
&\quad + \frac{C}{\lambda^2 \widetilde{F}_0^2}\int_\Omega \sum |f^i|^2 dx \\
&\leqslant C\Big(n, \frac{\Lambda}{\lambda}, |\Omega|\Big),
\end{aligned}
$$

where the constant C is independent of the lower bound for $|\Omega|$. It follows from the embedding theorem that

$$\|w\|_{L^{2^*}} \leqslant C,$$

which implies that, for $k > l$,

$$|A(k)|^{1/2^*} \log \frac{M+\varepsilon+\widetilde{F}_0}{M+\varepsilon+\widetilde{F}_0 - (k-l)} \leqslant C.$$

We take $k_0 - l = (1-\eta)(M+\varepsilon+\widetilde{F}_0)$, where η is a small positive constant to be determined. Then

$$|A(k_0)|^{1/2^*} \leqslant C\left[\log \frac{1}{\eta}\right]^{-1}.$$

It is clear that there exists $\eta > 0$ so that (4.15) is valid. By (4.17),

$$\operatorname{ess\,sup} u \leqslant \sup_{\partial\Omega} u^+ + (1-\eta)(M+\varepsilon+\widetilde{F}_0) + CF_0|\Omega|^{(1/n)-(1/p)}.$$

Since $M = \operatorname{ess\,sup} u - \sup_{\partial\Omega} u^+$ and ε is arbitrary, we conclude that

$$\operatorname{ess\,sup} u \leqslant \sup_{\partial\Omega} u^+ + CF_0|\Omega|^{(1/n)-(1/p)}.$$

The proof is complete. □

Remark. If the assumption (2.2) is replaced by the stronger assumption

$$\sum_i \|b^i\|_{L^p(\Omega)} + \sum_i \|d^i\|_{L^p(\Omega)} + \|c\|_{L^{p/2}(\Omega)} \leqslant \Lambda, \qquad (2.2)'$$

where $p > n$, then the constant C in the conclusion (4.9) of Theorem 4.2 depends only on $n, p, \Lambda/\lambda$ and Ω and is independent of b^i, d^i, c and the lower bound for $|\Omega|$.

In fact, for any $\varepsilon > 0$, we can prove that

$$\left|\int_\Omega \left\{d^j\varphi D_j\varphi + b^j\varphi D_j\varphi + c\varphi^2\right\}dx\right| \leqslant \Lambda\Big(\varepsilon\|\varphi\|^2_{H^1_0(\Omega)} + C_\varepsilon\|\varphi\|_{L^2(\Omega)}\Big). \tag{4.24}$$

For example, the first term on the left-hand side of (4.24) can be estimated as follows:

$$\begin{aligned}\left|\int_\Omega d^j\varphi D_j\varphi dx\right| &\leqslant \|d^j\|_{L^p}\|\varphi\|_{L^{2p/(p-2)}}\|D_j\varphi\|_{L^2} \\ &\leqslant \|d^j\|_{L^p}\|D_j\varphi\|_{L^2}(\varepsilon\|\varphi\|_{L^{2^*}} + C_\varepsilon\|\varphi\|_{L^2}).\end{aligned}$$

Using $(2.2)'$ and the embedding theorem, we then derive (4.24). Thus the constant C depends only on $n, p, \dfrac{\Lambda}{\lambda}$ and Ω. There is no change in other parts of the proof. The details are left as exercises for the reader.

Theorem 4.3. *Let the assumptions* (2.1), (2.2) *be in force, and let*

$$c - D_i d^i \geqslant 0 \qquad (\textit{in the sense of } \mathscr{D}'(\Omega)),$$

where Ω is an open bounded domain on which the Sobolev embedding theorem is valid. Then there exists a unique weak solution to the Dirichlet problem, and

$$\|u\|_{H^1_0(\Omega)} \leqslant C(\|T\|_{H^{-1}(\Omega)} + \|g\|_{H^1(\Omega)}). \tag{4.25}$$

Proof. Without loss of generality, we may assume that $g = 0$ (cf. the proof of Theorem 2.3). If $T = 0$, the zero solution is the only solution for (2.4), by the weak maximum principle (Theorem 4.1). By Theorem 3.2, there exists a unique weak solution $u \in H^1_0(\Omega)$ for (2.4), for any $T \in H^{-1}(\Omega)$. It follows that L is invertible, i.e.,

$$\|u\|_{H^1(\Omega)} \leqslant C\|Lu\|_{H^{-1}(\Omega)}, \quad \forall u \in H^1_0(\Omega).$$

From this inequality, (4.25) follows. □

5. Regularity for weak solutions

For simplicity, we consider only the following type of equation:

$$Lu = -D_j(a^{ij}D_iu) + b^iD_iu + cu = f. \tag{5.1}$$

Theorem 5.1. *Let the assumption* (2.2) *be in force, and let $a^{ij} \in W^{1,\infty}(\Omega)$, $b^i, c \in L^\infty(\Omega)$, $f \in L^2(\Omega)$. If $u \in H^1(\Omega)$ is a weak solution of* (5.1), *then for any $\Omega' \subset\subset \Omega$ (i.e., $\overline{\Omega'}$ is a compact subset of Ω) we have $u \in H^2(\Omega')$, and*

$$\|u\|_{H^2(\Omega')} \leqslant C(\|u\|_{H^1(\Omega)} + \|f\|_{L^2(\Omega)}), \tag{5.2}$$

where the constant C depends only on n, λ, $\|a^{ij}\|_{W^{1,\infty}}$, $\|b^i\|_{L^\infty(\Omega)}$, $\|c\|_{L^\infty(\Omega)}$ and $\operatorname{dist}\{\Omega', \partial\Omega\}$.

Proof. Set $q = f - b^i D_i u - cu$. Since u is a weak solution of (5.1), it satisfies

$$\int_\Omega a^{ij} D_i u D_j \varphi dx = \int_\Omega q\varphi dx, \qquad \forall \varphi \in H_0^1(\Omega). \tag{5.3}$$

Let τ_h be the translation operator in the x_1 direction, $\tau_h u(x) = u(x + he_1)$ (e_1 is the unit vector in the x_1 direction), and Δ_h the difference operator $\Delta_h = \frac{1}{h}(\tau_h - I)$. Let $v \in H_0^1(\Omega)$ be such that its support spt $v \subset\subset \Omega$. For $h < \frac{1}{2}\operatorname{dist}\{\text{spt } v, \partial\Omega\}$, we choose the test function $\varphi = \Delta_{-h} v$ in (5.3) and obtain

$$\int_\Omega \Delta_h(a^{ij} D_i u) D_j v dx = -\int_\Omega q \Delta_{-h} v dx.$$

Using the definition of q and the equality

$$\Delta_h(a^{ij} D_i u) = \tau_h a^{ij} \Delta_h D_i u + D_i u \Delta_h a^{ij},$$

we get

$$\begin{aligned} \int_\Omega \tau_h a^{ij} D_i \Delta_h u D_j v dx &= -\int_\Omega \left(\Delta_h a^{ij} D_i u D_j v + q\Delta_{-h} v\right) dx \\ &\leqslant C(\|u\|_{H^1(\Omega)} + \|f\|_{L^2(\Omega)})\|Dv\|_{L^2(\Omega)}. \end{aligned}$$

Take $\eta \in C_0^\infty(\Omega)$ such that $\eta(x) = 1$ for $x \in \Omega'$ and choose $v = \eta^2 \Delta_h u$. Then

$$\begin{aligned} &\int_\Omega \eta^2 \tau_h a^{ij} D_i \Delta_h u D_j \Delta_h u dx \\ &\leqslant -2\int_\Omega \eta \tau_h a^{ij} D_i \Delta_h u (D_j \eta) \Delta_h u dx \\ &\quad + C(\|u\|_{H^1(\Omega)} + \|f\|_{L^2(\Omega)})(\|\eta D\Delta_h u\|_{L^2(\Omega)} + 2\|\Delta_h u D\eta\|_{L^2(\Omega)}). \end{aligned}$$

Using the positive definiteness condition (2.2) and Cauchy's inequality, we obtain

$$\lambda \int_\Omega |\eta \Delta_h Du|^2 dx \leqslant C\int_\Omega |D\eta|^2 |\Delta_h u|^2 dx + C(\|u\|^2_{H^1(\Omega)} + \|f\|^2_{L^2(\Omega)}).$$

Hence

$$\|\eta \Delta_h Du\|^2_{L^2(\Omega)} \leqslant C(\|u\|^2_{H^1(\Omega)} + \|f\|^2_{L^2(\Omega)}). \tag{5.4}$$

By choosing appropriate η, using also Proposition 1.8 in Appendix 1, we derive that $D_1 Du \in L^2(\Omega')$. Since we can consider the difference operator in any direction, u belongs to $H^2(\Omega')$ and satisfies (5.2). □

The above regularity theorem for weak solutions is called *interior regularity*. If appropriate conditions are imposed on the boundary and boundary values, we can show that $u \in H^2(\Omega)$; this kind of result is called *global regularity*.

Definition 5.1. The boundary of Ω is said to belong to C^k, if for any $x^0 \in \partial\Omega$, there exist a neighborhood V of x^0 and a one-to-one map $\phi : V \to B_1$, such that

(1) $B_1 = \phi(V), \quad B_1^+ = \phi(V \cap \Omega)$,

(2) $\partial B_1^+ \cap B_1 = \phi(V \cap \partial\Omega)$,

(3) $\phi \in C^k(\overline{V}), \quad \phi^{-1} \in C^k(\overline{B}_1)$,

where B_1 is the n-dimensional unit ball centered at the origin and B_1^+ is the half ball.

Theorem 5.2. *Assume, in addition to the assumptions of Theorem 5.1, that* $\partial\Omega \in C^2$, $g \in H^2(\Omega)$. *Then a weak solution* u *of* (5.1) *satisfying* $u - g \in H_0^1(\Omega)$ *belongs to* $H^2(\Omega)$, *and*

$$\|u\|_{H^2(\Omega)} \leqslant C(\|u\|_{L^2(\Omega)} + \|f\|_{L^2(\Omega)} + \|g\|_{H^2(\Omega)}), \tag{5.5}$$

where the constant C *depends only on* n, λ, $\|a^{ij}\|_{W^{1,\infty}}$, $\|b^i\|_{L^\infty(\Omega)}$, $\|c\|_{L^\infty(\Omega)}$ *and* $\partial\Omega$.

Proof. First we assume that $g = 0$. For $x^0 \in \partial\Omega$, there exists $\phi : V \to B_1$ as in Definition 5.1. Introduce the new variable $y = \phi(x)$ and denote by J the Jacobian $\dfrac{\partial x}{\partial y}$. Letting $x = \phi^{-1}(y)$ in the integral identity (5.3), we obtain, for any $\varphi \in C_0^\infty(V \cap \Omega)$,

$$\int_{B_1^+} \widetilde{a}^{kl} \widetilde{D}_k u \widetilde{D}_l \varphi dy = \int_{B_1^+} \widetilde{q} \varphi dy, \tag{5.6}$$

where $\widetilde{D}_k = \dfrac{\partial}{\partial y_k}$, $B_1^+ = B_1 \cap \{y_n > 0\}$, and

$$\widetilde{a}^{kl} = J a^{ij} \frac{\partial y_k}{\partial x_i} \frac{\partial y_k}{\partial x_j}, \quad \widetilde{q} = Jq.$$

We now define the difference operator Δ_h in the x_k direction $(1 \leqslant k \leqslant n-1)$ and the cutoff function $\eta \in C_0^\infty(B_1)$ as before. Using a similar computation as in Theorem 5.1, we get $\widetilde{D}_k \widetilde{D} u \in L^2(B_{1/2}^+)$, and

$$\|\widetilde{D}_k \widetilde{D} u\|_{L^2(B_{1/2}^+)} \leqslant C(\|u\|_{H^1(\Omega)} + \|f\|_{L^2(\Omega)}), \tag{5.7}$$

where $1 \leqslant k \leqslant n-1$. By (5.6), for any $\varphi \in C_0^\infty(B_{1/2}^+)$,

$$\int_{B_{1/2}^+} \widetilde{a}^{nn} \widetilde{D}_n u \widetilde{D}_n \varphi dy = \int_{B_{1/2}^+} \Big[\widetilde{q} - \sum_{k+l<n} \widetilde{D}_l(\widetilde{a}^{kl} \widetilde{D}_k u)\Big] \varphi dy.$$

It follows from the definition of the generalized derivatives that

$$\widetilde{D}_n[a^{nn} \widetilde{D}_n u] = \widetilde{q} - \sum_{k+l<n} \widetilde{D}_l(\widetilde{a}^{kl} \widetilde{D}_k u) \quad \text{in } B_{1/2}^+,$$

from which we conclude that $\widetilde{D}_{nn} u \in H^2(B_{1/2}^+)$, and $\widetilde{D}_{nn} u$ satisfies an estimate similar to (5.7). Combining all these estimates, we obtain

$$\|\widetilde{D}^2 u\|_{L^2(B_{1/2}^+)} \leqslant C(\|u\|_{H^1(\Omega)} + \|f\|_{L^2(\Omega)}). \tag{5.8}$$

Letting $V' = \phi^{-1}(B_{1/2}^+)$ and changing the variable from y back to x, we derive

$$\|u\|_{H^2(V')} \leqslant C(\|u\|_{H^1(\Omega)} + \|f\|_{L^2(\Omega)}). \tag{5.9}$$

For any point x^0 on the boundary $\partial\Omega$, there exists a neighborhood V' on which (5.9) is valid. In the interior of Ω, we use Theorem 5.1 for the interior estimates. Using a finite cover for Ω, we conclude that

$$\|u\|_{H^2(\Omega)} \leqslant C(\|u\|_{H^1(\Omega)} + \|f\|_{L^2(\Omega)}). \tag{5.10}$$

Now suppose that $g \neq 0$. Let $w = u - g$. Then w satisfies

$$a(w, \varphi) = (f, \varphi)_0 - a(g, \varphi) = (f - Lg, \varphi)_0, \quad \forall \varphi \in H_0^1(\Omega).$$

Clearly $w \in H_0^1(\Omega)$. Using the result for $g = 0$, we obtain $w \in H^2(\Omega)$, and

$$\begin{aligned} \|w\|_{H^2(\Omega)} &\leqslant C(\|u\|_{H^1(\Omega)} + \|f - Lg\|_{L^2(\Omega)}) \\ &\leqslant C(\|u\|_{H^1(\Omega)} + \|f\|_{L^2(\Omega)} + \|g\|_{H^2(\Omega)}). \end{aligned}$$

(5.5) now follows from the interpolation theorem. □

From this theorem, we can also get higher regularity for the solution.

Theorem 5.3. *Let the assumption* (2.2) *be in force, and* $a^{ij} \in W^{k+1,\infty}(\Omega)$, $b^i, c \in W^{k,\infty}(\Omega)$, $f \in W^{k,2}(\Omega)$, *where* k *is a nonnegative integer. If* $u \in H^1(\Omega)$ *is a weak solution of* (5.1), *then* $u \in W_{loc}^{k+2,2}(\Omega)$.

Theorem 5.4. *In addition to the assumptions of Theorem 5.3, assume that* $\partial\Omega \in C^{k+2}$, $g \in W^{k+2,2}(\Omega)$. *If* $u \in H^1(\Omega)$ *is a weak solution of* (5.1) *satisfying* $u - g \in H_0^1(\Omega)$, *then* $u \in W^{k+2,2}(\Omega)$.

If a^{ij}, b^i, c and f are infinitely many times differentiable, then for any k we have $u \in W_{loc}^{k+2,2}(\Omega)$, and therefore, by the Sobolev embedding theorem, $u \in C^\infty(\Omega)$.

CHAPTER 2

Schauder Theory

The regularity theory for classical solutions of a second order elliptic equation was first established by Schauder. As far as classical solutions are concerned, the Schauder theory is quite complete. It provides a foundation for the study of nonlinear elliptic equations.

Here we use a new method introduced by Trudinger; his proof avoided the tedious computations of potentials.

1. Hölder spaces

In the process of studying potential equations, it is very inconvenient to discuss continuous differentiability alone — the results are not rich and plentiful. It is therefore necessary to introduce the concept of Hölder continuity, which can be viewed as fractional differentiation in a certain sense.

Definition 1.1. Let $\Omega \subset \mathbb{R}^n$, let $u(x)$ be defined on Ω, and let $x_0 \in \Omega$. If, for some $0 < \alpha < 1$,

$$H^{\alpha}_{x_0}[u;\Omega] = \sup_{x\in\Omega} \frac{u(x) - u(x_0)}{|x - x_0|^{\alpha}} < \infty, \tag{1.1}$$

then u is said to be *Hölder continuous at x_0 with respect to Ω, with exponent α*. If $\alpha = 1$ in the above definition, then u is said to be *Lipschitz continuous at x_0*.

We denote by $C^k(\overline{\Omega}) = C^{k,0}(\overline{\Omega})$ the space of functions which are k-times continuously differentiable on $\overline{\Omega}$. For Hölder continuity, we introduce the corresponding class of spaces, usually referred to as Hölder spaces.

For $0 < \alpha \leqslant 1$, introduce the semi-norms

$$[u]_{0,0;\Omega} = [u]_{0;\Omega} = \sup_{\Omega} |u(x)|, \tag{1.2}$$

$$[u]_{0,\alpha;\Omega} = [u]_{\alpha;\Omega} = \sup_{x_0\in\Omega} H^{\alpha}_{x_0}[u;\Omega], \tag{1.3}$$

$$[u]_{k,0;\Omega} = [u]_{k;\Omega} = \sum_{|\nu|=k} [D^{\nu}u]_{0;\Omega}, \tag{1.4}$$

$$[u]_{k,\alpha;\Omega} = \sum_{|\nu|=k} [D^{\nu}u]_{\alpha;\Omega}, \tag{1.5}$$

where k is a positive integer, $\nu = (\nu_1, \nu_2, \cdots, \nu_n)$ is a multi-index, $\nu_i \geqslant 0$ $(i = 1, 2, \cdots, n)$, $|\nu| = \sum_{i=1}^{n} \nu_i$, and

$$D^\nu u = \frac{\partial^{|\nu|} u}{\partial x_1^{\nu_1} \cdots \partial x_n^{\nu_n}}.$$

For simplicity, we often use $[D^k u]_{\alpha;\Omega}$ to represent $\sum_{|\nu|=k} [D^\nu u]_{\alpha;\Omega}$.

Definition 1.2. We use $C^{k,\alpha}(\overline{\Omega})$ $(0 < \alpha \leqslant 1)$ to denote the space consisting of all functions in $C^k(\overline{\Omega})$ satisfying $[u]_{k,\alpha;\Omega} < \infty$. Introduce the norm on $C^k(\overline{\Omega})$

$$|u|_{k;\Omega} = \sum_{m=0}^{k} [u]_{m;\Omega}, \tag{1.6}$$

and the norm on $C^{k,\alpha}(\overline{\Omega})$

$$|u|_{k,\alpha;\Omega} = |u|_{k;\Omega} + [u]_{k,\alpha;\Omega}. \tag{1.7}$$

It is not hard to verify that both $C^k(\overline{\Omega})$ and $C^{k,\alpha}(\overline{\Omega})$ are Banach spaces. We often drop the subscript Ω in the norms if there is no confusion.

For the product of two functions, the Hölder norm satisfies

Lemma 1.1. *For $u, v \in C^\alpha(\overline{\Omega})$ $(0 < \alpha \leqslant 1)$,*

$$[uv]_\alpha \leqslant [u]_0 [v]_\alpha + [u]_\alpha [v]_0 \leqslant |u|_\alpha |v|_\alpha. \tag{1.8}$$

The proof is left to the reader.

One of the most important properties in Hölder spaces is the interpolation inequality, which makes it possible to study only the most important term in deriving an a priori estimate, and thus simplifies the proof. Here we shall use the compactness argument for its proof.

Theorem 1.2. *Let Ω be a bounded domain, and $u \in C^{2,\alpha}(\overline{\Omega})$ $(0 < \alpha \leqslant 1)$. Then for any $\varepsilon > 0$,*

$$[u]_2 \leqslant \varepsilon [u]_{2,\alpha} + C_\varepsilon |u|_0, \tag{1.9}$$

$$[u]_1 \leqslant \varepsilon [u]_{2,\alpha} + C_\varepsilon |u|_0, \tag{1.10}$$

where C_ε depends on n, α, Ω, in addition to the dependence on ε. All the norms and semi-norms are defined on $\overline{\Omega}$.

Proof. We prove only (1.9); the proof for (1.10) is similar. If there is no constant C_ε such that (1.9) is valid for all functions in $C^{2,\alpha}(\overline{\Omega})$, then for any $N > 0$, there exists u_N such that

$$[u_N]_2 > \varepsilon [u_N]_{2,\alpha} + N |u_N|_0 \qquad (N = 1, 2, \cdots). \tag{1.11}$$

Using the homogeneity of the inequality, we can assume without loss of generality that $|u_N|_2 = 1$, for otherwise we can replace u_N by $u_N/|u_N|_2$. From (1.11),

$$[u_N]_{2,\alpha} < \frac{1}{\varepsilon}, \quad |u_N|_0 < \frac{1}{N} \qquad (N = 1, 2, \cdots). \tag{1.12}$$

Since u_N is uniformly bounded in $C^{2,\alpha}(\overline{\Omega})$, there exists a subsequence u_{N_k} which converges to some u in $C^2(\overline{\Omega})$, by the Ascoli-Arzelà theorem. However, the second inequality in (1.12) implies that u_N converges to 0 uniformly on $\overline{\Omega}$. It follows that $u = 0$, which contradicts $|u_N|_2 = 1$. The proof of (1.9) is complete. □

Definition 1.3. A domain Ω is said to satisfy a *cone property*, if there is a finite cone V such that for any $x \in \Omega$, there exists a cone congruent to V with vertex x and contained completely in Ω.

Theorem 1.3. *Suppose that Ω satisfies a cone property with h the height of the cone. Then for any $0 < \varepsilon \leqslant h$, we have*

$$[u]_2 \leqslant \varepsilon^\alpha [u]_{2,\alpha} + \frac{C}{\varepsilon^2}|u|_0, \tag{1.13}$$

$$[u]_1 \leqslant \varepsilon^{1+\alpha} [u]_{2,\alpha} + \frac{C}{\varepsilon}|u|_0, \tag{1.14}$$

where the constant C depends only on n and the solid angle of opening of the cone.

Proof. Let V_1 be a cone with the same solid angle of opening and height 1. Then Theorem 1.2 implies that

$$[u]_{2;V_1} \leqslant [u]_{2,\alpha;V_1} + C|u|_{0;V_1} \quad \text{for } u \in C^{2,\alpha}(\overline{V}_1), \tag{1.15}$$

where C depends only on the solid angle of opening. For $u \in C^{2,\alpha}(\overline{V}_\varepsilon)$ (V_ε shares the same solid angle of opening as V_1, with height ε and vertex at the origin), we use the change of variables $y = x/\varepsilon$, $\tilde{u}(y) = u(\varepsilon y) = u(x)$. Then $\tilde{u} \in C^{2,\alpha}(V_1)$, and (1.15) is valid for $\tilde{u}$. If we rewrite this inequality in terms of the original variables, then

$$[u]_{2;V_\varepsilon} \leqslant \varepsilon^\alpha [u]_{2,\alpha;V_\varepsilon} + \frac{C}{\varepsilon^2}|u|_{0;V_\varepsilon}.$$

For any $x \in \Omega$ and $\varepsilon \leqslant h$, there exists a cone V_ε with vertex x and $V_\varepsilon \subset \Omega$. Thus

$$[u]_{2;V_\varepsilon} \leqslant \varepsilon^\alpha [u]_{2,\alpha;V_\varepsilon} + \frac{C}{\varepsilon^2}|u|_{0;V_\varepsilon} \leqslant \varepsilon^\alpha [u]_{2,\alpha;\Omega} + \frac{C}{\varepsilon^2}|u|_{0;\Omega}.$$

Since x is arbitrary,

$$[u]_{2;\Omega} \leqslant \varepsilon^\alpha [u]_{2,\alpha;\Omega} + \frac{C}{\varepsilon^2}|u|_{0;\Omega}.$$

The proof of (1.14) is similar. □

2. Mollifiers

An equivalent Hölder norm of a function can be introduced using derivatives of its mollified function. This will reduce the Hölder estimates for solutions of elliptic equations to estimating the derivatives, and therefore will simplify the proof.

Let $\rho \in C_0^\infty(\mathbb{R}^n)$, $\rho(x) \geqslant 0$ with support $\mathrm{spt}\{\rho\} \subset B_1(0)$ and

$$\int_{\mathbb{R}^n} \rho(x)dx = 1. \tag{2.1}$$

Such a function is called a *mollifier*. One can take, for example,

$$\rho(x) = \begin{cases} k\exp\left\{\dfrac{1}{|x|^2-1}\right\} & \text{for } |x|<1, \\ 0 \quad \text{for } |x| \geqslant 0, \end{cases}$$

where k is chosen so that (2.1) is valid.

Definition 2.1. For $u \in L_{loc}(\mathbb{R}^n)$ and $\rho(x)$ a mollifier, the function

$$\widetilde{u}(x,\tau) = \tau^{-n}\int_{\mathbb{R}^n} \rho\Big(\frac{x-y}{\tau}\Big)u(y)dy \tag{2.2}$$

is said to be a *mollified function* of $u(x)$.

Lemma 2.1. *Let $u \in C(\mathbb{R}^n)$. Then $\widetilde{u}(x,\tau)$ converges uniformly on any compact set to $u(x)$ as $\tau \to 0$, and*

$$\sup|\widetilde{u}| \leqslant \sup|u|, \tag{2.3}$$

$$|D^k\widetilde{u}(x,\tau)| \leqslant C\tau^{-k}\sup_{B_\tau(x)}|u| \qquad (k=0,1,2,\cdots), \tag{2.4}$$

where D denotes the gradient in (x,τ) and C depends only on n, k and the mollifier ρ.

Proof. By (2.1),

$$\begin{aligned} \widetilde{u}(x,\tau)-u(x) &= \tau^{-n}\int_{\mathbb{R}^n}\rho\Big(\frac{x-y}{\tau}\Big)(u(y)-u(x))dy \\ &= \int_{B_1(0)}\rho(\eta)(u(x-\tau\eta)-u(x))d\eta, \end{aligned} \tag{2.5}$$

from which it is not difficult to derive that $\widetilde{u}(x,\tau)$ converges uniformly on any compact set to $u(x)$ as $\tau \to 0$. To prove (2.4), we first notice that, by induction,

$$D^k\widetilde{u}(x,\tau) = \tau^{-n-k}\int_{\mathbb{R}^n} P_k\Big(\frac{x-y}{\tau}\Big)u(y)dy,$$

where $P_k \in C_0^\infty(\mathbb{R}^n)$, $\mathrm{spt}\, P_k \subset \overline{B_1(0)}$. Letting $z=(x-y)/\tau$ in the above equality, we get

$$\begin{aligned} |D^k\widetilde{u}(x,\tau)| &\leqslant \tau^{-k}\left|\int_{\mathbb{R}^n}P_k(z)u(x-\tau z)dz\right| \\ &\leqslant \tau^{-k}\sup_{B_\tau(x)}|u|\int_{\mathbb{R}^n}|P_k(z)|dz \leqslant C\tau^{-k}\sup_{B_\tau(x)}|u|. \qquad \square \end{aligned}$$

Next we shall discuss the relationship between the derivatives of $\widetilde{u}(x,\tau)$ and the Hölder norm of $u(x)$.

Lemma 2.2. *Let $u \in C^{\alpha}_{loc}(\mathbb{R}^n)$ $(0 < \alpha \leqslant 1)$. Then*

$$|\widetilde{u}(x,\tau) - u(x)| \leqslant \tau^{\alpha} H_x^{\alpha}[u; B_\tau(x)], \tag{2.6}$$

$$|D^k \widetilde{u}(x,\tau)| \leqslant C\tau^{\alpha-k} H_x^{\alpha}[u; B_\tau(x)] \quad \text{for } k = 1,2,\cdots, \tag{2.7}$$

where C depends only on n, α, k and the mollifier ρ.

Proof. From (2.5) we easily derive (2.6). To prove (2.7), we denote by $\overline{\beta} = (\beta_0, \beta)$ the $n+1$ dimensional multi-index with $|\overline{\beta}| = k$, $D^{\overline{\beta}} = D_\tau^{\beta_0} D_x^{\beta}$. If $\beta = 0$, then we differentiate (2.5) with respect to τ to obtain

$$\begin{aligned} D^{\overline{\beta}}\widetilde{u}(x,\tau) &= \tau^{-n-k}\int_{\mathbb{R}^n} P_{\overline{\beta}}\Big(\frac{x-y}{\tau}\Big)(u(y)-u(x))dy \\ &= \tau^{-k}\int_{\mathbb{R}^n} P_{\overline{\beta}}(z)(u(x-\tau z)-u(x))dx, \end{aligned}$$

where $P_{\overline{\beta}}$ is a function with its support in $B_1(0)$. From this equality it is not difficult to derive (2.7). If $\beta \neq 0$, then

$$\begin{aligned} D^{\overline{\beta}}\widetilde{u}(x,\tau) &= \tau^{-n}\int_{\mathbb{R}^n} D^{\overline{\beta}}\Big[\rho\Big(\frac{x-y}{\tau}\Big)\Big]u(y)dy \\ &= \tau^{-n}\int_{\mathbb{R}^n} D^{\overline{\beta}}\Big[\rho\Big(\frac{x-y}{\tau}\Big)\Big](u(y)-u(x))dy \\ &\quad +(-1)^{|\beta|}\tau^{-n}u(x)\int_{\mathbb{R}^n} D_\tau^{\beta_0} D_y^{\beta}\Big[\rho\Big(\frac{x-y}{\tau}\Big)\Big]dy. \end{aligned}$$

The second term on the right-hand side of the above formula vanishes, by the divergence theorem. By estimating the first term, we derive (2.7). □

Lemma 2.3. *Let $u \in C(\mathbb{R}^n)$. If*

$$\sup_{y\in B_R(x),\, 0<\tau\leqslant R} \tau^{1-\alpha}|D\widetilde{u}(y,\tau)| < \infty$$

for some $0 < \alpha \leqslant 1$, $R > 0$, then u is Hölder continuous at x with respect to $B_R(x)$, with exponent α. Furthermore,

$$H_x^{\alpha}[u; B_R(x)] \leqslant C \sup_{y\in B_R(x),\, 0<\tau\leqslant R} \tau^{1-\alpha}|D\widetilde{u}(y,\tau)|, \tag{2.8}$$

where C depends only on n, α and the mollifier ρ.

Proof. For $|x-y| < R$, $0 < \tau \leqslant R$, we have

$$\begin{aligned} |u(x)-u(y)| &\leqslant |\widetilde{u}(x,\tau)-u(x)| + |\widetilde{u}(x,\tau)-\widetilde{u}(y,\tau)| \\ &\quad +|\widetilde{u}(y,\tau)-u(y)|. \end{aligned}$$

Notice that for $0 < \tau \leqslant R$,

$$\begin{aligned}|\widetilde{u}(x,\tau) - u(x)| &= \tau\left|\int_0^1 D_\tau \widetilde{u}(x,\eta\tau)d\eta\right| \\ &\leqslant \tau^\alpha \int_0^1 \frac{(\tau\eta)^{1-\alpha}|D_\tau\widetilde{u}(x,\eta\tau)|}{\eta^{1-\alpha}}d\eta \\ &\leqslant \sup_{0<\tau\leqslant R} \tau^{1-\alpha}|D_\tau\widetilde{u}(x,\tau)| \cdot \tau^\alpha \int_0^1 \frac{1}{\eta^{1-\alpha}}d\eta.\end{aligned}$$

Therefore

$$\begin{aligned}|u(x) - u(y)| &\leqslant \frac{2}{\alpha}\tau^\alpha \sup_{0<\tau\leqslant R,\ z\in B_R(x)} \tau^{1-\alpha}|D_\tau\widetilde{u}(z,\tau)| \\ &\quad +|D_x\widetilde{u}(x^*,\tau)||x-y|,\end{aligned}$$

where x^* is a point on the line segment connecting x and y. Now we take $\tau = |x-y|$; then for $y \in B_R(x)$,

$$\frac{|u(x)-u(y)|}{|x-y|^\alpha} \leqslant \left(\frac{2}{\alpha}+1\right) \sup_{0<\tau\leqslant R,\ z\in B_R(x)} \tau^{1-\alpha}|D\widetilde{u}(z,\tau)|.$$

The lemma is proved. □

Corollary 2.4. (1) *There exists a constant C, depending only on n, α and the mollifier ρ, such that*

$$\frac{1}{C}[u]_\alpha \leqslant \sup_{\tau>0,\ x\in\mathbb{R}^n} \tau^{1-\alpha}|D\widetilde{u}(x,\tau)| \leqslant C[u]_\alpha, \tag{2.9}$$

for $u \in C^\alpha(\mathbb{R}^n)$ $(0 < \alpha \leqslant 1)$.

(2) *For any n dimensional multi-index β, there exists a constant $C = C(n,\alpha,\rho)$ such that*

$$[D^{\beta+1}u]_\alpha \leqslant \sup_{\tau>0}[DD_x^\beta\widetilde{u}(x,\tau)]_\alpha^x \leqslant C[D^{\beta+1}u]_\alpha, \tag{2.10}$$

where $D^{\beta+1} = DD^\beta$, and $[\cdot]_\alpha^x$ denotes the Hölder norm in the x direction.

Proof. (2.9) is a direct corollary of Lemmas 2.2 and 2.3.

The first inequality in (2.10) is obvious. In fact, for any $\lambda > 0$,

$$\frac{|D_x^{\beta+1}\widetilde{u}(x,\lambda) - D_x^{\beta+1}\widetilde{u}(y,\lambda)|}{|x-y|^\alpha} \leqslant \sup_{\tau>0}[D_x^{\beta+1}\widetilde{u}]_\alpha^x .$$

Letting $\lambda \to 0$, we obtain

$$[D^{\beta+1}u]_\alpha \leqslant \sup_{\tau>0}[D_x^{\beta+1}\widetilde{u}]_\alpha^x .$$

Now we prove the second inequality in (2.10). Let $y = x+h$, $A_h u = u(x+h)$. Then

$$\begin{aligned}|DD_x^\beta\widetilde{u}(x,\tau) - DD_x^\beta\widetilde{u}(y,\tau)| &= |DD_x^\beta\widetilde{(u - A_h u)}(x,\tau)| \\ &= |\widetilde{DD_x^\beta(u - A_h u)}(x,\tau)| \leqslant C[D^{\beta+1}(u - A_h u)]_0,\end{aligned}$$

where in deriving the last inequality we made use of (2.7) (the case $k = 1$, $\alpha = 1$) of Lemma 2.2. (2.10) follows immediately. □

3. $C^{2,\alpha}$ estimates for solutions of potential equations

In §2, Hölder norm estimates of solutions are reduced to estimating the derivatives of their mollified functions. Therefore we need to establish the derivative estimates for solutions of the potential equations.

Lemma 3.1. *Let $u \in C^\infty(\mathbb{R}^n)$ satisfy*

$$-\Delta u = f,$$

where Δ is the Laplacian operator. Then for any $R > 0$, we have

$$|D_i u(x)| \leqslant \frac{n}{R} \operatorname*{osc}_{B_R(x)} u + R \sup_{B_R(x)} |f| \qquad (i = 1, 2, \cdots, n), \tag{3.1}$$

where $\operatorname*{osc}_{\Omega} u$ denotes the oscillation of u on Ω.

Proof. Assume without loss of generality that x is the origin. Set $F_0 = \sup_{B_R(x)} |f|$. Then by the divergence formula

$$\begin{aligned}
\int_{B_\rho} \Delta(D_i u)dx &= \int_{\partial B_\rho} \frac{\partial D_i u}{\partial r} dS \\
&= \rho^{n-1} \int_{|\omega|=1} \frac{\partial D_i u}{\partial \rho}(\rho\omega)d\omega \\
&= \rho^{n-1} \frac{\partial}{\partial \rho}\left[\rho^{1-n} \int_{\partial B_\rho} D_i u dS\right].
\end{aligned}$$

On the other hand,

$$\begin{aligned}
\int_{B_\rho} \Delta(D_i u)dx &= \int_{\partial B_\rho} \Delta u \cos(r, x_i)dS \\
&= -\int_{\partial B_\rho} f \cos(r, x_i)dS.
\end{aligned}$$

Combining the above two equalities, we have

$$\pm\frac{\partial}{\partial \rho}\left[\rho^{1-n} \int_{\partial B_\rho} D_i u dS\right] \leqslant n\omega_n F_0,$$

where ω_n is the volume of an n-dimensional unit ball. Integrating the above inequality over $\rho \in [0, r]$, we obtain

$$\pm\left[r^{1-n} \int_{\partial B_r} D_i u dS - n\omega_n D_i u(0)\right] \leqslant n\omega_n r F_0.$$

Multiplying the above inequality by r^{n-1} and integrating r over $[0, R]$, we get

$$\pm\left[\int_{B_R} D_i u dx - \omega_n R^n D_i u(0)\right] \leqslant \omega_n R^{n+1} F_0.$$

Rewriting this inequality, we have

$$\begin{aligned}|D_iu(0)| &\leqslant RF_0 + \left|\frac{1}{\omega_n R^n}\int_{B_R} D_i u dx\right| \\ &\leqslant RF_0 + \frac{1}{\omega_n R^n}\left|\int_{\partial B_R}(u(x)-u(0))\cos(r,x_i)dS\right| \\ &\leqslant RF_0 + \frac{n}{R}\operatorname*{osc}_{B_R} u.\end{aligned}$$

The proof of the lemma is complete. □

Theorem 3.2. *Let* $u \in C_0^{2,\alpha}(\mathbb{R}^n)$ $(0 < \alpha < 1)$ *satisfy*

$$-\Delta u = f. \tag{3.2}$$

Then

$$[D^2u]_\alpha \leqslant C[f]_\alpha, \tag{3.3}$$

where C *depends only on* n, α.

Proof. Let $g(x) = f(x) - f(x_0)$ in $B_R(x_0)$. We can rewrite the equation (3.2) as

$$-\Delta u(x) - f(x_0) = g(x), \tag{3.4}$$

where g satisfies

$$\sup_{B_R(x_0)} |g(x)| \leqslant R^\alpha [f]_\alpha. \tag{3.5}$$

Mollifying (3.4), we obtain

$$-\Delta\widetilde{u}(x,\tau) - f(x_0) = \widetilde{g}(x,\tau), \qquad \tau > 0.$$

Differentiating the equation twice with respect to x, τ, we get

$$-\Delta D^2\widetilde{u}(x,\tau) = D^2\widetilde{g}(x,\tau).$$

Restricting D^2 to be DD_x and applying Lemma 3.1, we derive

$$\begin{aligned}|DD_x^2\widetilde{u}(x_0,\tau)| &\leqslant C\left\{\frac{1}{R}\operatorname*{osc}_{B_R(x_0)} DD_x\widetilde{u}(x,\tau) + R\sup_{B_R(x_0)}|D^2\widetilde{g}|\right\} \\ &\leqslant C\left\{\frac{1}{R^{1-\alpha}}[DD_x\widetilde{u}]_\alpha^x + R\sup_{B_R(x_0)}|D^2\widetilde{g}|\right\}.\end{aligned}$$

We take $\tau \leqslant R$. Applying Lemma 2.1 and Corollary 2.4 (2), we obtain

$$\tau^{1-\alpha}|DD_x^2\widetilde{u}(x_0,\tau)| \leqslant C\left(\frac{\tau^{1-\alpha}}{R^{1-\alpha}}[D^2u]_\alpha + R\tau^{-1-\alpha}\sup_{B_{R+\tau}(x_0)}|g|\right).$$

Take $R = N\tau$, where N is to be determined. By (3.5),

$$\tau^{1-\alpha}|DD_x^2\widetilde{u}(x_0,\tau)| \leqslant C\left\{N^{\alpha-1}[D^2u]_\alpha + N^{1+\alpha}[f]_\alpha\right\}.$$

It follows from Corollary 2.4 (1) that

$$\begin{aligned}[D^2u]_\alpha &\leqslant C \sup_{\tau>0, x_0\in\mathbb{R}^n} \tau^{1-\alpha}|DD_x^2\widetilde{u}(x_0,\tau)| \\ &\leqslant C\Big\{N^{\alpha-1}[D^2u]_\alpha + N^{1+\alpha}[f]_\alpha\Big\}.\end{aligned}$$

If we take N to be sufficiently large so that $CN^{\alpha-1} = 1/2$, then the desired estimate (3.3) follows. □

We can easily generalize this result to elliptic equations with *constant* coefficients

$$-a^{ij}D_{ij}u = f, \tag{3.6}$$

where the summation convention is followed. Assume that the coefficient matrix (a^{ij}) satisfies

$$\lambda|\xi|^2 \leqslant a^{ij}\xi_i\xi_j \leqslant \Lambda|\xi|^2, \quad \forall \xi \in \mathbb{R}^n, \tag{3.7}$$

where $\Lambda \geqslant \lambda > 0$.

Theorem 3.3. *Let $u \in C_0^{2,\alpha}(\mathbb{R}^n)$ $(0 < \alpha < 1)$ satisfy* (3.6). *Assume that the constant coefficient matrix satisfies* (3.7). *Then*

$$[D^2u]_\alpha \leqslant C\lambda^{-1}[f]_\alpha, \tag{3.8}$$

where C depends only on n, α and Λ/λ.

Proof. Assume without loss of generality that $\lambda = 1$. Introduce a change of variables $y = Bx$ and $\overline{u}(y) = u(x)$, $\overline{f}(y) = f(x)$. Then a simple calculation shows that $\overline{u}(y)$ satisfies the equation

$$-\overline{a}^{ij}D_{ij}\overline{u} = \overline{f}, \quad \overline{A} = (\overline{a}^{ij}) = B^TAB.$$

There exists a nondegenerate B such that $\overline{A}$ is the identity matrix, and thus

$$-\Delta_y\overline{u} = \overline{f}.$$

By Theorem 3.2,

$$[D^2\overline{u}]_\alpha \leqslant C[\overline{f}]_\alpha.$$

(3.8) will follow after changing the variables back to x. The details are left to the reader. □

In order to deal with estimates of solutions of the Dirichlet problem near the boundary, we consider the Dirichlet problem for the potential equation in a half space $\mathbb{R}_+^n = \mathbb{R}^n \cap \{x_n > 0\}$.

Theorem 3.4. *Let $u \in C_0^{2,\alpha}(\overline{\mathbb{R}_+^n})$ $(0 < \alpha < 1)$ satisfy*

$$\begin{cases} -\Delta u = f & \text{in } \mathbb{R}_+^n, \\ u = 0 & \text{on } \partial\mathbb{R}_+^n. \end{cases} \tag{3.9}$$

Then there exists a constant C, depending only on n, α, such that

$$[D^2u]_\alpha \leqslant C[f]_\alpha. \tag{3.10}$$

Proof. For each $x_0 \in \mathbb{R}^n_+$, we rewrite the equation as

$$-\Delta u - f(x_0) = g(x),$$

where $g(x) = f(x) - f(x_0)$. We now extend u, g and $f(x_0)$ to $x_n < 0$ using odd extension, i.e.,

$$v(x) = \begin{cases} u(x', x_n), & x_n \geqslant 0, \\ -u(x', -x_n), & x_n < 0, \end{cases}$$

$$h(x) = \begin{cases} g(x', x_n), & x_n \geqslant 0, \\ -g(x', -x_n), & x_n < 0, \end{cases}$$

$$f_0(x) = \begin{cases} f(x_0), & x_n \geqslant 0, \\ -f(x_0), & x_n < 0, \end{cases}$$

where $x' = (x_1, \cdots, x_{n-1})$. It is not difficult to verify that $v \in W^{2,\infty}(\mathbb{R}^n)$, $D_{x'}v \in C^{1,\alpha}(\mathbb{R}^n)$, and

$$-\Delta v - f_0(x) = h(x) \quad \text{on } \mathbb{R}^n \setminus \{x_n = 0\}.$$

Mollifying the equation, we get

$$-\Delta \widetilde{v}(x, \tau) - \widetilde{f}_0(x, \tau) = \widetilde{h}(x, \tau), \tag{3.11}$$

where

$$\widetilde{v}(x, \tau) = \tau^{-n} \int_{-\infty}^{\infty} dy_n \int_{\mathbb{R}^{n-1}} \rho_1\Big(\frac{x_n - y_n}{\tau}\Big) \rho_{n-1}\Big(\frac{x' - y'}{\tau}\Big) v(y) dy'$$

and ρ_1, ρ_{n-1} are the 1 dimensional and $n-1$ dimensional mollifiers, respectively.

Applying the differentiation $DD_{x'}$ to (3.11), we obtain

$$-\Delta DD_{x'}\widetilde{v}(x, \tau) = DD_{x'}\widetilde{h}(x, \tau).$$

Applying Lemma 3.1, using also an argument similar to that in Theorem 3.2, we derive

$$\begin{aligned} |D_x DD_{x'}\widetilde{v}(x_0, \tau)| &\leqslant \frac{n}{R} \operatorname*{osc}_{B_R(x_0)} DD_{x'}\widetilde{v}(x, \tau) + R \sup_{B_R(x_0)} |DD_{x'}\widetilde{h}(x, \tau)| \\ &\leqslant C\Big[R^{\alpha-1}[DD_{x'}v]_\alpha + R\tau^{-2} \sup_{B_{R+\tau}(x_0)} |h|\Big]. \end{aligned}$$

Take $R = N\tau$ as in Theorem 3.2. Applying Lemma 2.3, we derive, for sufficiently large N,

$$[DD_{x'}u]_\alpha \leqslant C[f]_\alpha. \tag{3.12}$$

Using the equation for u, we have

$$D_{nn}u = -\sum_{i=1}^{n-1} D_{ii}u - f.$$

By (3.12),

$$[D_{nn}u]_\alpha \leqslant C[f]_\alpha.$$

The proof is complete. □

The above result can also be extended to elliptic equations with constant coefficients.

Theorem 3.5. *Suppose that the constant coefficient matrix* (a^{ij}) *satisfies* (3.7), *and that, for some* $0 < \alpha < 1$, $u \in C_0^{2,\alpha}(\overline{\mathbb{R}^n_+})$ *satisfies* (3.6) *and the boundary condition*

$$u\Big|_{\partial\mathbb{R}^n_+} = 0.$$

Then we have the estimate

$$[D^2u]_\alpha \leqslant C\lambda^{-1}[f]_\alpha,$$

where C *depends only on* n, α *and* Λ/λ.

4. Interior Schauder estimates

The following lemma is very useful in studying any interior estimates.

Lemma 4.1. *Let* $\varphi(t)$ *be a bounded nonnegative function defined on the interval* $[T_0, T_1]$, *where* $T_1 > T_0 \geqslant 0$. *Suppose that for any* $T_0 \leqslant t < s \leqslant T_1$, φ *satisfies*

$$\varphi(t) \leqslant \theta\varphi(s) + \frac{A}{(s-t)^\alpha} + B, \tag{4.1}$$

where θ, A, B *and* α *are nonnegative constants,* $\theta < 1$. *Then*

$$\varphi(\rho) \leqslant C\left[\frac{A}{(R-\rho)^\alpha} + B\right], \quad \forall T_0 \leqslant \rho < R \leqslant T_1, \tag{4.2}$$

where C *depends only on* α, θ.

Proof. Let $t_0 = \rho$, $t_{i+1} = t_i + (1-\tau)\tau^i(R-\rho)$ $(i = 0, 1, 2, \cdots)$, where $0 < \tau < 1$ is to be determined. From (4.1),

$$\varphi(t_i) \leqslant \theta\varphi(t_{i+1}) + \frac{A}{[(1-\tau)\tau^i(R-\rho)]^\alpha} + B \quad (i = 0, 1, 2, \cdots).$$

By induction,

$$\varphi(t_0) \leqslant \theta^k\varphi(t_k) + \left[\frac{A}{(1-\tau)^\alpha(R-\rho)^\alpha} + B\right]\sum_{i=0}^{k-1}\theta^i\tau^{-i\alpha}.$$

We choose τ such that $\theta\tau^{-\alpha} < 1$. By letting $k \to \infty$, we get (4.2). □

This lemma can be interpreted as asserting that if (4.1) is valid for any $t < s$ for some $\theta < 1$, then we can get rid of the term $\theta\varphi(s)$.

Let Ω be a bounded open domain. Now we consider the general linear second order elliptic equation:

$$Lu = -a^{ij}D_{ij}u + b^iD_iu + cu = f \quad \text{in } \Omega. \tag{4.3}$$

Assume that there exist $\Lambda \geqslant \lambda > 0$ such that

$$\lambda|\xi|^2 \leqslant a^{ij}(x)\xi_i\xi_j \leqslant \Lambda|\xi|^2, \quad \forall x \in \Omega, \xi \in \mathbb{R}^n, \tag{4.4}$$

$a^{ij}, b^i, c \in C^\alpha(\overline{\Omega})$ $(0 < \alpha < 1)$ and

$$\frac{1}{\lambda}\left\{\sum_{i,j}|a^{ij}|_{\alpha;\Omega} + \sum_i|b^i|_{\alpha;\Omega} + |c|_{\alpha;\Omega}\right\} \leqslant \Lambda_\alpha. \tag{4.5}$$

Lemma 4.2. *Suppose that the coefficients in the equation (4.3) satisfy the assumptions (4.4) and (4.5). Then there exists $R_0 \leqslant 1$, depending only on $n, \alpha, \Lambda/\lambda$ and Λ_α, such that for any $0 < R \leqslant R_0$ with $B_R \subset \Omega$ and any solution $u \in C_0^{2,\alpha}(B_R)$ $(0 < \alpha < 1)$ of (4.3), the following estimate is valid:*

$$[D^2 u]_{\alpha;B_R} \leqslant C\left\{\frac{1}{\lambda}[f]_{\alpha;B_R} + R^{-2-\alpha}|u|_{\alpha;B_R}\right\}, \tag{4.6}$$

where the constant C depends only on $n, \alpha, \Lambda/\lambda$ and Λ_α.

Proof. Assume without loss of generality that $\lambda = 1$. Let B_R be a ball centered at x_0. We shall use the method of freezing the coefficients by rewriting (4.3) as

$$-a^{ij}(x_0)D_{ij}u = \widetilde{f},$$

where

$$\widetilde{f} = f + (a^{ij}(x) - a^{ij}(x_0))D_{ij}u - b^i D_i u - cu.$$

By Theorem 3.3,

$$[D^2 u]_\alpha \leqslant C[\widetilde{f}]_\alpha \leqslant C\Big\{[f]_\alpha + R^\alpha [D^2 u]_\alpha + |u|_2\Big\},$$

where the norms are on B_R. Using the interpolation inequalities (1.13) and (1.14) (take $\varepsilon = R/2$), we obtain

$$[D^2 u]_\alpha \leqslant CR^\alpha [D^2 u]_\alpha + C\Big\{[f]_\alpha + R^{-2}|u|_0\Big\}.$$

If we take $R_0 = \left(\dfrac{1}{2C}\right)^{1/\alpha}$, then for $0 < R \leqslant R_0$,

$$[D^2 u]_\alpha \leqslant C\Big\{[f]_\alpha + R^{-2}|u|_0\Big\}.$$

The lemma is proved. □

In Lemma 4.2, the assumption that u is compactly supported in B_R is very restrictive. We can use Lemma 4.1 to get rid of this assumption and to obtain $C^{2,\alpha}$ interior estimates for solutions of (4.3).

Theorem 4.3 (Interior Schauder estimates). *Suppose that the coefficients in the equation (4.3) satisfy the assumptions (4.4) and (4.5). Let $u \in C^{2,\alpha}(\Omega)$ $(0 < \alpha < 1)$ be a solution of (4.3). Then for $\Omega' \subset\subset \Omega$, we have*

$$|u|_{2,\alpha;\Omega'} \leqslant C\left(\frac{1}{\lambda}|f|_{\alpha;\Omega} + |u|_{0;\Omega}\right), \tag{4.7}$$

where C depends only on $n, \alpha, \Lambda/\lambda, \Lambda_\alpha$ and $\operatorname{dist}\{\Omega', \partial\Omega\}$.

Proof. Assume without loss of generality that $\lambda = 1$. Let R_0 be the constant in Lemma 4.2 and $\overline{R}_0 = \min\Big\{R_0, \frac{1}{2}\operatorname{dist}\{\Omega', \partial\Omega\}\Big\}$. For any $x_0 \in \Omega'$ and $0 < R \leqslant \overline{R}_0$, we denote $B_R = B_R(x_0)$. Let $\zeta(x)$ be a cutoff function such that

$$\begin{aligned} &\zeta \in C_0^\infty(B_R), \qquad \zeta(x) = 1 \quad \text{for } x \in B_{\tau R}, \\ &[D^k\zeta]_0 + (1-\tau)^\alpha R^\alpha [D^k \zeta]_\alpha \leqslant \frac{C}{(1-\tau)^k R^k}, \end{aligned} \tag{4.8}$$

where $C = C(n,k)$, $0 < \tau < 1$. Such a function can be constructed by using a mollifier. Let $v = \zeta u$. Then $v \in C_0^{2,\alpha}(B_R)$, and

$$Lv = \zeta f + [-a^{ij}D_{ij}\zeta + b^i D_i\zeta]u - 2a^{ij}D_i\zeta D_j u.$$

In order to apply Lemma 4.2, we need to estimate the Hölder norm of the right-hand side of the above equation. We first estimate one of the terms:

$$\begin{aligned}[a^{ij}D_i\zeta D_j u]_\alpha &\leqslant \Lambda_\alpha\Big\{[D\zeta]_\alpha[u]_1 + [D\zeta]_0[u]_{1,\alpha} + [D\zeta]_0[u]_1\Big\}\\ &\leqslant C\Big\{\frac{[u]_1}{(1-\tau)^{1+\alpha}R^{1+\alpha}} + \frac{R^{1-\alpha}[u]_2}{(1-\tau)R}\Big\}.\end{aligned}$$

Using the interpolation inequality (Theorem 1.3), we obtain, for any $\varepsilon \in (0,1)$,

$$[a^{ij}D_i\zeta D_j u]_\alpha \leqslant C\Big\{\varepsilon[D^2u]_\alpha + C_\varepsilon\frac{|u|_0}{(1-\tau)^{(2/\alpha)+1}R^{2+\alpha}}\Big\}.$$

Similarly, by Lemmas 4.2 and 1.1, Theorem 1.3 and (4.8),

$$\begin{aligned}[D^2v]_{\alpha;B_R} &\leqslant C\Big\{\frac{1}{(1-\tau)^\alpha R^\alpha}[f]_{0;B_R} + [f]_{\alpha;B_R} + \varepsilon[D^2u]_{\alpha;B_R}\\ &\qquad + \frac{C_\varepsilon}{(1-\tau)^{(2/\alpha)+1}R^{2+\alpha}}|u|_{0;B_R}\Big\}.\end{aligned}$$

Notice that $\zeta(x) = 1$ for $x \in B_{\tau R}$. Using the product rule for derivatives and the interpolation inequalities (1.13), (1.14), we obtain

$$\begin{aligned}[D^2u]_{\alpha;B_{\tau R}} &\leqslant C\Big\{[f]_{\alpha;B_R} + \varepsilon[D^2u]_{\alpha;B_R}\\ &\qquad + \frac{C_\varepsilon}{(1-\tau)^{(2/\alpha)+1}R^{(2/\alpha)+1}}\Big[R^{(2/\alpha)-1-\alpha}|u|_{0;B_R} + R^{(2/\alpha)+1-\alpha}[f]_{0;B_R}\Big]\Big\}.\end{aligned}$$

Let $\varphi(s) = [D^2u]_{\alpha;B_s}$ and $\tau R = s$, $R = t$ in the above inequality. Then for $0 \leqslant s < t \leqslant \overline{R}_0$,

$$\begin{aligned}\varphi(s) &\leqslant C\Big\{\varepsilon\varphi(t) + [f]_{\alpha;B_R}\\ &\qquad + \frac{C_\varepsilon}{(t-s)^{(2/\alpha)+1}}\Big[\overline{R}_0^{(2/\alpha)-1-\alpha}|u|_{0;B_{\overline{R}_0}} + \overline{R}_0^{(2/\alpha)+1-\alpha}[f]_{0;B_{\overline{R}_0}}\Big]\Big\}.\end{aligned}$$

We take ε so that $C\varepsilon = 1/2$. Then by Lemma 4.1,

$$\begin{aligned}[D^2u]_{\alpha;B_\rho} &\leqslant C\Big\{[f]_{\alpha;B_R} + \frac{C_\varepsilon}{(R-\rho)^{(2/\alpha)+1}}\Big[\overline{R}_0^{(2/\alpha)-1-\alpha}|u|_{0;B_{\overline{R}_0}}\\ &\qquad + \overline{R}_0^{(2/\alpha)+1-\alpha}[f]_{0;B_{\overline{R}_0}}\Big]\Big\} \qquad \forall 0 < \rho < R \leqslant \overline{R}_0.\end{aligned}$$

Finally, taking $R = \overline{R}_0$, $\rho = \overline{R}_0/2$, and using the interpolation inequality, we get

$$|u|_{2,\alpha;B_{\overline{R}_0/2}} \leqslant C\{|f|_{\alpha;\Omega} + |u|_{0;\Omega}\}. \tag{4.9}$$

Now we conclude the proof by covering Ω' with finitely many balls of radius $\overline{R}_0/2$ and applying (4.9) on each of these balls. □

Remark. Suppose that the coefficients satisfy (4.4), $a^{ij}, b^i, c, f \in C^{k,\alpha}(\Omega)$ ($k \geqslant 0, 0 < \alpha < 1, i,j = 1,2,\cdots,n$). Using finite differences, we can prove that any solution u of (4.3) belongs to $C^{k+2,\alpha}(\Omega)$, and for any $\Omega' \subset\subset \Omega'' \subset\subset \Omega$,

$$|u|_{k+2,\alpha;\Omega'} \leqslant C(|f|_{k,\alpha;\Omega''} + |u|_{0;\Omega''}),$$

where C depends only on $n, k, \Lambda/\lambda$, $\operatorname{dist}\{\Omega', \partial\Omega''\}$, and the $C^{k,\alpha}(\overline{\Omega''})$ norm of the coefficients a^{ij}, b^i, c.

5. Global Schauder estimates

Following the procedure of interior estimates, using half-balls in place of balls, we can derive estimates near the boundary.

Lemma 5.1. *Suppose that Ω has a flat part of the boundary S, $\Omega \subset \mathbb{R}^n \cap \{x_n > 0\}$, and $S \subset \partial\mathbb{R}^n_+$. Let the assumptions (4.4) and (4.5) be in force. Then there exist R_0 and C, depending only on $n, \alpha, \Lambda/\lambda$ and Λ_α, such that for any $0 < R \leqslant R_0$ with $B_R^+ \subset \Omega$ (B_R is a ball with its center on S) and any solution $u \in C^{2,\alpha}(B_R^+)$ of (4.3) which vanishes in a neighborhood of $\partial B_R^+ \cap \{x_n > 0\}$ and vanishes on $S \cap \partial B_R^+$, the following estimate is valid:*

$$[D^2 u]_{\alpha;B_R^+} \leqslant C\left\{\frac{1}{\lambda}[f]_{\alpha;B_R^+} + R^{-2-\alpha}|u|_{0;B_R^+}\right\}. \tag{5.1}$$

The proof is exactly the same as that of Lemma 4.2.

Lemma 5.2. *Let Ω be as in Lemma 5.1. Let the assumptions (4.4) and (4.5) be in force. Suppose that $u \in C^{2,\alpha}(\Omega \cup S)$ is a solution of the equation (4.3) satisfying $u = 0$ on S. Then for any $\Omega' \subset\subset \Omega \cup S$, we have*

$$|u|_{2,\alpha;\Omega'} \leqslant C\left\{\frac{1}{\lambda}|f|_{\alpha;\Omega} + |u|_{0;\Omega}\right\}, \tag{5.2}$$

where C depends only on $n, \alpha, \Lambda/\lambda$, Λ_α and $\operatorname{dist}\{\Omega', \partial\Omega \setminus S\}$.

Proof. Assume without loss of generality that $\lambda = 1$. Let R_0 be the constant in Lemma 5.1, and set $\overline{R}_0 = \min\left\{R_0, \frac{1}{2}\operatorname{dist}\{\Omega', \partial\Omega \setminus S\}\right\}$, $\Omega'' = \Omega' \cap \left\{x_n > \frac{1}{4}\overline{R}_0\right\}$. Since $\Omega'' \subset\subset \Omega$, the interior Schauder estimate gives

$$|u|_{2,\alpha;\Omega''} \leqslant C\{|f|_{\alpha;\Omega} + |u|_{0;\Omega}\}. \tag{5.3}$$

Now we let $B_{\overline{R}_0}$ be a ball with its center inside $\overline{\Omega}' \cap S$. By following exactly the same proof as in Theorem 4.3, we obtain

$$|u|_{2,\alpha;B^+_{\overline{R}_0/2}} \leqslant C(|f|_{\alpha;\Omega} + |u|_{0;\Omega}).$$

We can cover $\overline{\Omega'} \setminus \overline{\Omega''}$ with finitely many balls $B^+_{\overline{R}_0/2}$ with their centers inside $\overline{\Omega'} \cap S$; thus

$$|u|_{2,\alpha;\Omega' \setminus \Omega''} \leqslant C(|f|_{\alpha;\Omega} + |u|_{0;\Omega}). \tag{5.4}$$

Estimates (5.3) and (5.4) imply (5.2). □

In order to apply the above result to a general domain, the boundary of the domain must have a certain smoothness. Analogously to Definition 5.1 of Chapter 1, we can introduce the class $C^{k,\alpha}$ of boundaries for $\partial\Omega$. We shall not repeat the definition here.

Theorem 5.3 (Global Schauder estimates). *Let the assumptions* (4.4), (4.5) *be in force, and* $\partial\Omega \in C^{2,\alpha}$ $(0 < \alpha < 1)$. *Suppose that* $u \in C^{2,\alpha}(\overline{\Omega})$ *is a solution of the equation* (4.3) *satisfying the boundary condition* $u\Big|_{\partial\Omega} = 0$. *Then*

$$|u|_{2,\alpha;\Omega} \leqslant C\left(\frac{1}{\lambda}|f|_{\alpha;\Omega} + |u|_{0;\Omega}\right), \tag{5.5}$$

where C *depends only on* $n, \alpha, \Lambda/\lambda, \Lambda_\alpha$ *and* Ω.

Proof. Assume without loss of generality that $\lambda = 1$. Let $x_0 \in \partial\Omega$, and suppose that ϕ is the smooth function defined in Definition 5.1 of Chapter 1 such that

$$|\phi|_{2,\alpha;V\cap\Omega} \leqslant K, \qquad |\phi^{-1}|_{2,\alpha;B_1^+} \leqslant K. \tag{5.6}$$

Under the change of variables $y = \phi(x)$, $\overline{u}(y) = u(x)$, the function $\overline{u}$ satisfies the equation

$$\overline{a}^{rs} D_{rs}\overline{u} + \overline{b}^r D_r\overline{u} + \overline{c}\,\overline{u} = \overline{f}, \tag{5.7}$$

where

$$\begin{aligned} &\overline{a}^{rs}(y) = a^{ij}\frac{\partial y_r}{\partial x_i}\frac{\partial y_s}{\partial x_j}, \quad \overline{b}^r(y) = a^{ij}\frac{\partial^2 y_r}{\partial x_i \partial x_j} + b^i\frac{\partial y_r}{\partial x_i}, \\ &\overline{c}(y) = c \circ \phi^{-1}(y). \end{aligned}$$

Applying Lemma 5.2 to the equation (5.7), we get

$$|\overline{u}|_{2,\alpha;B_{1/2}^+} \leqslant C(|\overline{u}|_{0;B_1^+} + |\overline{f}|_{\alpha;B_1^+}).$$

Rewriting this inequality in terms of the variable x, we have

$$|u|_{2,\alpha;\phi^{-1}(B_{1/2}^+)} \leqslant C(|u|_{0;\Omega} + |f|_{\alpha;\Omega}).$$

Next one can apply interior estimates and a finite covering technique to complete the proof for (5.5). The above proof is only an outline; the detailed calculations and analysis are left to the reader. □

Remark 5.1. If, in Theorem 5.3, the zero boundary condition for u is replaced by

$$u = \varphi \quad \text{on } \partial\Omega, \tag{5.8}$$

where $\varphi \in C^{2,\alpha}(\overline{\Omega})$, then the following Schauder estimate is valid:

$$|u|_{2,\alpha;\Omega} \leqslant C(|f|_{\alpha;\Omega} + |\varphi|_{2,\alpha;\Omega} + |u|_{0;\Omega}). \tag{5.9}$$

In fact, we can simply apply Theorem 5.3 to the function $w = u - \varphi$.

Remark 5.2. Suppose that $u \in C^{2,\alpha}(\overline{\Omega})$ is a solution of (4.3) satisfying the boundary condition (5.8), where (a^{ij}) satisfies (4.4), $a^{ij}, b^i, c, f \in C^{k,\alpha}(\overline{\Omega})$, $\partial\Omega \in C^{k+2,\alpha}$ and $\varphi \in C^{k+2,\alpha}(\overline{\Omega})$. Using the finite difference technique, we can show that $u \in C^{k+2,\alpha}(\overline{\Omega})$ and

$$|u|_{k+2,\alpha} \leqslant C(|f|_{k,\alpha} + |\varphi|_{k+2,\alpha} + |u|_0),$$

where C depends only on $n, \alpha, k, \Lambda/\lambda, \Omega$ and the $C^{k,\alpha}(\overline{\Omega})$ norms of a^{ij}, b^i, c.

6. A maximum principle for classical solutions

The maximum principle is an important property for solutions of second order elliptic equations. Using it, we can estimate the maximum norm $|u|_{0;\Omega}$, which is needed in using the Schauder estimates. Furthermore, combining the maximum principle with the construction of auxiliary functions gives us an important tool in studying partial differential equations.

We shall assume throughout this section that Ω is a bounded domain in $\mathbb{R}^n$.

Theorem 6.1 (Weak maximum principle for classical solutions). *Suppose that L is defined in (4.3), with the coefficient matrix (a^{ij}) satisfying (4.4), b^i being bounded, and $c \geqslant 0$. If $u \in C^2(\Omega) \cap C(\overline{\Omega})$ satisfies $Lu \leqslant f$, then*

$$\sup_{\Omega} u \leqslant \sup_{\partial\Omega} u^+ + C|f|_{0;\Omega}, \tag{6.1}$$

where C depends only on n, $\frac{1}{\lambda}\sum_i |b|_{0;\Omega}$ and $\operatorname{diam}\Omega$.

Proof. The proof is divided into two steps.

(1) First, we assume that $c(x) \geqslant c_0 > 0$. Let $v = u - \sup\limits_{\partial\Omega} u^+$. Then v satisfies

$$\begin{aligned} Lv &\leqslant f - c\sup_{\partial\Omega} u^+ \leqslant f \quad \text{in } \Omega, \\ v &\leqslant 0 \quad \text{on } \partial\Omega. \end{aligned} \tag{6.2}$$

If v attains its positive maximum at some interior point $x_0 \in \Omega$, then

$$(D^2 v(x_0)) \leqslant 0, \qquad Dv(x_0) = 0,$$

where $D^2 v$ is the Hessian matrix of v. Thus

$$[-a^{ij} D_{ij} v + b^i D_i v]_{x_0} \geqslant 0.$$

It follows from (6.2) that

$$c(x_0)v(x_0) \leqslant |f|_{0;\Omega},$$

and therefore

$$\sup_{\Omega} v \leqslant \frac{|f|_{0;\Omega}}{c_0},$$

i.e.,

$$\sup_{\Omega} u \leqslant \sup_{\partial\Omega} u^+ + \frac{1}{c_0}|f|_{0;\Omega}. \tag{6.3}$$

(2) In the general case $c(x) \geqslant 0$, we let $v = zw$, where $z > 0$ is to be determined. We shall choose the auxiliary function z to reduce the problem to the special case studied in Step (1). A direct calculation shows that w satisfies

$$\begin{aligned}(6.4)\qquad & -a^{ij}D_{ij}w + \Big(b^i - \frac{2}{z}a^{ij}\frac{\partial z}{\partial x_j}\Big)D_i w \\ & + \Big[c + \frac{1}{z}(b^i D_i z - a^{ij}D_{ij}z)\Big]w \leqslant \frac{f}{z}.\end{aligned}$$

We choose z so that

$$z > 0, \quad b^i D_i z - a^{ij} D_{ij} z > 0 \quad \text{in } \overline{\Omega}.$$

Then the differential inequality (6.4) for w belongs to the special case studied in Step (1). We assume without loss of generality that Ω is contained in the strip $\{0 < x_1 < d\}$, and take

$$z = e^{2\alpha d} - e^{\alpha x_1}.$$

If α is sufficiently large, then

$$\begin{aligned}-a^{ij}D_{ij}z + b^i D_i z &= (a^{11}\alpha^2 - b^1\alpha)e^{\alpha x_1} \\ &\geqslant \lambda\Big(\alpha^2 - \frac{|b^1|_{0;\Omega}}{\lambda}\alpha\Big) > 0.\end{aligned}$$

Since $w\Big|_{\partial\Omega} \leqslant 0$, the result from Step (1) implies that

$$\sup_{\Omega} w \leqslant C|f|_{0;\Omega}.$$

(6.1) follows immediately. □

Theorem 6.2. *Under the assumptions of Theorem 6.1, if* $u \in C^2(\Omega) \cap C(\overline{\Omega})$ *is a solution of the equation* (4.3), *then*

$$(6.5)\qquad |u|_{0;\Omega} \leqslant \sup_{\partial\Omega} |u| + C|f|_{0;\Omega},$$

where C *depends on the same quantities as those in Theorem 6.1.*

Proof. We can apply Theorem 6.1 to both u and $-u$. □

Remark. If $f \equiv 0$, then (6.1) implies that "the nonnegative maximum can always be attained on $\partial\Omega$". This conclusion is weaker than the statement "if u is not a constant, then u cannot attain its nonnegative maximum in the interior of Ω". Thus Theorem 6.1 is called the *weak maximum principle.*

7. Solvability of the Dirichlet problem

Consider the Dirichlet problem

$$(7.1)\qquad -a^{ij}D_{ij}u + b^i D_i u + cu = f \quad \text{in } \Omega,$$

$$(7.2)\qquad u = \varphi \quad \text{on } \partial\Omega.$$

Lemma 7.1. *Let* $\partial\Omega \in C^{[n/2]+4}$. *Suppose that the coefficients in the equation* (7.1) *satisfy* (4.4) *and* (4.5), $c \geqslant 0$, $f \in C^\alpha(\overline{\Omega})$, *and* $\varphi \in C^{2,\alpha}(\overline{\Omega})$, *where* $0 < \alpha < 1$. *Then the problem* (7.1), (7.2) *admits a unique solution* $u \in C^{2,\alpha}(\overline{\Omega})$.

Proof. Assume without loss of generality that $\varphi \equiv 0$. Let a^{ij}_N, b^i_N, c_N, $f_N \in C^\infty(\overline{\Omega})$ $(N = 1, 2, \cdots)$ be sequences of functions which converge to a^{ij}, b^i, c, f $(i, j = 1, 2, \cdots, n)$ uniformly on $\overline{\Omega}$, and

$$\begin{aligned}
&\frac{\lambda}{2}|\xi|^2 \leqslant a^{ij}_N \xi_i \xi_j \leqslant 2\Lambda|\xi|^2, \quad \forall x \in \Omega, \xi \in \mathbb{R}^n,\\
&c_N \geqslant 0, \quad \|f_N\|_{\alpha;\Omega} \leqslant 2\|f\|_{\alpha;\Omega},\\
&\frac{1}{\lambda}\Big\{\sum_{i,j}|a^{ij}_N|_\alpha + \sum_i |b^i_N|_\alpha + |c_N|_\alpha\Big\} \leqslant 2\Lambda_\alpha.
\end{aligned}$$

We now consider the approximation problem

$$\left\{\begin{array}{ll} -a^{ij}_N D_{ij}u_N + b^i_N D_i u_N + c_N u_N = f_N & \text{in } \Omega, \\ u_N = \varphi \quad \text{on } \partial\Omega. \end{array}\right. \tag{7.3}$$

Theorem 5.4 in Chapter 1 implies that there exists a solution $u_N \in W^{[n/2]+4,2}(\Omega) \cap W^{1,2}_0(\Omega)$ for problem (7.3). By the Sobolev embedding theorem, $u_N \in C^{2,\alpha}(\overline{\Omega})$. From the global Schauder estimate (Theorem 5.3) and Theorem 6.2, we obtain

$$|u_N|_{2,\alpha;\Omega} \leqslant C|f_N|_{\alpha;\Omega} \leqslant 2C|f|_{\alpha;\Omega},$$

where C is independent of N. By the Ascoli-Arzelà theorem, there exists a subsequence $\{u_{N_k}\}$ which converges in $C^2(\overline{\Omega})$ to some function u. It is easy to verify that $u \in C^{2,\alpha}(\overline{\Omega})$ and satisfies (7.1) and (7.2). □

This lemma requires a very high regularity on $\partial\Omega$; therefore improvements are necessary. We now consider a general domain.

Definition 7.1. A domain $\Omega \subset \mathbb{R}^n$ is said to satisfy an *exterior sphere condition*, if for any $x_0 \in \partial\Omega$ there exists a sphere $B_\rho(y) \subset \mathbb{R}^n \setminus \overline{\Omega}$ such that $\overline{B_\rho(y)} \cap \overline{\Omega} = \{x_0\}$.

Theorem 7.2. *Let Ω be a bounded domain satisfying the exterior sphere condition. Suppose that the coefficients in the equation* (7.1) *satisfy* (4.4) *and* (4.5), *$c \geqslant 0$, $f \in C^\alpha(\overline{\Omega})$, and $\varphi \in C(\overline{\Omega})$, where $0 < \alpha < 1$. Then the problem* (7.1), (7.2) *admits a unique solution $u \in C^{2,\alpha}(\Omega) \cap C(\overline{\Omega})$.*

Proof. Take a sequence of domains Ω_N such that $\Omega_N \subset \Omega$, $\partial\Omega_N \in C^{[n/2]+4}$, and $\sup\limits_{x \in \partial\Omega_N} \operatorname{dist}\{x, \partial\Omega\} \leqslant 1/N$. Take also a sequence of functions $\varphi_N \in C^{2,\alpha}(\overline{\Omega})$ such that $|\varphi_N - \varphi|_{0;\Omega} \leqslant 1/N$. By Lemma 7.1, the Dirichlet problem

$$\left\{\begin{array}{ll} -a^{ij} D_{ij}u_N + b^i D_i u_N + c u_N = f & \text{in } \Omega_N, \\ u_N = \varphi_N \quad \text{on } \partial\Omega_N \end{array}\right. \tag{7.4}$$

has a unique solution $u_N \in C^{2,\alpha}(\overline{\Omega}_N)$. For any $\Omega' \subset\subset \Omega$, we can apply the interior Schauder estimate to derive, for sufficiently large N,

$$\begin{aligned}
|u_N|_{2,\alpha;\Omega'} &\leqslant C\{|f|_{\alpha;\Omega} + |u_N|_{0;\Omega}\}\\
&\leqslant C\Big\{|f|_{\alpha;\Omega} + |\varphi|_{0;\Omega} + \frac{1}{N}\Big\},
\end{aligned}$$

where C is independent of N, and the second inequality is derived from Theorem 6.2. By the Ascoli-Arzelà theorem, there exists a subsequence $\{u_{N_k}\}$ which converges in $C^2(\overline{\Omega'})$ to some function $u \in C^{2,\alpha}(\overline{\Omega'})$, for any $\Omega' \subset\subset \Omega$. By taking the limit in (7.4), we easily verify that u satisfies (7.1). It remains to prove that $u \in C(\overline{\Omega})$ satisfies the boundary condition (7.2). We use the barrier function method. For $x_0 \in \partial\Omega$, let $B_\rho(y)$ be the exterior sphere as described in Definition 7.1. A *barrier function* at x_0 is a function $\omega(x)$ with the following properties:

(1) $\omega(x_0) = 0$, $\omega(x) > 0$ for $x \in \overline{\Omega} \setminus \{x_0\}$;

(2) $\omega \in C^2(\overline{\Omega})$, $L\omega > 0$.

We shall construct a barrier function as follows:

$$\omega(x) = e^{-\beta\rho^2} - e^{-\beta|x-y|^2}, \tag{7.5}$$

where β is to be determined. It is obvious that property (1) is satisfied. Now we compute (notice that $c \geqslant 0$)

$$\begin{aligned} L\omega &\geqslant [4a^{ij}\beta^2(x_i - y_i)(x_j - y_j) - 2a^{ij}\delta_{ij}\beta + 2\beta b^i(x_i - y_i)]e^{-\beta|x-y|^2} \\ &\geqslant [4\lambda\beta^2\rho^2 - 2n\Lambda\beta - 2\beta\lambda\Lambda_\alpha\rho]e^{-\beta(d+\rho)^2}, \end{aligned}$$

where $d = \operatorname{diam}\Omega$. If β is large enough, then

$$L\omega \geqslant \theta > 0, \tag{7.6}$$

for some positive constant θ. For any $\varepsilon > 0$, there exists a neighborhood $\mathcal{N}(x_0)$ of x_0 such that

$$|\varphi(x) - \varphi(x_0)| < \varepsilon \quad \text{for } x \in \mathcal{N}(x_0) \cap \Omega.$$

Since $\omega(x)$ is bounded from below by a positive constant on $\Omega \setminus \mathcal{N}(x_0)$, we can choose C large enough (depending on ε) so that

$$C\omega(x) + \varphi(x_0) + \varepsilon > \varphi(x) > -C\omega(x) + \varphi(x_0) - \varepsilon, \quad x \in \overline{\Omega}.$$

If N is sufficiently large, then

$$C\omega(x) + \varphi(x_0) + \varepsilon \geqslant \varphi_N(x) \geqslant -C\omega(x) + \varphi(x_0) - \varepsilon, \quad x \in \overline{\Omega}.$$

From (7.6), we derive, for sufficiently large C,

$$L\Big(C\omega(x) + \varphi(x_0) + \varepsilon\Big) \geqslant Lu_N \geqslant L\Big(-C\omega(x) + \varphi(x_0) - \varepsilon\Big), \quad x \in \Omega_N.$$

Applying the weak maximum principle, we obtain

$$C\omega(x) + \varphi(x_0) + \varepsilon \geqslant u_N(x) \geqslant -C\omega(x) + \varphi(x_0) - \varepsilon, \quad x \in \Omega_N.$$

Letting $N \to \infty$, we get

$$C\omega(x) + \varphi(x_0) + \varepsilon \geqslant u(x) \geqslant -C\omega(x) + \varphi(x_0) - \varepsilon, \quad x \in \Omega.$$

Letting $x \to x_0$, we conclude that

$$\varphi(x_0) + \varepsilon \geqslant \limsup_{x \to x_0} u(x) \geqslant \liminf_{x \to x_0} u(x) \geqslant \varphi(x_0) - \varepsilon.$$

Since ε is arbitrary,

$$\lim_{x \to x_0} u(x) = \varphi(x_0).$$

The point $x_0 \in \partial\Omega$ is arbitrary. Therefore $u(x)$ extends continuously to $\partial\Omega$ and satisfies (7.2). □

The method in the above proof is very important. The application of the interior Schauder estimates and the techniques for utilizing barrier functions for the boundary conditions are both significant in general.

Theorem 7.3. *Let $\partial\Omega \in C^{2,\alpha}$ $(0 < \alpha < 1)$. Suppose that the coefficients in the equation (7.1) satisfy (4.4) and (4.5), $c \geqslant 0$, $f \in C^{\alpha}(\overline{\Omega})$, and $\varphi \in C^{2,\alpha}(\overline{\Omega})$. Then the problem (7.1), (7.2) admits a unique solution $u \in C^{2,\alpha}(\overline{\Omega})$.*

Proof. Assume without loss of generality that $\varphi \equiv 0$. Since $\partial\Omega \in C^{2,\alpha}$, the exterior sphere condition is satisfied. Therefore there exists a unique solution $u \in C^{2,\alpha}(\Omega) \cap C(\overline{\Omega})$ to the problem (7.1), (7.2). It suffices to prove that u is in $C^{2,\alpha}$ in every neighborhood of $x_0 \in \partial\Omega$. Let ϕ be the map defined in Definition 5.1 in Chapter 1, $\phi \in C^{2,\alpha}(\overline{V}), \phi^{-1} \in C^{2,\alpha}(\overline{B}_1)$. We assume without loss of generality that $u \in C^{2,\alpha}(B_1^+) \cap C(\overline{B_1^+})$ and satisfies (7.1) in B_1^+. We want to show that $u \in C^{2,\alpha}(\overline{B_{1/2}^+})$. Mollifying the singularities of ∂B_1^+ if necessary, we may assume that ∂B_1^+ is sufficiently smooth. Take $\varphi_N \in C^{2,\alpha}(\overline{B_1^+})$ such that $\varphi_N\Big|_{\partial B_1^+ \setminus \partial B_1} = 0$, and

$$|\varphi_N - u|_{0;B_1^+} \to 0 \quad \text{as } N \to \infty.$$

The boundary value problem

$$\text{(7.7)} \qquad \begin{cases} -a^{ij}D_{ij}u_N + b^i D_i u_N + cu_N = f & \text{in } B_1^+, \\ u_N = \varphi_N & \text{on } \partial B_1^+ \end{cases}$$

has a solution $u_N \in C^{2,\alpha}(\overline{B_1^+})$, by Theorem 7.1. Using the interior Schauder estimate and the maximum principle, we see that a subsequence $\{u_{N_k}\}$ of $\{u_N\}$ converges uniformly on B_1^+ to some function $\widetilde{u}$, and for any $\Omega' \subset\subset B_1^+$, $\{u_{N_k}\}$ converges in $C^2(\overline{\Omega'})$ to $\widetilde{u}$. Letting $N = N_k \to \infty$ in (7.7), we obtain

$$\begin{aligned} &-a^{ij}D_{ij}\widetilde{u} + b^i D_i \widetilde{u} + c\widetilde{u} = f \quad \text{in } B_1^+, \\ &\widetilde{u} = u \quad \text{on } \partial B_1^+. \end{aligned}$$

It follows that $\widetilde{u} \equiv u$ in B_1^+. Applying Lemma 5.2 to (7.7), we get

$$[D^2 u_N]_{\alpha;B_{1/2}^+} \leqslant C\{|f|_{\alpha;B_1^+} + |u_N|_{0;B_1^+}\} \leqslant C\{|f|_{\alpha;B_1^+} + |\varphi_N|_{0;B_1^+}\},$$

where C is independent of N. Letting $N \to \infty$, we derive

$$[D^2 u]_{\alpha;B_{1/2}^+} \leqslant C\{|f|_{\alpha;B_1^+} + |u|_{0;B_1^+}\}.$$

Hence $u \in C^{2,\alpha}(\overline{B_{1/2}^+})$. This completes the proof. □

Remark. The proof of Theorem 7.3 implies the following: Suppose that the coefficients in the equation (7.1) satisfy (4.4) and (4.5), $c \geqslant 0$, $f \in C^{\alpha}(\overline{\Omega})$, Ω satisfies the exterior sphere condition with part of its boundary $S \in C^{2,\alpha}$, and $\varphi \in C(\overline{\Omega}) \cap C^{2,\alpha}(\Omega \cup S)$. Then the Dirichlet problem (7.1), (7.2) admits a unique solution $u \in C^{2,\alpha}(\Omega) \cap C(\overline{\Omega})$; furthermore, for any $\Omega' \subset\subset \Omega \cup S$, $u \in C^{2,\alpha}(\overline{\Omega'})$.

CHAPTER 3

L^p Theory

In this chapter, we shall give interior and global L^p estimates for second order elliptic equations with continuous principal coefficients. Using this estimate, we shall establish the existence of $W^{2,p}$ solutions. The results in this chapter are very important in studying nonlinear elliptic and parabolic equations.

The assumption of the continuity of the principal coefficients is essential. When applying the result of this chapter to nonlinear equations, one should pay close attention to make sure that this assumption is satisfied.

1. The Marcinkiewicz interpolation theorem

Definition 1.1. For $f \in L^1(\Omega)$ and $t \geqslant 0$, we set

$$A_t(f) = \left\{x \in \Omega \middle| |f(x)| > t\right\}. \tag{1.1}$$

The function

$$\lambda_f(t) = |A_t(f)| = \text{meas}\left\{x \in \Omega \middle| |f(x)| > t\right\} \tag{1.2}$$

is called the *distribution function* of f.

Lemma 1.1. *Let $f \in L^p(\Omega)$ $(1 \leqslant p < \infty)$. Then*

$$\int_\Omega |f|^p dx = p \int_0^\infty t^{p-1} |A_t(f)| dt. \tag{1.3}$$

Proof. By Fubini's theorem,

$$\begin{aligned} \int_\Omega |f|^p dx &= \int_\Omega dx \int_0^{|f(x)|} pt^{p-1} dt \\ &= \int_\Omega dx \int_0^\infty pt^{p-1} \chi_{A_t} dt = p \int_0^\infty t^{p-1} \int_\Omega \chi_{A_t} dx dt, \end{aligned}$$

where χ_{A_t} is the characteristic function on the set $A_t(f)$. (1.3) follows immediately. $\square$

Definition 1.2 (Marcinkiewicz spaces). Let $p \geqslant 1$. A measurable function f is said to belong to the *weak L^p space* (denoted by $L^p_w(\Omega)$), if

$$\|f\|_{L^p_w(\Omega)} = \inf\left\{A \middle| \lambda_f(t) \leqslant t^{-p} A^p, \ \forall t > 0\right\} < \infty. \tag{1.4}$$

It should be noted that $\|\cdot\|_{L^p_w(\Omega)}$ is not a norm. Furthermore, $L^\infty_w(\Omega) = L^\infty(\Omega)$. In fact, $\|f\|_\infty = \inf\{t|\lambda_f(t) = 0\}$, and therefore $\lambda_f(t) = 0$ for $t \geqslant \|f\|_\infty$. It is also clear that $\lambda_f(t) > 0$ for $t < \|f\|_\infty$. Thus the definition (1.4) implies that $\|f\|_{L^\infty_w} = \|f\|_{L^\infty}$.

One can also prove that

$$L^p(\Omega) \subsetneqq L^p_w(\Omega) \subset L^q(\Omega), \quad \forall 1 \leqslant q < p. \tag{1.5}$$

In fact, for $f \in L^p(\Omega)$,

$$t^p \operatorname{meas} A_t(f) \leqslant \int_{A_t} |f(x)|^p dx \leqslant \int_\Omega |f(x)|^p dx.$$

By (1.4),

$$\|f\|_{L^p_w(\Omega)} \leqslant \|f\|_{L^p}.$$

Next if $f \in L^p_w(\Omega)$, then by Lemma 1.1,

$$\begin{aligned}
\int_\Omega |f|^q dx &= q\int_0^\infty t^{q-1}|A_t(f)|dt \\
&\leqslant q\int_0^1 t^{q-1}|A_t(f)|dt + q\int_1^\infty t^{q-1}|A_t(f)|dt \\
&\leqslant q|\Omega| + \|f\|_{L^p_w(\Omega)}\, q\int_1^\infty t^{q-1-p}dt \;<\; \infty.
\end{aligned}$$

Definition 1.3. A map $T : L^p(\Omega) \to L^q(\Omega)$ is said to be a *quasilinear map* if there exists a $Q > 0$ such that

$$|T(f+g)(x)| \leqslant Q(|Tf(x)| + |Tg(x)|), \quad \forall f, g \in L^p(\Omega), \text{ a.e. } x \in \Omega. \tag{1.6}$$

A quasilinear map T is said to be of *strong type* (p, q) if there exists a $C > 0$ such that

$$\|Tf\|_{L^q(\Omega)} \leqslant C\|f\|_{L^p(\Omega)}, \quad \forall f \in L^p(\Omega).$$

In this case, we denote

$$\|T\|_{(p,q)} = \sup_{f\in L^p(\Omega), f\neq 0} \frac{\|Tf\|_{L^q(\Omega)}}{\|f\|_{L^p(\Omega)}}.$$

T is said to be of *weak type* (p, q) if

$$\|Tf\|_{L^q_w(\Omega)} \leqslant C\|f\|_{L^p(\Omega)}, \quad \forall f \in L^p(\Omega).$$

From (1.5) it is clear that for $1 \leqslant p, q \leqslant \infty$, a quasilinear map of strong type (p, q) must also be of weak type (p, q).

Theorem 1.2 (Marcinkiewicz interpolation theorem). *Let $1 \leqslant p < q \leqslant \infty$. Suppose that a quasilinear map T is of both weak type (p, p) and weak type (q, q), i.e.,*

$$\|Tf\|_{L^p_w} \leqslant B_p\|f\|_{L^p}, \quad \forall f \in L^p(\Omega), \tag{1.7}$$

$$\|Tf\|_{L^q_w} \leqslant B_q\|f\|_{L^q}, \quad \forall f \in L^q(\Omega). \tag{1.8}$$

Then for any $r \in (p, q)$, T is of strong type (r, r), and

$$\|T\|_{(r,r)} \leqslant CB_p^\theta B_q^{1-\theta},$$

where C depends only on r, p, q and the constant Q in (1.6), and $\theta = \dfrac{p\,(q-r)}{r\,(q-p)}$.

Proof. Let $f \in L^r(\Omega)$. Decompose f as $f = f_1 + f_2$, where

$$f_2(x) = \begin{cases} 0 & \text{for } |f(x)| > \gamma s, \\ f(x) & \text{for } |f(x)| \leqslant \gamma s, \end{cases}$$

where γ is a constant to be determined, and $s \geqslant 0$. It is easy to verify that $f_2 \in L^q(\Omega)$, $f_1 \in L^p(\Omega)$. Since T is a quasilinear map,

$$\begin{aligned} &\text{meas}\left\{x \in \Omega \middle| |Tf(x)| > s\right\} \\ \leqslant\; &\text{meas}\left\{x \in \Omega \middle| |Tf_1(x)| > \frac{s}{2Q}\right\} \\ &+\text{meas}\left\{x \in \Omega \middle| |Tf_2(x)| > \frac{s}{2Q}\right\}. \end{aligned} \tag{1.9}$$

If $q < \infty$, then from (1.7), (1.8) and (1.9) we derive

$$\lambda_{Tf}(s) \leqslant \frac{(2QB_p)^p \|f_1\|_p^p}{s^p} + \frac{(2QB_q)^q \|f_2\|_q^q}{s^q}.$$

By Lemma 1.1,

$$\begin{aligned} \int_\Omega |Tf|^r dx &= \int_0^\infty r s^{r-1} \lambda_{Tf}(s) ds \\ &\leqslant (2QB_p)^p \int_0^\infty r s^{r-p-1} ds \int_{|f|>\gamma s} |f|^p dx \\ &\quad + (2QB_q)^q \int_0^\infty r s^{r-q-1} ds \int_{|f|\leqslant \gamma s} |f|^q dx \\ &= (2QB_p)^p r \int_\Omega |f|^p dx \int_0^{|f|/\gamma} s^{r-p-1} ds \\ &\quad + (2QB_q)^q r \int_\Omega |f|^q dx \int_{|f|/\gamma}^\infty \frac{1}{s^{q-r+1}} ds \\ &= \left[\frac{(2QB_p)^p r}{r-p} \gamma^{p-r} + \frac{(2QB_q)^q r}{q-r} \gamma^{q-r}\right] \int_\Omega |f|^r dx. \end{aligned}$$

We take $\gamma = (B_p^p B_q^{-q})^{1/(q-p)}$. Then

$$\|T\|_{(r,r)} \leqslant C B_p^\theta B_q^{1-\theta},$$

where C depends only on p, q, r and Q.

In the case $q = \infty$, we can take γ to be large enough so that the second term on the right-hand side of (1.9) vanishes. In fact, (1.8) implies that $\|Tf_2\|_\infty \leqslant B_\infty \|f_2\|_\infty \leqslant B_\infty \gamma s$; thus we can choose $\gamma = 1/(2QB_\infty)$.

Using an argument similar to that for the case $q < \infty$, we obtain

$$\|T\|_{(r,r)} \leqslant C B_p^{p/r} B_\infty^{(r-p)/r}.$$

The proof is complete. $\square$

2. A decomposition lemma

Lemma 2.1. *For $f \in L^1(\mathbb{R}^n)$, $f \geqslant 0$ and fixed $\alpha > 0$, there exist two sets F and Ω such that*

(i) $\mathbb{R}^n = F \cup \Omega$, $F \cap \Omega = \emptyset$,

(ii) $f(x) \leqslant \alpha$, *a.e.* $x \in F$,

(iii) $\Omega = \bigcup_{k=1}^{\infty} Q_k$, *where* $\{Q_k\}$ *are nonoverlapping (disjoint interior) cubes with their sides parallel to the coordinate axes such that*

$$\alpha < \fint_{Q_k} f(x)dx \leqslant 2^n\alpha \qquad (k = 1, 2, \cdots), \tag{2.1}$$

where

$$\fint_{Q_k} f(x)dx = \frac{1}{|Q_k|}\int_{Q_k} f(x)dx.$$

Proof. Since $\int_{\mathbb{R}^n} f(x)dx$ is finite, we can decompose $\mathbb{R}^n$ into congruent cubes, with the side so large that for any such cube Q',

$$\fint_{Q'} f dx \leqslant \alpha.$$

We divide each Q' into 2^n equal cubes Q''. There are only two possibilities:

Case 1: $\fint_{Q''} f(x)dx \leqslant \alpha$,

Case 2: $\fint_{Q''} f(x)dx > \alpha$.

We take those Q'' satisfying Case 2 into our family of $\{Q_k\}$ as stated in the lemma. For such a Q'', (2.1) is valid, since

$$\alpha < \fint_{Q''} f(x)dx \leqslant \frac{1}{2^{-n}|Q'|}\int_{Q'} f(x)dx \leqslant 2^n\alpha.$$

We will further divide those cubes satisfying Case 1. This process will continue until Case 2 appears. Let Ω be the union of all such cubes satisfying Case 2, and let $F = \mathbb{R}^n \setminus \Omega$. With our construction process, it is obvious that conclusions (i) and (iii) hold. Now we prove (ii). For any $x \in F$, there exists cubes $\{\widetilde{Q}_l\}$ such that

$$x \in \widetilde{Q}_l, \quad |\widetilde{Q}_l| \to 0 \quad \text{as } l \to \infty.$$

Furthermore, for each $\widetilde{Q}_l$, Case 1 is valid. Since f is integrable,

$$f(x) = \lim_{l\to\infty} \frac{1}{|\widetilde{Q}_l|}\int_{\widetilde{Q}_l} f(y)dy, \quad \text{a.e. } x \in F,$$

from which it follows that

$$f(x) \leqslant \alpha, \quad \text{a.e. } x \in F.$$

The proof is complete. □

This lemma implies that

$$|\Omega| \leqslant \frac{1}{\alpha}\|f\|_{L^1(\mathbb{R}^n)}.$$

In fact, by (2.1)

$$|\Omega| = \sum_{k=1}^{\infty} |Q_k| < \sum_{k=1}^{\infty} \frac{1}{\alpha} \int_{Q_k} f(x)dx \leqslant \frac{1}{\alpha} \|f\|_{L^1(\mathbb{R}^n)}.$$

This decomposition lemma was discovered by Calderón and Zygmund. It is very important in the study of singular integral operators and has become a fundamental tool in measure theory (cf. §2 of Chapter 6).

3. Estimates for solutions of potential equations

It is well known that the fundamental solution of the Laplace equation is given by

$$\Gamma(x) = \frac{1}{n(n-2)\omega_n} \frac{1}{|x|^{n-2}} \qquad (n \geqslant 3). \tag{3.1}$$

For $f \in C_0^\infty(\mathbb{R}^n)$, we consider the Newtonian potential with density f:

$$w(x) = \int_{\mathbb{R}^n} \Gamma(x-\xi) f(\xi) d\xi. \tag{3.2}$$

Lemma 3.1. *If $f \in C_0^\infty(\mathbb{R}^n)$, then the Newtonian potential $w \in C^\infty(\mathbb{R}^n)$ and satisfies the equation*

$$-\Delta w = f, \quad \forall x \in \mathbb{R}^n. \tag{3.3}$$

Proof. We rewrite (3.2) as

$$w(x) = \int_{\mathbb{R}^n} \Gamma(\xi) f(x-\xi) d\xi.$$

Then it is obvious that $w \in C^\infty(\mathbb{R}^n)$. Using integration by parts, we obtain

$$\begin{aligned} \Delta w(x) &= \int_{\mathbb{R}^n} \Gamma(\xi) \Delta f(x-\xi) d\xi \\ &= -\int_{\mathbb{R}^n} \frac{\partial}{\partial \xi_i} \Gamma(\xi) \frac{\partial}{\partial \xi_i} f(x-\xi) d\xi \\ &= -\lim_{\varepsilon \to 0^+} \int_{|\xi| \geqslant \varepsilon} D_i \Gamma(\xi) \frac{\partial}{\partial \xi_i} f(x-\xi) d\xi. \end{aligned}$$

Using integration by parts again, we get

$$\Delta w(x) = \lim_{\varepsilon \to 0^+} \int_{|\xi| = \varepsilon} D_i \Gamma(\xi) f(x-\xi) \frac{\xi_i}{|\xi|} dS,$$

where we made use of the fact that $\Delta\Gamma(\xi) = 0$ for $|\xi| \neq 0$. From the definition of $\Gamma(\xi)$ in (3.1), it is not difficult to compute the limit and derive (3.3). □

We now write (3.2) as $w = Nf$, where N is a linear operator from $C_0^\infty(\mathbb{R}^n)$ to $C^\infty(\mathbb{R}^n)$. For fixed i, j $(1 \leqslant i, j \leqslant n)$, we define

$$Tf = D_{ij} Nf, \tag{3.4}$$

where D_{ij} is the differential operator $\partial^2/\partial x_i \partial x_j$. It is clear that T is also a linear operator from $C_0^\infty(\mathbb{R}^n)$ to $C^\infty(\mathbb{R}^n)$.

Lemma 3.2. *T is a bounded linear operator from $L^2(\mathbb{R}^n)$ to $L^2(\mathbb{R}^n)$, and*

$$\|T\|_{(2,2)} \leqslant 1. \tag{3.5}$$

Proof. First we assume that $f \in C_0^\infty(\mathbb{R}^n)$. By (3.3), for any $B_R = B_R(0)$,

$$\int_{B_R} f^2 dx = \int_{B_R} (\Delta w)^2 dx = \sum_{ij} \int_{B_R} D_{ii} w D_{jj} w dx.$$

Using integration by parts twice in the above equality, we obtain

$$\begin{aligned} \int_{B_R} f^2 dx &= \int_{B_R} \sum_{ij} (D_{ij} w)^2 dx \\ &\quad + \int_{\partial B_R} \sum_{ij} D_i w \Big(D_{jj} w \frac{x_i}{R} - D_{ij} w \frac{x_j}{R} \Big) dS. \end{aligned} \tag{3.6}$$

Now we suppose that $\operatorname{spt} f \subset B_{R_0}$. Then for $R > 2R_0$ and $x \in \partial B_R$,

$$|D_i w| \leqslant \int_{B_{R_0}} |D_i \Gamma(x - \xi)| |f(\xi)| d\xi \leqslant \frac{C}{R^{n-1}},$$

$$|D^2 w| \leqslant \int_{B_{R_0}} |D^2 \Gamma(x - \xi)| |f(\xi)| d\xi \leqslant \frac{C}{R^n}.$$

Letting $R \to \infty$ in (3.6), we obtain

$$\int_{\mathbb{R}^n} \sum_{ij} (D_{ij} w)^2 dx = \int_{\mathbb{R}^n} f^2 dx. \tag{3.7}$$

This equality implies that

$$\|Tf\|_{L^2} \leqslant \|f\|_{L^2}, \quad \forall f \in C_0^\infty(\mathbb{R}^n).$$

Since $C_0^\infty(\mathbb{R}^n)$ is dense in $L^2(\mathbb{R}^n)$, T can be uniquely extended to a bounded linear operator on $L^2(\mathbb{R}^n)$, i.e., T is of strong type $(2, 2)$. □

Lemma 3.3. *For the fundamental solution $\Gamma(x)$, the following estimate holds:*

$$J \triangleq \sup_{\xi \neq 0,\ 1 \leqslant i,j \leqslant n} \int_{|x| \geqslant 2|\xi|} |D_{ij} \Gamma(x - \xi) - D_{ij} \Gamma(x)| dx < \infty, \tag{3.8}$$

where J depends only on n.

Proof. By the mean value theorem,

$$\int_{|x| \geqslant 2|\xi|} |D_{ij} \Gamma(x - \xi) - D_{ij} \Gamma(x)| dx \leqslant \int_{|x| \geqslant 2|\xi|} \sum_{k=1}^n |D_{ijk} \Gamma(x - \lambda \xi)|\, |\xi_k| dx,$$

where $0 < \lambda < 1$. Using the formula for $\Gamma(x)$, noticing also that $|x - \lambda \xi| \geqslant \frac{1}{2}|x|$ for $|x| \geqslant 2|\xi|$, we obtain

$$\begin{aligned} &\int_{|x| \geqslant 2|\xi|} |D_{ij} \Gamma(x - \xi) - D_{ij} \Gamma(x)| dx \\ &\leqslant \int_{|x| \geqslant 2|\xi|} \frac{C}{|x - \lambda \xi|^{n+1}} |\xi| dx \\ &\leqslant C \int_{|x| \geqslant 2|\xi|} \frac{|\xi|}{|x|^{n+1}} dx \leqslant C. \end{aligned}$$

□

Lemma 3.4. *T is of weak type $(1,1)$.*

Proof. First we assume that $f \in C_0^\infty(\mathbb{R}^n)$. Applying the Calderón–Zygmund decomposition lemma to $|f(x)|$, we get two sets F and Ω such that

$$\begin{aligned}
&\mathbb{R}^n = F \cup \Omega, \quad F \cap \Omega = \emptyset,\\
&|f(x)| \leqslant \alpha, \quad \text{a.e. } x \in F,\\
&\Omega = \bigcup_{k=1}^{\infty} Q_k, \quad |\Omega| = \sum_{k=1}^{\infty} |Q_k| \leqslant \frac{1}{\alpha}\|f\|_{L^1},\\
&⨍_{Q_k} |f| dx \leqslant 2^n \alpha.
\end{aligned}$$

Define

$$g(x) = \begin{cases} f(x), & x \in F, \\ ⨍_{Q_k} f(\xi) d\xi, & x \in Q_k \qquad (k = 1, 2, \cdots), \end{cases} \tag{3.9}$$

and

$$b(x) = f(x) - g(x).$$

It is easy to verify that g and b satisfy

$$|g(x)| \leqslant 2^n \alpha, \qquad \text{a.e. } x \in \mathbb{R}^n, \tag{3.10}$$

$$\|g\|_{L^1} \leqslant \|f\|_{L^1}, \qquad \|b\|_{L^1} \leqslant 2\|f\|_{L^1}, \tag{3.11}$$

$$b(x) = 0 \text{ for } x \in F, \qquad ⨍_{Q_k} b(x) dx = 0 \quad (k = 1, 2, \cdots). \tag{3.12}$$

Since $f \in C_0^\infty(\mathbb{R}^n)$, both g and b belong to $L^2(\mathbb{R}^n)$. Clearly, $Tf = Tg + Tb$. For any $\alpha > 0$,

$$\lambda_{Tf}(\alpha) \leqslant \lambda_{Tg}\Big(\frac{\alpha}{2}\Big) + \lambda_{Tb}\Big(\frac{\alpha}{2}\Big). \tag{3.13}$$

By Lemma 3.2, T is of strong type $(2,2)$. Therefore it must also be of weak type $(2,2)$. Using also (3.10), (3.11), we have

$$\lambda_{Tg}\Big(\frac{\alpha}{2}\Big) \leqslant \frac{4}{\alpha^2}\|g\|_{L^2}^2 \leqslant \frac{4}{\alpha^2}\|g\|_{L^\infty}\|g\|_{L^1} \leqslant \frac{2^{n+2}}{\alpha}\|f\|_{L^1}. \tag{3.14}$$

Next, we estimate $\lambda_{Tb}\Big(\frac{\alpha}{2}\Big)$. Let Q_k^* be a cube with the same center as Q_k and its side length $2\sqrt{n}$ times that of Q_k. Set

$$\Omega^* = \bigcup_{k=1}^{\infty} Q_k^*, \quad F^* = \mathbb{R}^n \setminus \Omega^*. \tag{3.15}$$

Obviously,

$$|\Omega^*| \leqslant (2\sqrt{n})^n |\Omega| \leqslant \frac{(2\sqrt{n})^n}{\alpha}\|f\|_{L^1}. \tag{3.16}$$

Set

$$b_k(x) = b(x)\chi_{Q_k} = \begin{cases} b(x), & x \in Q_k, \\ 0, & \text{elsewhere.} \end{cases}$$

Then by (3.12)

$$\int_{Q_k} b_k(x)dx = 0.$$

We now take a sequence of functions $b_k^l \in C_0^\infty(Q_k)$ $(l = 1, 2, \cdots)$ such that

$$\|b_k^l - b_k\|_{L^2} \to 0 \text{ as } l \to \infty, \qquad \int_{Q_k} b_k^l(x)dx = 0. \tag{3.17}$$

By the definition of T, for $x \in (Q_k^*)^c = \mathbb{R}^n \setminus Q_k^*$,

$$\begin{aligned} Tb_k^l(x) &= D_{ij}\int_{Q_k} \Gamma(x-\xi)b_k^l(\xi)d\xi \\ &= \int_{Q_k} [D_{ij}\Gamma(x-\xi) - D_{ij}\Gamma(x-x^k)]b_k^l(\xi)d\xi, \end{aligned}$$

where x^k is the center of Q_k. In deriving the above equality, we made use of (3.17). Integrating x over $(Q_k^*)^c$, using also Lemma 3.3, we obtain

$$\begin{aligned} &\int_{(Q_k^*)^c} |Tb_k^l(x)|dx \\ &\leqslant \sup_{\xi\in Q_k}\int_{(Q_k^*)^c} |D_{ij}\Gamma(x-\xi) - D_{ij}\Gamma(x-x^k)|dx \cdot \int_{Q_k} |b_k^l(\xi)|d\xi \\ &\leqslant J\int_{Q_k} |b_k^l(\xi)|d\xi. \end{aligned}$$

Clearly, as $l \to \infty$, b_k^l and Tb_k^l converge in $L^2(\mathbb{R}^n)$ to b_k and Tb_k respectively. By Fatou's lemma, we can let $l \to \infty$ in the above inequality to derive

$$\int_{(Q_k^*)^c} |Tb_k(x)|dx \leqslant J\int_{Q_k} |b_k(\xi)|d\xi = J\int_{Q_k} |b(\xi)|d\xi.$$

Therefore

$$\int_{F^*} |Tb(x)|dx \leqslant \sum_{k=1}^{\infty}\int_{(Q_k^*)^c} |Tb_k(x)|dx \leqslant J\|b\|_{L^1} \leqslant 2J\|f\|_{L^1},$$

which implies that

$$\text{meas}\left\{x \in F^* \middle| |Tb(x)| > \frac{\alpha}{2}\right\} \leqslant \frac{4}{\alpha}J\|f\|_{L^1}.$$

By (3.16),

$$\begin{aligned} \lambda_{Tb}\Big(\frac{\alpha}{2}\Big) &\leqslant |\Omega^*| + \text{meas}\left\{x \in F^* \middle| |Tb(x)| > \frac{\alpha}{2}\right\} \\ &\leqslant [4J + (2\sqrt{n})^n]\frac{\|f\|_{L^1}}{\alpha}. \end{aligned} \tag{3.18}$$

Substituting (3.14), (3.18) into (3.13), we get

$$\lambda_{Tf}(\alpha) \leqslant [2^{n+2} + 4J + (2\sqrt{n})^n]\frac{\|f\|_{L^1}}{\alpha}, \quad \forall f \in C_0^\infty(\mathbb{R}^n). \tag{3.19}$$

Since $C_0^\infty(\mathbb{R}^n)$ is dense in $L^1(\mathbb{R}^n)$, the above inequality holds also for any $f \in L^1(\mathbb{R}^n)$. (3.19) implies that T is of weak type $(1,1)$. □

Theorem 3.5. *For $1 < p < \infty$, T is of strong type (p,p).*

Proof. By Lemmas 3.2 and 3.4, T is both of strong type $(2,2)$ and of weak type $(1,1)$. If follows from the Marcinkiewicz interpolation theorem that T is of strong type (p,p), for any $1 < p \leqslant 2$.

For $2 < p < \infty$, we set $p' = p/(p-1)$. For any $f, h \in C_0^\infty(\mathbb{R}^n)$

$$\begin{aligned}\int_{\mathbb{R}^n} Tf(x)h(x)dx &= \int_{\mathbb{R}^n} h(x)D_{ij}\left[\int_{\mathbb{R}^n}\Gamma(x-\xi)f(\xi)d\xi\right]dx\\ &= \int_{\mathbb{R}^n} D_{ij}h(x)dx\int_{\mathbb{R}^n}\Gamma(x-\xi)f(\xi)d\xi\\ &= \int_{\mathbb{R}^n} f(\xi)d\xi\int_{\mathbb{R}^n}\Gamma(x-\xi)D_{ij}h(x)dx\\ &= \int_{\mathbb{R}^n} f(\xi)D_{ij}\left[\int_{\mathbb{R}^n}\Gamma(x-\xi)h(x)dx\right]d\xi\\ &\leqslant \|f\|_{L^p}\|Th\|_{L^{p'}}.\end{aligned}$$

Notice that $1 < p' < 2$. Thus the previous result (the case $1 < p \leqslant 2$) implies that

$$\int_{\mathbb{R}^n} Tf(x)h(x)dx \leqslant C\|f\|_{L^p}\|h\|_{L^{p'}}, \quad \forall f, h \in C_0^\infty(\mathbb{R}^n).$$

Therefore

$$\|Tf\|_{L^p} \leqslant C\|f\|_{L^p}, \quad \forall f \in C_0^\infty(\mathbb{R}^n),$$

which implies that T is of strong type (p,p). □

The above method is due primarily to Calderón and Zygmund. The simplified proof for potential integrals was given by Gilbarg and Trudinger.

Theorem 3.6. *Suppose that $u \in W_0^{2,p}(B_R)$ satisfies*

$$-\Delta u = f.$$

Then for $1 < p < \infty$, there exists a constant C, depending only on n, p, such that

$$\|D^2u\|_{L^p(B_R)} \leqslant C\|f\|_{L^p(B_R)}.$$

Proof. Assume without loss of generality that $u \in C_0^\infty(\mathbb{R}^n)$. By Lemma 3.1,

$$\begin{aligned}u(x) &= \int_{\mathbb{R}^n}\Gamma(x-\xi)(-\Delta u(\xi))d\xi\\ &= \int_{\mathbb{R}^n}\Gamma(x-\xi)f(\xi)d\xi.\end{aligned}$$

Applying Theorem 3.5, we obtain the theorem immediately. □

4. Interior $W^{2,p}$ estimates

Consider the boundary value problem

$$Lu = -a^{ij}D_{ij}u + b^iD_iu + cu = f \quad \text{in } \Omega, \tag{4.1}$$

$$u = 0 \quad \text{on } \partial\Omega. \tag{4.2}$$

Assume that the coefficients in (4.1) satisfy

$$a^{ij}\xi_i\xi_j \geqslant \lambda|\xi|^2, \quad \forall x \in \Omega, \xi \in \mathbb{R}^n, \quad \lambda > 0, \tag{4.3}$$

$$\sum_{ij}\|a^{ij}\|_{L^\infty(\Omega)} + \sum_i \|b^i\|_{L^\infty(\Omega)} + \|c\|_{L^\infty(\Omega)} \leqslant \Lambda, \tag{4.4}$$

$$a^{ij} \in C(\overline{\Omega}) \quad (i, j = 1, 2, \cdots, n). \tag{4.5}$$

The method given here is similar to that for the interior Schauder estimates.

Lemma 4.1. *Let the assumptions (4.3)–(4.5) be in force. Then there exists $R_0 > 0$, depending only on $n, p, \Lambda/\lambda$ and the modulus of continuity of a^{ij}, such that for any $0 < R \leqslant R_0$ and any $u \in W_0^{2,p}(B_R)$ $(1 < p < \infty)$ satisfying (4.1) almost everywhere, the following estimate holds:*

$$\|D^2u\|_{L^p(B_R)} \leqslant C\Big\{\frac{1}{\lambda}\|f\|_{L^p(B_R)} + R^{-2}\|u\|_{L^p(B_R)}\Big\}, \tag{4.6}$$

where C depends only on $n, p, \Lambda/\lambda$.

Proof. Assume without loss of generality that $\lambda = 1$. Let B_R be a ball centered at x_0. We freeze the coefficients at x_0 and rewrite (4.1) as

$$-a^{ij}(x_0)D_{ij}u = \widetilde{f}, \tag{4.7}$$

where

$$\widetilde{f} = f + [a^{ij}(x) - a^{ij}(x_0)]D_{ij}u - b^iD_iu - cu.$$

For the constant coefficients elliptic equation (4.7), the result of Theorem 3.6 holds. Therefore,

$$\|D^2u\|_{L^p} \leqslant C\|\widetilde{f}\|_{L^p}, \tag{4.8}$$

where C depends only on n, p and Λ/λ, $\lambda = 1$. Denote the modulus of continuity of a^{ij} by

$$\omega(R) = \sup_{|x-y|\leqslant R,\ 1\leqslant i,j\leqslant n} |a^{ij}(x) - a^{ij}(y)|. \tag{4.9}$$

By the assumption (4.5), $\omega(R) \to 0$ as $R \to 0$. Thus (4.8) implies that

$$\|D^2u\|_{L^p} \leqslant C\{\|f\|_{L^p} + \omega(R)\|D^2u\|_{L^p} + \|u\|_{W^{1,p}}\}.$$

We take R_0 so that $C\omega(R) \leqslant 1/2$ for $0 < R \leqslant R_0$. Then

$$\|D^2u\|_{L^p} \leqslant C\{\|f\|_{L^p} + \|u\|_{W^{1,p}}\} \quad \text{for } 0 < R \leqslant R_0.$$

Applying the interpolation inequality in Sobolev spaces, we immediately obtain the lemma. □

Theorem 4.2. *Let the assumptions (4.3)–(4.5) be in force. Suppose that $u \in W^{2,p}_{loc}(\Omega)$ satisfies (4.1) almost everywhere. Then for any $\Omega' \subset\subset \Omega$,*

$$\|u\|_{W^{2,p}(\Omega')} \leqslant C\Big\{\frac{1}{\lambda}\|f\|_{L^p(\Omega)} + \|u\|_{L^p(\Omega)}\Big\},$$

where C depends only on $n, p, \Lambda/\lambda$, $\operatorname{dist}\{\Omega', \partial\Omega\}$ and the modulus of continuity of a^{ij}.

Proof. Assume without loss of generality that $\lambda = 1$. Let R_0 be the constant in Lemma 4.1 and $\overline{R}_0 = \min\Big\{R_0, \frac{1}{2}\operatorname{dist}\{\Omega', \partial\Omega\}\Big\}$. For any $x_0 \in \Omega'$ and $\overline{R}_0/2 \leqslant \rho < R \leqslant \overline{R}_0$, we let $\zeta(x)$ be a cutoff function on $B_R(x_0)$ such that

$$\begin{aligned} &\zeta \in C_0^\infty(B_R(x_0)), \qquad \zeta(x) = 1 \quad \text{for } x \in B_\rho(x_0), \\ &|D^k\zeta| \leqslant \frac{C}{(R-\rho)^k} \quad (k = 1, 2). \end{aligned} \tag{4.10}$$

The function $v = \zeta u$ satisfies the equation

$$-a^{ij}D_{ij}v + b^iD_iv + cv = \widetilde{f},$$

where

$$\widetilde{f} = \zeta f + [-a^{ij}D_{ij}\zeta + b^iD_i\zeta]u - 2a^{ij}D_i\zeta D_ju.$$

By Lemma 4.1,

$$\begin{aligned} \|D^2v\|_{L^p(B_R(x_0))} &\leqslant C\Big[\|\widetilde{f}\|_{L^p(B_R(x_0))} + \frac{1}{R^2}\|v\|_{L^p(B_R(x_0))}\Big] \\ &\leqslant C\Big[\|f\|_{L^p(B_R(x_0))} + \frac{1}{(R-\rho)}\|Du\|_{L^p(B_R(x_0))} + \frac{1}{(R-\rho)^2}\|u\|_{L^p(B_R(x_0))}\Big]. \end{aligned}$$

Denote $B_R = B_R(x_0)$. Using the interpolation inequalities, we obtain

$$\|D^2u\|_{L^p(B_\rho)} \leqslant C\Big[\varepsilon\|D^2u\|_{L^p(B_R)} + \|f\|_{L^p(B_R)} + \frac{C_\varepsilon}{(R-\rho)^2}\|u\|_{L^p(B_R)}\Big].$$

By Lemma 4.1 in Chapter 2,

$$\|D^2u\|_{L^p(B_\rho)} \leqslant C\Big[\|f\|_{L^p(B_R)} + \frac{1}{(R-\rho)^2}\|u\|_{L^p(B_R)}\Big] \qquad \text{for } \frac{\overline{R}_0}{2} \leqslant \rho < R \leqslant \overline{R}_0.$$

We take $R = \overline{R}_0$, $\rho = \overline{R}_0/2$, and then cover Ω' with finitely many balls of radius $\overline{R}_0/2$ to obtain the theorem. □

5. Global $W^{2,p}$ estimates

The method for the global $W^{2,p}$ estimates is similar to that for the Schauder estimates. Therefore we shall only outline the proof in terms of several lemmas.

Lemma 5.1. *Let $u \in W^{2,p}(B_R^+) \cap W_0^{1,p}(B_R^+)$, where $B_R^+ = B_R(0) \cap \{x_n > 0\}$. Suppose that u vanishes in a neighborhood of $\partial B_R^+ \cap \{x_n > 0\}$, and satisfies*

$$-\Delta u = f, \quad a.e.\ x \in B_R^+. \tag{5.1}$$

Then

$$\|D^2u\|_{L^p(B_R^+)} \leqslant C\|f\|_{L^p(B_R^+)}, \tag{5.2}$$

where C depends only on n, p.

Proof. Let

$$\widetilde{u} = \begin{cases} u(x', x_n) & \text{for } x_n \geqslant 0, \\ -u(x', -x_n) & \text{for } x_n < 0. \end{cases}$$

Then $\widetilde{u} \in W_0^{2,p}(B_R)$ and satisfies the equation

$$-\Delta\widetilde{u} = \widetilde{f}$$

almost everywhere. By Theorem 3.6,

$$\|D^2\widetilde{u}\|_{L^p(B_R^+)} \leqslant C\|\widetilde{f}\|_{L^p(B_R^+)} \leqslant 2C\|f\|_{L^p(B_R^+)},$$

from which (5.2) follows. □

Lemma 5.2. *Let the assumptions* (4.3)–(4.5) *be in force. Suppose that Ω has a flat part of the boundary S, $\Omega \subset \mathbb{R}^n_+$, and $S \subset \partial\mathbb{R}^n_+$. Then there exist R_0 and C, depending only on $n, p, \Lambda/\lambda$ and the modulus of continuity of a^{ij}, such that for any $0 < R \leqslant R_0$ with $B_R^+ \subset \Omega$ (B_R is a ball with its center on S) and any solution $u \in W^{2,p}(B_R^+) \cap W_0^{1,p}(B_R^+)$ of* (4.1) *which vanishes in a neighborhood of $\partial B_R^+ \cap \{x_n > 0\}$, the following estimate is valid:*

$$\|D^2u\|_{L^p(B_R^+)} \leqslant C\Big\{\frac{1}{\lambda}\|f\|_{L^p(B_R^+)} + R^{-2}\|u\|_{L^p(B_R^+)}\Big\}. \tag{5.3}$$

Lemma 5.3. *Under the assumptions of Lemma 5.2, if $u \in W^{2,p}(\Omega)$ satisfies* (4.1) *almost everywhere in Ω and $u = 0$ on S, then for any $\Omega' \subset\subset \Omega \cup S$, we have*

$$\|u\|_{W^{2,p}(\Omega')} \leqslant C\Big\{\frac{1}{\lambda}\|f\|_{L^p(\Omega)} + \|u\|_{L^p(\Omega)}\Big\}, \tag{5.4}$$

where C depends only on $n, p, \Lambda/\lambda, \operatorname{dist}\{\Omega', \partial\Omega \setminus S\}$ and the modulus of continuity of a^{ij}.

Finally, we have the global $W^{2,p}$ estimates.

Theorem 5.4. *Let $\partial\Omega \in C^{1,1}$, and let the assumptions* (4.3)–(4.5) *be in force. If $u \in W^{2,p}(\Omega) \cap W_0^{1,p}(\Omega)$ satisfies* (4.1) *almost everywhere in Ω, then*

$$\|u\|_{W^{2,p}(\Omega)} \leqslant C\Big\{\frac{1}{\lambda}\|f\|_{L^p(\Omega)} + \|u\|_{L^p(\Omega)}\Big\}, \tag{5.5}$$

where C depends only on $n, p, \Lambda/\lambda, \Omega$ and the modulus of continuity of a^{ij}.

We want to emphasize that $W^{2,p}$ estimates depend on the modulus of continuity of a^{ij}; one must be very careful when applying them.

Remark. Consider now the nonhomogeneous boundary condition

$$u = \varphi \quad \text{on } \partial\Omega, \tag{5.6}$$

where $\varphi \in W^{2,p}(\Omega)$. If $u \in W^{2,p}(\Omega)$, $u - \varphi \in W_0^{1,p}(\Omega)$ and satisfies (4.1) almost everywhere, then u is said to be a solution (or *strong solution*) to the Dirichlet problem (4.1), (5.6). Set

$$\|\varphi\|_{W^{2-1/p,p}(\partial\Omega)} = \inf\left\{\|\Phi\|_{W^{2,p}(\Omega)}\Big|\Phi \in W^{2,p}(\Omega),\ \Phi - \varphi \in W_0^{1,p}(\Omega)\right\}.$$

Then a strong solution $u \in W^{2,p}(\Omega)$ of (4.1), (5.6) satisfies the estimate

$$\|u\|_{W^{2,p}(\Omega)} \leqslant C\left\{\|f\|_{L^p(\Omega)} + \|\varphi\|_{W^{2-1/p,p}(\partial\Omega)} + \|u\|_{L^p(\Omega)}\right\}.$$

6. Existence of $W^{2,p}$ solutions

We need the maximum principle for strong solutions. Based on the concepts of normal mappings, the proof was first given by Alexandroff, Bakelman and Pucci. The proof is long and therefore is postponed to Chapter 6. Here we just state the result.

Theorem 6.1. *Let the assumptions* (4.3), (4.4) *be in force, and* $c \geqslant 0$. *If* $u \in C(\overline{\Omega}) \cap W_{loc}^{2,n}(\Omega)$ *satisfies* (4.1) *almost everywhere on* Ω, *then*

$$\sup_{\Omega}|u| \leqslant \sup_{\partial\Omega}|u| + \frac{C}{\lambda}\|f\|_{L^n(\Omega)}, \tag{6.1}$$

where C *depends only on* $n, \Lambda/\lambda$ *and* $\operatorname{diam}\Omega$.

This theorem, together with the $W^{2,p}$ estimate, gives us the solvability of the Dirichlet problem (4.1), (5.6) in $W^{2,p}(\Omega)$ $(p \geqslant n)$.

Lemma 6.2. *Let* $\partial\Omega \in C^{2,\alpha}$ $(\alpha > 0)$. *Suppose that the coefficients in* (4.1) *satisfy* (4.3)–(4.5), $c \geqslant 0$, $S \subset \partial\Omega$ *is open relative to* $\partial\Omega$, $\varphi \in W_{loc}^{2,p}(\Omega) \cap C(\overline{\Omega})$, $\varphi = 0$ *on* S, *and* $f \in L^p(\Omega)$ $(p \geqslant n)$. *Under these assumptions, problem* (4.1), (5.6) *admits a solution* $u \in W_{loc}^{2,p}(\Omega) \cap C(\overline{\Omega})$; *furthermore, for any* $\Omega' \subset\subset \Omega \cup S$, $u \in W^{2,p}(\Omega')$.

Proof. Step 1. First, we assume that $S = \partial\Omega$. Take approximating sequences $a_N^{ij}, b_N^i, c_N, f_N \in C^\alpha(\overline{\Omega})$ such that a_N^{ij}, b_N^i, c_N satisfy (4.3)–(4.5), a_N^{ij} have their moduli of continuity uniformly in N, and

$$\begin{aligned} &f_N \to f \quad (L^p(\Omega)), \qquad a_N^{ij} \to a^{ij} \quad (C(\overline{\Omega})), \\ &b_N^i \xrightarrow{w*} b^i, \quad c_N \xrightarrow{w*} c \quad (L^\infty(\Omega)), \end{aligned} \tag{6.2}$$

where $\xrightarrow{w*}$ means weak star convergence. Consider the approximating problem

$$\begin{aligned} &-a_N^{ij}D_{ij}u_N + b_N^iD_iu_N + c_Nu_N = f_N \quad \text{in } \Omega, \\ &u_N = 0 \quad \text{on } \partial\Omega. \end{aligned} \tag{6.3}$$

By §7 of Chapter 2, (6.3) admits a solution $u_N \in C^{2,\alpha}(\overline{\Omega})$. Clearly $u_N \in W^{2,p}(\Omega) \cap W_0^{1,p}(\Omega)$. By the $W^{2,p}$ estimate and (6.1),

$$\|u_N\|_{W^{2,p}(\Omega)} \leqslant \frac{C}{\lambda}\|f_N\|_{L^p(\Omega)},$$

where C is independent of N. By the weak compactness in $W^{2,p}(\Omega)$, $\{u_N\}$ has a subsequence which converges weakly to some $u \in W^{2,p}(\Omega)$. It is easy to verify that $u \in W^{2,p}(\Omega) \cap W_0^{1,p}(\Omega)$ satisfies (4.1) almost everywhere.

Step 2. Let S be a nontrivial subset of $\partial\Omega$. We take a sequence of functions $\varphi_N \in C^2(\overline{\Omega})$ such that $\varphi_N = 0$ on S and $\varphi_N \to \varphi$ $(C(\overline{\Omega}))$. The equation (4.1) with the boundary condition

$$u\Big|_{\partial\Omega} = \varphi_N \tag{6.4}$$

admits a solution $u_N \in W^{2,p}(\Omega)$, $u_N - \varphi_N \in W_0^{1,p}(\Omega)$, by Step 1 (consider the equation for $u - \varphi_N$). By the $W^{2,p}$ local estimates near the boundary (Lemma 5.3),

$$\|u_N\|_{W^{2,p}(\Omega')} \leqslant C(\|f\|_{L^p(\Omega)} + \|u_N\|_{L^p(\Omega)}), \tag{6.5}$$

where $\Omega' \subset\subset \Omega \cup S$, and C is independent of N. By (6.1),

$$\sup |u_N - u_{N'}| \leqslant \sup |\varphi_N - \varphi_{N'}| \quad \text{for any } N, N'.$$

It follows that u_N converges uniformly to some function $u \in C(\overline{\Omega})$. By the interior $W^{2,p}$ estimates and diagonal process, there exists a subsequence of u_N which converges weakly in $W^{2,p}_{loc}(\Omega)$ to some function u. It follows that $u \in W^{2,p}_{loc}(\Omega) \cap C(\overline{\Omega})$ is a solution. By (6.5), $u \in W^{2,p}(\Omega')$ for any $\Omega' \subset\subset \Omega \cup S$. The proof is complete. □

Now we consider a class of elliptic operators

$$\mathscr{L} = \{L | L = -a^{ij}D_{ij} + b^i D_i + c \text{ satisfies (4.3)–(4.5), } c \geqslant 0\}, \tag{6.6}$$

where we rewrite (4.5) as

$$\sup_{1 \leqslant i,j \leqslant n} \Big(\sup_{|x-y| \leqslant R} |a^{ij}(x) - a^{ij}(y)| \Big) \leqslant \omega(R), \qquad \omega(R) \to 0 \text{ as } R \to 0.$$

Theorem 6.3. *Let Ω be a bounded domain with $\partial\Omega \in C^{1,1}$. Suppose that for any $L \in \mathscr{L}$, the Dirichlet problem* (4.1), (4.2) *admits at most one solution in $W^{2,p}(\Omega)$, $1 < p < \infty$. Then for any $L \in \mathscr{L}$, the solution $u \in W^{2,p}(\Omega)$ of* (4.1), (4.2) *satisfies the estimate*

$$\|u\|_{W^{2,p}(\Omega)} \leqslant \frac{C}{\lambda} \|f\|_{L^p(\Omega)}, \tag{6.7}$$

where C depends only on $n, p, \Lambda/\lambda, \Omega$ and $\omega(R)$.

Furthermore, if $f \in L^p(\Omega)$ $(1 < p < \infty)$, then the Dirichlet problem (4.1), (4.2) *admits a solution $u \in W^{2,p}(\Omega) \cap W_0^{1,p}(\Omega)$.*

Proof. Assume without loss of generality that $\lambda = 1$. If (6.7) does not hold, then for any N, there exist $L_N = -a_N^{ij}D_{ij} + b_N^i D_i + c_N \in \mathscr{L}$, $f_N \in L^p(\Omega)$, $u_N \in W^{2,p}(\Omega) \cap W_0^{1,p}(\Omega)$ such that $L_N u_N = f_N$, $\|u_N\|_{L^p(\Omega)} = 1$, and

$$\|u_N\|_{W^{2,p}(\Omega)} \geqslant N \|f_N\|_{L^p(\Omega)}.$$

By the $W^{2,p}$ estimates,

$$\|u_N\|_{W^{2,p}(\Omega)} \leqslant C\{\|f_N\|_{L^p(\Omega)} + \|u_N\|_{L^p(\Omega)}\} \leqslant \frac{C}{N}\|u_N\|_{W^{2,p}(\Omega)} + C,$$

where C is independent of N. It follows that

$$\|u_N\|_{W^{2,p}(\Omega)} \leqslant C \quad \text{when } N \geqslant 2C.$$

Therefore a subsequence of $\{u_N\}$ converges weakly in $W^{2,p}(\Omega)$ to some function $u \in W^{2,p}(\Omega)$. Similarly, there are further subsequences of a_N^{ij}, b_N^i, c_N and f_N which satisfy (6.2). It is not difficult to verify that $u \in W^{2,p}(\Omega) \cap W_0^{1,p}(\Omega)$ satisfies $Lu = 0$, $\|u\|_{L^p} = 1$. This contradicts the uniqueness assumptions. The proof for (6.7) is complete.

Next, we prove the existence of a $W^{2,p}$ solution. Just as in the first step in the proof of Lemma 6.2, we consider the approximation (6.3). Since $\partial\Omega \in C^{1,1}$, the exterior sphere condition is satisfied. Therefore a solution $u_N \in C^{2,\alpha}(\Omega) \cap C(\overline{\Omega})$ exists. We want to prove that $u_N \in W^{2,p}(\Omega)$; once this is proved, we can apply the global $W^{2,p}$ estimates and obtain the theorem immediately. Thus it suffices to prove that u_N belongs to $W^{2,p}$ in a neighborhood of every boundary point. Let $x_0 \in \partial\Omega$. Since $\partial\Omega \in C^{1,1}$, there exist a neighborhood V of x_0 and a map ϕ as in Definition 5.1. Mollifying the singularities of ∂B_1^+, we may assume without loss of generality that ∂B_1^+ is sufficiently smooth. Let $y = \phi(x)$ and $\widetilde{u}_N(y) = u_N \circ \phi^{-1}(y)$. Then u_N satisfies

$$-\widetilde{a}_N^{rs} D_{rs}\widetilde{u}_N + \widetilde{b}_N^r D_r\widetilde{u}_N + \widetilde{c}_N\widetilde{u}_N = \widetilde{f}_N,$$

where

$$\widetilde{a}_N^{rs}(y) = a_N^{ij}\frac{\partial y_r}{\partial x_i}\frac{\partial y_s}{\partial x_j}, \quad \widetilde{b}_N^r(y) = a_N^{ij}\frac{\partial^2 y_r}{\partial x_i \partial x_j} + b_N^i\frac{\partial y_r}{\partial x_i},$$

$$\widetilde{c}_N = c_N \circ \phi^{-1}(y), \quad \widetilde{f}_N = f_N \circ \phi^{-1}(y).$$

Since f_N is continuous on $\overline{\Omega}$, $\widetilde{f}_N \in L^q(B_1^+)$, where $q = \max\{n, p\}$. By Theorem 6.1 and Lemma 6.2, $\widetilde{u}_N \in W^{2,q}(B_{1/2}^+)$. Changing the variables back to x, we conclude that u_N belongs to $W^{2,p}$ in a neighborhood of x_0. The proof is complete. □

Although Theorem 6.3 gives the existence of a solution in $W^{2,p}$, it is proved under the uniqueness assumption. Thus by Theorem 6.1, the solvability problem is complete for $p \geqslant n$. We next give the uniqueness in $W^{2,p}$ for $p < n$.

Theorem 6.4. *Suppose that $\Omega \in C^{1,1}$, $L \in \mathscr{L}$. Then for $1 < p < \infty$, the Dirichlet problem* (4.1), (4.2) *admits at most one solution.*

Proof. Step 1. We first prove that there exists $\sigma > 0$ such that the zero solution is the only solution in $W^{2,p}(\Omega) \cap W_0^{1,p}(\Omega)$ for the equation

$$Lu + \sigma u = 0. \tag{6.8}$$

Assume without loss of generality that $\lambda = 1$. We use the method of dimension increment. Denote a point in $n+1$ dimensional space $\mathbb{R}^{n+1}$ by $(x_1, x_2, \cdots, x_n, t)$, and set

$$\widetilde{\Omega} = \Omega \times (-1, 1) \subset \mathbb{R}^{n+1}.$$

Consider the operator

$$\widetilde{L} = L - D_{tt}.$$

If u is a solution of (6.8), then it is easy to verify that $v(x,t) = [\cos(\sigma^{1/2}t)]u(x)$ is a solution of $\widetilde{L}v = 0$. Now let $\widetilde{\Omega}' = \Omega \times (-1/2, 1/2)$. By Lemma 5.3,

$$\|v\|_{W^{2,p}(\widetilde{\Omega}')} \leqslant C\|v\|_{L^p(\widetilde{\Omega})} \leqslant C\|u\|_{L^p(\Omega)},$$

where C depends only on $n, p, \Lambda/\lambda, \Omega$ and the modulus of continuity of a^{ij}; it is independent of σ. In particular,

$$\|D_{tt}v\|_{L^p(\widetilde{\Omega}')} \leqslant C\|u\|_{L^p(\Omega)},$$

i.e.,

$$\sigma\|u\|_{L^p(\Omega)}\left(\int_{-1/2}^{1/2} |\cos\sigma^{1/2}t|^p dt\right)^{1/p} \leqslant C\|u\|_{L^p(\Omega)}.$$

We introduce a change of integration variable $\tau = \sigma^{1/2}t$ ($\sigma \geqslant 1$). Then

$$\sigma^{1-\frac{1}{2p}}\|u\|_{L^p(\Omega)}\left(\int_{-1/2}^{1/2} |\cos\tau|^p d\tau\right)^{1/p} \leqslant C\|u\|_{L^p(\Omega)}.$$

If σ is sufficiently large (depending only on $n, p, \Lambda/\lambda, \Omega$ and modulus of continuity of a^{ij}), then $u = 0$.

We now consider a class of operators $\widetilde{\mathscr{L}} = \{L + \sigma \mid L \in \mathscr{L}\}$. Analogously to Theorem 6.3, if $f \in L^p(\Omega)$, then the Dirichlet problem

$$Lu + \sigma u = f, \quad \text{a.e. } x \in \Omega,$$
$$u \in W^{2,p}(\Omega) \cap W_0^{1,p}(\Omega)$$

admits a solution.

Step 2. Suppose that $u \in W^{2,p}(\Omega) \cap W_0^{1,p}(\Omega)$ ($p < n$) is a solution of $Lu = 0$. We want to show that $u = 0$. By the Sobolev embedding theorem, $u \in L^q(\Omega)$, where $\frac{1}{q} = \frac{1}{p} - \frac{1}{n}$. Let σ be the constant determined in Step 1. Clearly,

$$Lu + \sigma u = f,$$

where $f = \sigma u \in L^q(\Omega)$. From Step 1 we derive that $u \in W^{2,q}(\Omega)$. If q is still less than n, we shall continue the above process to derive $u \in W^{2,q_2}(\Omega)$, where $\frac{1}{q_2} = \frac{1}{q} - \frac{1}{n} = \frac{1}{p} - \frac{2}{n}$. After finitely many iterations, we conclude that $u \in W^{2,n}(\Omega)$. Therefore, $u = 0$, by Theorem 6.1. □

CHAPTER 4

De Giorgi-Nash-Moser Estimates

In both the Schauder and $W^{2,p}$ estimates, the leading order coefficients are assumed to be continuous. Because of this assumption, it is often difficult to apply these estimates directly to a nonlinear equation. In 1957, De Giorgi established Hölder norm estimates for solutions of elliptic equations with measurable coefficients. In 1958, Nash independently obtained a similar estimate for parabolic equations. This was a breakthrough for the study of quasilinear elliptic and parabolic equations. In 1960, Moser gave a simplified proof for this estimate and also established the Harnack inequality; his method has become a classical method for studying regularity of weak solutions.

1. Local properties of weak solutions

In this section we study the following equation of divergence form:

$$-D_j(a^{ij}D_iu)+b^iD_iu+cu=0, \tag{1.1}$$

where we assume that

$$\lambda|\xi|^2\leqslant a^{ij}\xi_i\xi_j\leqslant\Lambda|\xi|^2,\quad \forall x\in\Omega,\ \xi\in\mathbb{R}^n, \tag{1.2}$$

$$\sum_{i,j}\|a^{ij}\|_{L^\infty(\Omega)}+\sum_i\|b^i\|_{L^\infty(\Omega)}+\|c\|_{L^\infty(\Omega)}\leqslant\Lambda. \tag{1.3}$$

For convenience in computation and for clarity in presentation of the method, we consider only a special case:

$$-D_j(a^{ij}(x)D_iu)=0. \tag{1.1$'$}$$

We also assume that $n\geqslant 3$. The method, however, applies to the equation (1.1) and the case $n=2$.

Lemma 1.1. *Suppose $\Phi(s)\in C^{0,1}_{loc}(\mathbb{R})$ is a convex function ($\Phi(\sigma s_1+(1-\sigma)s_2)\leqslant\sigma\Phi(s_1)+(1-\sigma)\Phi(s_2)$ for $0<\sigma<1$).*

(1) *If u is a weak subsolution of (1.1)$'$, $\Phi'(s)\geqslant 0$ and $\Phi(u)\in H^1(\Omega)$, then $v=\Phi(u)$ is also a weak subsolution of (1.1)$'$.*

(2) *If u is a weak supersolution of (1.1)$'$, $\Phi'(s)\leqslant 0$ and $\Phi(u)\in H^1(\Omega)$, then $v=\Phi(u)$ is a weak subsolution of (1.1)$'$.*

Proof. We prove only (1). First assume that $\Phi \in C^2_{loc}(\mathbb{R})$. Let $v = \Phi(u)$. By assumption, $\Phi'(s) \geqslant 0$, and u is a weak subsolution of $(1.1)'$. Thus for any $\varphi \in C_0^\infty(\Omega)$, $\varphi \geqslant 0$,

$$\begin{aligned}\int_\Omega a^{ij} D_i v D_j \varphi dx &= \int_\Omega \Phi'(u) a^{ij} D_i u D_j \varphi dx \\ &= \int_\Omega a^{ij} D_i u D_j [\Phi'(u)\varphi] dx \\ &\qquad - \int_\Omega a^{ij} D_i u D_j u \Phi''(u) \varphi dx \\ &\leqslant 0.\end{aligned}$$

It follows that v is a weak subsolution of $(1.1)'$.

For a general convex $\Phi(s)$, we use its mollified function $\Phi_\tau(s)$. We still have $\Phi'_\tau(s) \geqslant 0$ and $\Phi_\tau(s)$ is convex. Thus $v_\tau = \Phi_\tau(u)$ is a weak subsolution of $(1.1)'$, i.e., for any $\varphi \in C_0^\infty(\Omega)$, $\varphi \geqslant 0$,

$$0 \geqslant \int_\Omega a^{ij} D_i v_\tau D_j \varphi dx = \int_\Omega \Phi'_\tau(u) a^{ij} D_i u D_j \varphi dx.$$

By the dominate convergence theorem, as $\tau \to 0$,

$$0 \geqslant \int_\Omega a^{ij} D_i v D_j \varphi dx.$$

The proof is complete. $\square$

As a special case, if u is a weak subsolution of $(1.1)'$, so is u^+.

Lemma 1.2 (Local maximum principle). *Let the assumption* (1.2) *be in force. If $v \in W^{1,2}(B_R)$ is a bounded weak subsolution of* $(1.1)'$, *then for any $p > 0$ and $0 < \theta < 1$,*

$$\operatorname*{ess\,sup}_{B_{\theta R}} v \leqslant C\left(\fint_{B_R} (v^+)^p dx \right)^{1/p}, \tag{1.4}$$

where C depends only on $n, \Lambda/\lambda, p$ and $(1-\theta)^{-1}$; and

$$\fint_{B_R} (v^+)^p dx = \frac{1}{|B_R|} \int_{B_R} (v^+)^p dx.$$

Proof. First we assume that $p \geqslant 2$. By Lemma 1.1, v^+ is also a weak subsolution of $(1.1)'$. Therefore we can assume without loss of generality that $v \geqslant 0$. v satisfies

$$\int_{B_R} a^{ij} D_i v D_j \varphi dx \leqslant 0, \quad \forall \varphi \in W_0^{1,2}(B_R),\ \varphi \geqslant 0.$$

Let $\zeta \in C_0^\infty(B_R)$, and take $\varphi = \zeta^2 v^{p-1}$. Then

$$\int_{B_R} a^{ij} D_i v D_j (\zeta^2 v^{p-1}) dx \leqslant 0,$$

which implies that

$$(p-1)\int_{B_R}(a^{ij}D_ivD_jv)v^{p-2}\zeta^2dx \leqslant -2\int_{B_R}a^{ij}v^{p-1}\zeta D_ivD_j\zeta dx.$$

By Cauchy's inequality

$$(p-1)^2\int_{B_R}\zeta^2v^{p-2}|Dv|^2dx \leqslant C\int_{B_R}|D\zeta|^2v^pdx,$$

where C depends only on $n, \Lambda/\lambda$. Since $v^{p-2}|Dv|^2 = \dfrac{4}{p^2}|D(v^{p/2})|^2$,

$$\int_{B_R}\zeta^2|D(v^{p/2})|^2dx \leqslant C\int_{B_R}|D\zeta|^2v^pdx.$$

It easily follows that

$$\int_{B_R}|D(\zeta v^{p/2})|^2dx \leqslant C\int_{B_R}|D\zeta|^2v^pdx.$$

By the Sobolev embedding theorem

$$\left(\int_{B_R}(\zeta v^{p/2})^{2^*}dx\right)^{2/2^*} \leqslant C\int_{B_R}|D\zeta|^2v^pdx, \tag{1.5}$$

where $2^* = 2n/(n-2)$; C depends only on $n, \Lambda/\lambda$ and is independent of p. Set

$$R_k = R\left(\theta + \frac{1-\theta}{2^k}\right) \qquad (k = 0, 1, 2, \cdots), \tag{1.6}$$

where $\theta \in (0,1)$. Choose $\zeta_k \in C_0^\infty(B_{R_k})$ such that $0 \leqslant \zeta_k \leqslant 1$, $\zeta_k(x) = 1$ on $B_{R_{k+1}}$, and

$$|D\zeta_k| \leqslant \frac{2}{R_k - R_{k+1}} = \frac{2^{k+1}}{(1-\theta)R}.$$

By letting $\zeta = \zeta_k$ and $B_R = B_{R_k}$ in (1.5), we get

$$\left(\int_{B_{R_{k+1}}}v^{np/(n-2)}\right)^{(n-2)/n} \leqslant \frac{C4^k}{(1-\theta)^2R^2}\int_{B_{R_k}}v^pdx. \tag{1.7}$$

Set $p_k = p\left(\dfrac{n}{n-2}\right)^k$ $(k = 0, 1, 2, \cdots)$. Choosing $p = p_k$ in (1.7) and taking p_k-th root, we obtain

$$\|v\|_{L^{p_{k+1}}(B_{R_{k+1}})} \leqslant \left[\frac{C4^k}{(1-\theta)^2R^2}\right]^{1/p_k}\|v\|_{L^{p_k}(B_{R_k})}.$$

After iteration, we find that

$$\|v\|_{L^{p_{k+1}}(B_{R_{k+1}})} \leqslant \left[\frac{C}{(1-\theta)^2R^2}\right]^{\sum 1/p_k} 4^{\sum k/p_k}\|v\|_{L^p(B_R)}.$$

Notice that $\sum \dfrac{k}{p_k}$ is convergent and $\displaystyle\sum_{k=0}^{\infty}\frac{1}{p_k} = \frac{n}{2p}$. Therefore we can rewrite the above inequality as

$$\|v\|_{L^{p_{k+1}}(B_{R_{k+1}})} \leqslant \frac{C}{[(1-\theta)R]^{n/p}}\|v\|_{L^p(B_R)}.$$

Letting $k \to \infty$, we find that

$$\operatorname*{ess\,sup}_{B_{\theta R}} v \leqslant \frac{C}{[(1-\theta)R]^{n/p}} \|v\|_{L^p(B_R)}. \tag{1.8}$$

The lemma is proved for $p \geqslant 2$.

Now assume that $0 < p < 2$. We choose $p = 2$ in (1.8); then

$$\operatorname*{ess\,sup}_{B_{\theta R}} v \leqslant \frac{C}{[(1-\theta)R]^{n/2}} \Big(\operatorname*{ess\,sup}_{B_R} v\Big)^{1-p/2} \Big(\int_{B_R} v^p dx\Big)^{1/2}.$$

By Young's inequality,

$$\operatorname*{ess\,sup}_{B_{\theta R}} v \leqslant \frac{1}{2} \operatorname*{ess\,sup}_{B_R} v + \frac{C}{[(1-\theta)R]^{n/p}} \|v\|_{L^p(B_R)}.$$

Let $\varphi(s) = \operatorname*{ess\,sup}_{B_s} v$. We choose $s = \theta R$ and $t = R$ in the above inequality. Then

$$\varphi(s) \leqslant \frac{1}{2}\varphi(t) + \frac{C}{(t-s)^{n/p}} \|v\|_{L^p(B_R)}, \quad \forall 0 < s < t \leqslant R.$$

By Lemma 4.1 in Chapter 2,

$$\varphi(\theta R) \leqslant \frac{C}{[(1-\theta)R]^{n/p}} \|v\|_{L^p(B_R)}.$$

The proof is complete. □

The proof of the above lemma is called the Moser iteration. Both the Moser and De Giorgi iterations (§4, Chapter 1) are important tools in a priori estimates.

Remark. In fact, the boundedness assumption on the weak solution in the above lemma can be removed; see [GT, Chapter 8] for details.

Lemma 1.3 (Weak Harnack inequality)**.** *Let the assumption* (1.2) *be in force on* $B_{\sigma R}$ ($\sigma > 1$). *Suppose that* v *is a nonnegative, bounded weak supersolution of* (1.1)$'$. *Then there exist* $p_0 > 0$, $C > 0$ *such that*

$$\operatorname*{ess\,inf}_{B_{\theta R}} v \geqslant \frac{1}{C} \Big(\fint_{B_R} v^{p_0} dx \Big)^{1/p_0}, \quad \forall 0 < \theta < 1, \tag{1.9}$$

where p_0, C *depend only on* $n, \Lambda/\lambda$, $(\sigma-1)^{-1}$ *and* $(1-\theta)^{-1}$.

Proof. Replacing v with $v + \varepsilon$ if necessary, we can assume without loss of generality that $v \geqslant \varepsilon > 0$. We also assume without loss of generality that $R = 1$. By Lemma 1.1 (2), v^{-1} is a bounded weak subsolution of (1.1)$'$. Lemma 1.2 implies that

$$\operatorname*{ess\,sup}_{B_\theta} v^{-p} \leqslant C \int_{B_1} v^{-p} dx.$$

It follows that

$$\begin{aligned} \operatorname*{ess\,inf}_{B_\theta} v &\geqslant C^{-1/p} \Big(\int_{B_1} v^{-p} dx \Big)^{-1/p} \\ &= C^{-1/p} \Big[\int_{B_1} v^{-p} dx \int_{B_1} v^p dx \Big]^{-1/p} \Big[\int_{B_1} v^p dx \Big]^{1/p}. \end{aligned}$$

It is clear that to complete the proof of (1.9), it suffices to establish that

$$\int_{B_1} v^{-p} dx \int_{B_1} v^p dx \leqslant C \tag{1.10}$$

for some $p > 0$. Let $w = \log v - \beta$, where β is to be determined. We want to establish, for some $p > 0$, that

$$\int_{B_1} e^{p|w|} dx \leqslant C, \tag{1.11}$$

which implies that

$$\int_{B_1} e^{p(\beta - \log v)} dx \leqslant C \quad \text{and} \quad \int_{B_1} e^{p(\log v - \beta)} dx \leqslant C.$$

The product of the above two inequalities gives (1.10). Now the problem reduces to establishing (1.11).

By assumption, v is a weak supersolution of (1.1)$'$ on B_σ. For any $\varphi \in W_0^{1,2}(\Omega)$, $\varphi \geqslant 0$, we choose the test function as $v^{-1}\varphi$. Then

$$\int_{B_\sigma} a^{ij} D_i v D_j (v^{-1}\varphi) dx \geqslant 0.$$

A simple calculation shows that

$$\int_{B_\sigma} a^{ij} D_i w D_j \varphi dx - \int_{B_\sigma} (a^{ij} D_i w D_j w) \varphi dx \geqslant 0, \ \forall \varphi \in W_0^{1,2}(\Omega), \ \varphi \geqslant 0. \tag{1.12}$$

Let $\overline{\sigma} = (1+\sigma)/2$. We take a cutoff function $\zeta \in C_0^\infty(B_\sigma)$ such that $\zeta = 1$ on $B_{\overline{\sigma}}$. We choose $\varphi = \zeta^2$ in (1.12) and use Cauchy's inequality to obtain

$$\int_{B_\sigma} \zeta^2 a^{ij} D_i w D_j w dx \leqslant 8 \int_{B_\sigma} a^{ij} D_i \zeta D_j \zeta dx.$$

Using the ellipticity condition (1.2), we derive

$$\int_{B_{\overline{\sigma}}} |Dw|^2 dx \leqslant C, \tag{1.13}$$

where C depends only on $n, \Lambda/\lambda$ and $(\sigma - 1)^{-1}$. If we take $\beta = \fint_{B_{\overline{\sigma}}} \log v dx$, then by Poincaré's inequality,

$$\int_{B_{\overline{\sigma}}} |w|^2 dx \leqslant C. \tag{1.14}$$

Next we want to estimate $\|w\|_{L^q(B_1)}$ for any integer $q \geqslant 2$. Although the calculations are tedious, the basic idea is the Moser iteration. We choose the test function $\varphi = \zeta^2 |w|^{2q}$ $(q \geqslant 1)$ in (1.12), where $\zeta \in C_0^\infty(B_{\overline{\sigma}})$. Then

$$\begin{aligned} \int_{B_\sigma} \zeta^2 |w|^{2q} a^{ij} D_i w D_j w dx \leqslant 2q \int_{B_\sigma} \zeta^2 |w|^{2q-1} a^{ij} D_i w D_j |w| dx \\ + \int_{B_\sigma} 2\zeta |w|^{2q} a^{ij} D_i w D_j \zeta dx. \end{aligned} \tag{1.15}$$

By Young's inequality,

$$2q|w|^{2q-1} \leqslant \frac{2q-1}{2q} |w|^{2q} + (2q)^{2q-1}.$$

Substituting this inequality into (1.15) (using also the inequality $a^{ij}\xi_i\eta_j \leqslant 2qa^{ij}\eta_i\eta_j + [1/(8q)]a^{ij}\xi_i\xi_j$), we get

$$\begin{aligned}&\frac{1}{2q}\int_{B_\sigma}\zeta^2|w|^{2q}a^{ij}D_iwD_jwdx\\ &\quad\leqslant (2q)^{2q-1}\int_{B_\sigma}a^{ij}D_iwD_jwdx + \frac{1}{4q}\int_{B_\sigma}\zeta^2|w|^{2q}a^{ij}D_iwD_jwdx\\ &\quad\quad +4q\int_{B_\sigma}a^{ij}|w|^{2q}D_i\zeta D_j\zeta dx.\end{aligned}$$

Using the ellipticity condition (1.2) and the estimate (1.13) in the above inequality, we obtain

$$\lambda\int_{B_\sigma}\zeta^2|w|^{2q}|Dw|^2dx \leqslant C\Lambda(2q)^{2q} + 16\Lambda q^2\int_{B_\sigma}|w|^{2q}|D\zeta|^2dx, \tag{1.16}$$

where C is independent of q. For $\delta \geqslant 1$, $\tau > 0$, $\delta+\tau \leqslant \overline{\sigma}$, we choose $\zeta(x)$ to be a cutoff function on $B_{\delta+\tau}$ such that

$$\begin{aligned}&\zeta \in C_0^\infty(B_{\delta+\tau}), \quad 0 \leqslant \zeta \leqslant 1,\\ &\zeta = 1 \quad \text{for } x \in B_\delta, \quad |D\zeta| \leqslant \frac{2}{\tau}.\end{aligned}$$

By Cauchy's inequality and Young's inequality,

$$\begin{aligned}|D(\zeta^2|w|^{2q})| &\leqslant 2q\zeta^2|w|^{2q-1}|D|w|\,| + 2\zeta|D\zeta|\,|w|^{2q}\\ &\leqslant \zeta^2|w|^{2q}|Dw|^2 + q^2\zeta^2|w|^{2q-2} + 4\tau^{-1}|w|^{2q}\\ &\leqslant \zeta^2|w|^{2q}|Dw|^2 + \zeta^2|w|^{2q} + \zeta^2q^{2q} + 4\tau^{-1}|w|^{2q}.\end{aligned}$$

Substituting this inequality into (1.16), we obtain

$$\int_{B_{\overline{\sigma}}}|D(\zeta^2|w|^{2q})|dx \leqslant C(2q)^{2q} + C\tau^{-2}q^2\int_{B_{\delta+\tau}}|w|^{2q}dx,$$

where C depends only on n, Λ/λ and $(\overline{\sigma}-1)^{-1}$. Denote $\kappa = n/(n-1)$. By the Sobolev embedding theorem,

$$\left(\int_{B_\delta}|w|^{2q\kappa}dx\right)^{1/\kappa} \leqslant C(2q)^{2q} + C\tau^{-2}q^2\int_{B_{\delta+\tau}}|w|^{2q}dx. \tag{1.17}$$

Let

$$q_i = \kappa^{i-1}, \quad \delta_0 = \overline{\sigma}, \quad \delta_i = \delta_{i-1} - \frac{\overline{\sigma}-1}{2^i} \quad (i = 1, 2, \cdots).$$

Then (1.17) gives

$$\left(\int_{B_{\delta_i}}|w|^{2\kappa^i}dx\right)^{1/\kappa} \leqslant C\kappa^{2(i-1)\kappa^{i-1}} + C(4\kappa)^i\int_{B_{\delta_{i-1}}}|w|^{2\kappa^{i-1}}dx.$$

Taking the $2k^{i-1}$-th root on both sides, for

$$I_j = \left(\int_{B_{\delta_j}}|w|^{2\kappa^j}dx\right)^{1/(2\kappa^j)}$$

we obtain the following inequality:

$$I_i \leqslant C^{1/\kappa^{i-1}}\kappa^{i-1} + C^{1/\kappa^{i-1}}(4\kappa)^{i/\kappa^{i-1}}I_{i-1} \qquad (i = 1, 2, \cdots).$$

By iteration,

$$I_j \leqslant C^{\sum 1/\kappa^{i-1}} (4\kappa)^{\sum i/\kappa^{i-1}} \left(\sum_{i=1}^{j} \kappa^{i-1} + I_0 \right) \qquad (j = 1, 2, \cdots).$$

Notice that $\sum_{i=1}^{j} \kappa^{i-1} \leqslant C\kappa^j$. Therefore

$$I_j \leqslant C\kappa^j + CI_0 \qquad (j = 1, 2, \cdots).$$

For each integer $q \geqslant 2$, there exists j such that $2\kappa^{j-1} \leqslant q \leqslant 2\kappa^j$. By Hölder's inequality and (1.14),

$$\|w\|_{L^q(B_1)} \leqslant CI_j \leqslant Cq + CI_0 \leqslant \widetilde{C}q, \quad q \geqslant 2.$$

Since $q^q \leqslant e^q q!$,

$$\int_{B_1} |w|^q dx \leqslant \widetilde{C}^q q^q \leqslant (\widetilde{C}e)^q q!, \quad q \geqslant 2.$$

Thus for $p = (2\widetilde{C}e)^{-1}$,

$$\int_{B_1} \frac{(p|w|)^q}{q!} dx \leqslant 2^{-q}, \quad q \geqslant 2.$$

Summing over all q's, we conclude that

$$\int_{B_1} e^{p|w|} dx \leqslant C,$$

which is just what we want to prove. □

Theorem 1.4 (Harnack inequality). *Let the assumption* (1.2) *be in force on* B_R. *Suppose that* u *is a nonnegative, bounded weak solution of* $(1.1)'$ *on* B_R. *Then for any* $0 < \theta < 1$,

$$\operatorname*{ess\,sup}_{B_{\theta R}} u \leqslant C \operatorname*{ess\,inf}_{B_{\theta R}} u, \tag{1.18}$$

where C *depends only on* $n, \Lambda/\lambda$ *and* $(1-\theta)^{-1}$.

Proof. Let $R_1 = \dfrac{(1+\theta)}{2} R$. By Lemma 1.2,

$$\operatorname*{ess\,sup}_{B_{\theta R}} u \leqslant C \left[\fint_{B_{R_1}} |u|^{p_0} dx \right]^{1/p_0},$$

where p_0 is the constant in Lemma 1.3. By Lemma 1.3,

$$\operatorname*{ess\,inf}_{B_{\theta R}} u \geqslant C^{-1} \left[\fint_{B_{R_1}} |u|^{p_0} dx \right]^{1/p_0}.$$

Combining the above two inequalities, we get (1.18). □

2. Interior Hölder continuity

The following auxiliary lemma is frequently used in studying Hölder continuity.

Lemma 2.1. *Let $\omega(P)$ be a nonnegative, nondecreasing function on $[0, R_0]$. Suppose that for some $0 < \theta, \eta < 1$, $0 < \alpha \leqslant 1$ and $K \geqslant 0$, ω satisfies*

$$\omega(\theta R) \leqslant \eta\omega(R) + KR^\alpha, \quad \forall 0 < R \leqslant R_0. \tag{2.1}$$

Then there exist $0 < \gamma \leqslant \alpha$, $C > 0$, such that

$$\omega(R) \leqslant C\left(\frac{R}{R_0}\right)^\gamma [\omega(R_0) + KR_0^\alpha], \tag{2.2}$$

where γ, C depend only on θ, η, α.

Proof. Let $\widetilde{R}_0 \in (\theta R_0, R_0]$ and $R_s = \theta^s \widetilde{R}_0$ $(s = 1, 2, \cdots)$. By (2.1),

$$\omega(R_{s+1}) \leqslant \eta\omega(R_s) + KR_s^\alpha \quad (s = 1, 2, \cdots). \tag{2.3}$$

Assume without loss of generality that $\theta^{-\alpha}\eta > 1$, for otherwise we may take η closer to 1. We iterate (2.3) to obtain

$$\begin{aligned} \omega(R_s) &\leqslant \eta^s\omega(\widetilde{R}_0) + \sum_{m=0}^{s-1} K\eta^m R_{s-m-1}^\alpha \\ &\leqslant \eta^s\omega(\widetilde{R}_0) + K\widetilde{R}_0^\alpha \theta^{\alpha(s-1)} \sum_{m=0}^{s-1} \eta^m \theta^{-\alpha m} \\ &\leqslant \eta^s\omega(\widetilde{R}_0) + K\widetilde{R}_0^\alpha \theta^{\alpha(s-1)} \frac{(\theta^{-\alpha}\eta)^s}{\theta^{-\alpha}\eta - 1} \\ &\leqslant \eta^s[\omega(R_0) + CKR_0^\alpha], \end{aligned}$$

where C depends only on θ, η, α. Since $s = \log_\theta \dfrac{R_s}{\widetilde{R}_0}$,

$$\omega(R_s) \leqslant \left(\frac{R_s}{\widetilde{R}_0}\right)^{\log_\theta \eta} [CKR_0^\alpha + \omega(R_0)].$$

Let $\gamma = \log\eta / \log\theta$. Then

$$\omega(R_s) \leqslant \left(\frac{R_s}{\widetilde{R}_0}\right)^{\gamma} [CKR_0^\alpha + \omega(R_0)].$$

If $\widetilde{R}_0$ runs over the interval $(\theta R_0, R_0]$, then R_s $(s = 0, 1, 2, \cdots)$ runs over the entire interval $[0, R_0]$. The proof is complete. □

Theorem 2.2. *Let the assumption (1.2) be in force. Suppose that u is a bounded weak solution of (1.1)′. Then there exist $C \geqslant 0$ and $0 < \gamma < 1$ such that for any $B_R(x) \subset \Omega$,*

$$\operatorname*{ess\,osc}_{B_R(x)} u \leqslant C\left(\frac{R}{d_x}\right)^\gamma \operatorname*{ess\,osc}_{B_{d_x}(x)} u, \tag{2.4}$$

where $d_x = \operatorname{dist}\{x, \partial\Omega\}$, C *and* γ *depend only on* n *and* Λ/λ, *and*

$$\operatorname*{ess\,osc}_A u = \operatorname*{ess\,sup}_A u - \operatorname*{ess\,inf}_A u.$$

Proof. Let $M(R) = \operatorname*{ess\,sup}_{B_R(x_0)} u$, $m(R) = \operatorname*{ess\,inf}_{B_R(x_0)} u$ and $\omega(R) = M(R) - m(R)$, where $0 < R \leqslant d_{x_0}$. Then the function $v = u - m(R)$ is a bounded, nonnegative weak solution of (1.1)$'$. By Theorem 1.4, for any $0 < \theta < 1$,

$$C \operatorname*{ess\,inf}_{B_{\theta R}} v \geqslant \operatorname*{ess\,sup}_{B_{\theta R}} v,$$

i.e.,

$$C[m(\theta R) - m(R)] \geqslant [M(\theta R) - m(R)].$$

Thus

$$\begin{aligned} \omega(\theta R) &= M(\theta R) - m(\theta R) \\ &\leqslant M(\theta R) - \left[m(R) + \frac{1}{C}\Big(M(\theta R) - m(R)\Big)\right] \\ &\leqslant \Big(1 - \frac{1}{C}\Big)M(\theta R) - m(R)\Big(1 - \frac{1}{C}\Big) \\ &\leqslant \Big(1 - \frac{1}{C}\Big)\omega(R). \end{aligned}$$

Applying Lemma 2.1, we get (2.4) immediately. □

In fact, the estimate of the form (2.4) implies the interior Hölder continuity.

Corollary. *Under the assumptions of Theorem 2.2, u is continuous in Ω (here we actually mean that u is equal almost everywhere to a function which is continuous in Ω). Furthermore, there exist $C > 0$ and $0 < \gamma < 1$ such that*

$$|u(x) - u(y)| \leqslant \left(\frac{|x-y|}{d_{x,y}}\right)^{\gamma} \|u\|_{L^\infty(\Omega)} \quad \text{for } x, y \in \Omega, \tag{2.5}$$

where $d_{xy} = \min\{d_x, d_y\}$, *and* C *depends only on* n *and* Λ/λ.

Proof. First we prove that u is continuous in Ω. For any $x \in \Omega$, $0 < \delta \leqslant d_x$, define

$$u_\delta(x) = \frac{1}{|B_\delta|}\int_{B_\delta(x)} u(z)dz.$$

For any δ and $\overline{\delta}$ with $0 < \delta \leqslant \overline{\delta} \leqslant d_x$, we apply (2.4) to obtain

$$\begin{aligned} |u_\delta(x) - u_{\overline{\delta}}(x)| &\leqslant \frac{1}{|B_{\overline{\delta}}|}\int_{B_{\overline{\delta}}(x)} |u(z) - u_\delta(x)|dz \\ &\leqslant \operatorname*{ess\,osc}_{B_{\overline{\delta}}(x)} u \leqslant C\left(\frac{\overline{\delta}}{d_x}\right)^{\gamma} \|u\|_{L^\infty(\Omega)}. \end{aligned}$$

Therefore $u_\delta(x)$ converges to a function $\widetilde{u}(x)$ uniformly on any compact subset of Ω as $\delta \to 0$. It is clear that $u_\delta(x)$ is continuous for fixed δ. Hence $\widetilde{u}(x)$ is continuous in Ω. By the Lebesgue differentiation theorem, $u_\delta(x) \to u(x)$ (a.e. $x \in \Omega$); it follows that $u(x) = \widetilde{u}(x)$ (a.e. $x \in \Omega$). We can identify $u(x)$ with $\widetilde{u}(x)$, and thus $u(x)$ itself is continuous in Ω.

We now prove (2.5). If $|x-y| \geqslant \frac{1}{2} d_{x,y}$, then clearly

$$|u(x)-u(y)| \leqslant 2\|u\|_{L^\infty}\left(\frac{2|x-y|}{d_{xy}}\right)^\gamma .$$

If $|x-y| \leqslant \frac{1}{2} d_{xy}$, then by (2.4) (assuming that $d_{x,y}=d_x$)

$$|u(x)-u(y)| \leqslant \operatorname*{osc}_{B_{2|x-y|}(x)} u \leqslant C\left(\frac{|x-y|}{d_x}\right)^\gamma \|u\|_{L^\infty}.$$

The proof is complete. □

Now we consider the nonhomogeneous equation

$$-D_j(a^{ij}D_iu)+b^iD_iu+cu=f+D_if^i. \tag{2.6}$$

For simplicity, we consider only

$$-D_j(a^{ij}D_iu)=f+D_if^i. \tag{2.6$'$}$$

Theorem 2.3. *Suppose that the coefficients of* (2.6)$'$ *satisfy* (1.2), $f \in L^{q_*}(\Omega)$, $f^i \in L^q(\Omega)$ *for some* $q>n$, *where* $q_*=nq/(n+q)$. *If* u *is a weak solution of* (2.6)$'$, *then there exist* $C>0$ *and* $0<\gamma<1$ *such that for* $B_R(x) \subset \Omega$,

$$\operatorname*{ess\,osc}_{B_R(x)} u \leqslant C\left(\frac{R}{d_x}\right)^\gamma \left[\operatorname*{ess\,osc}_{B_{d_x}(x)} u + d_x^\gamma\left(\|f\|_{L^{q_*}(\Omega)}+\sum_i \|f^i\|_{L^q(\Omega)}\right)\right], \tag{2.7}$$

where C *and* γ *depend only on* n, q *and* $\operatorname{diam}\Omega$.

Proof. For $B_R=B_R(x_0)\subset\Omega$, we consider the Dirichlet problem

$$\begin{cases} -D_j(a^{ij}D_iv)=f+D_if^i & \text{in } B_R,\\ v=0 \quad \text{on } \partial B_R. \end{cases}$$

By Chapter 1, for the above equation, a weak solution $v \in W_0^{1,2}(B_R)$ exists. Furthermore, by the maximum principle for weak solutions,

$$\|v\|_{L^\infty(B_R)} \leqslant CF_0|B_R(x_0)|^{(1/n)-(1/q)} \leqslant CR^{1-n/q}F_0, \tag{2.8}$$

where $F_0=\|f\|_{L^{q_*}(\Omega)}+\sum_i \|f^i\|_{L^q(\Omega)}$. Obviously, the function $w=u-v$ is a weak solution of

$$-D_j(a^{ij}D_iw)=0.$$

Denote by $\omega_u(R)$, $\omega_w(R)$, $\omega_v(R)$ the oscillation of u, w, v on $B_R(x_0)$, respectively. By Theorem 2.2, for fixed $0<\theta<1$,

$$\omega_w(\theta R) \leqslant \left(1-\frac{1}{C}\right)\omega_w(R), \quad 0<R\leqslant d_{x_0}. \tag{2.9}$$

Combining (2.8) and (2.9), we get

$$\begin{aligned}\omega_u(\theta R) &\leqslant \omega_w(\theta R) + \omega_v(\theta R)\\ &\leqslant \left(1 - \frac{1}{C}\right)\omega_w(R) + \omega_v(\theta R)\\ &\leqslant \left(1 - \frac{1}{C}\right)\omega_u(R) + 4\|v\|_{L^\infty(B_R)}\\ &\leqslant \left(1 - \frac{1}{C}\right)\omega_u(R) + CF_0R^{1-n/q}.\end{aligned}$$

By Lemma 2.1, there exist $C > 0$ and $0 < \gamma < 1$ such that

$$\omega_u(\theta R) \leqslant C\left(\frac{R}{R_0}\right)^\gamma [\omega_u(R_0) + F_0R_0^\gamma],$$

where $R_0 = d_{x_0}$. The proof is complete. □

3. Global Hölder continuity

In order to obtain Hölder estimates near the boundary, some assumptions on the boundary of Ω are necessary.

Definition 3.1. We say that Ω satisfies a *uniform exterior cone condition* if there exists a cone V_h of height h such that for any point $x_0 \in \partial\Omega$, there exists a cone congruent to V_h with vertex x_0 and lying outside Ω.

Let $x_0 \in \partial\Omega$. We set

$$u_M^+ = \begin{cases} \max\{u, M\} & \text{for } x \in B_R(x_0) \cap \Omega,\\ M & \text{for } x \in B_R(x_0) \setminus \Omega,\end{cases} \tag{3.1}$$

$$u_m^- = \begin{cases} \min\{u, m\} & \text{for } x \in B_R(x_0) \cap \Omega,\\ m & \text{for } x \in B_R(x_0) \setminus \Omega.\end{cases} \tag{3.2}$$

Lemma 3.1. *Suppose that v is a bounded weak subsolution of* (1.1)′, $x_0 \in \partial\Omega$, $M = \sup\limits_{\partial\Omega \cap B_R(x_0)} v^+$. *Then for any $p > 0$, $0 < \theta < 1$,*

$$\sup_{B_{\theta R}(x_0)} v_M^+ \leqslant C\left[⨍_{B_R(x_0)} (v_M^+)^p dx\right]^{1/p}, \tag{3.3}$$

where C depends only on $n, \Lambda/\lambda$, p and $(1-\theta)^{-1}$.

Proof. We choose the test function $\varphi = \zeta^2[v^{p-1} - M^{p-1}]^+$, where ζ is a cutoff function on $B_R(x_0)$. Clearly, $\varphi \in W_0^{1,2}(\Omega)$ and $\varphi \geqslant 0$. The remaining proof is similar to that of Lemma 1.2. □

Lemma 3.2. *Suppose that v is a bounded, nonnegative weak supersolution of* (1.1)′, *Ω satisfies a uniform exterior cone condition, and (a^{ij}) satisfies* (1.2). *For*

$x_0 \in \partial\Omega$, *we set* $m = \inf_{\partial\Omega\cap B_R(x_0)} v$. *Then there exists* $p_0 > 0$ *such that for* $0 < \theta < 1$,

$$\inf_{B_{\theta R}(x_0)} v_m^- \geqslant C^{-1}\left[\fint_{B_R(x_0)} (v_m^-)^{p_0} dx\right]^{1/p_0}, \quad \forall 0 < R \leqslant h, \tag{3.4}$$

where h *is the height of the exterior cone;* p_0 *and* C *depend only on* $n, \Lambda/\lambda, (1-\theta)^{-1}$ *and* h.

Proof. From Lemma 1.1, we derive that v_m^- is a weak supersolution of (1.1)′ and $(v_m^-)^{-p}$ is a weak subsolution of (1.1)′. Using translation and scaling if necessary, we may assume without loss of generality that x_0 is the origin and $R = 1$. From Lemma 3.1,

$$\begin{aligned} \inf_{B_\theta} v_m^- &\geqslant C^{-1}\left[\int_{B_1} (v_m^-)^p dx\right]^{-1/p} \\ &= \frac{1}{C}\left[\int_{B_1} (v_m^-)^{-p} dx \int_{B_1} (v_m^-)^p dx\right]^{-1/p}\left[\int_{B_1} (v_m^-)^p\right]^{1/p}. \end{aligned}$$

Similarly, it suffices to prove that, for some $p > 0$,

$$\int_{B_1} e^{p|w|} dx \leqslant C,$$

where $w = \log v_m^- - \beta$. If we choose $\beta = \log m$, then w vanishes on $B_1 \setminus \Omega$. Since Ω satisfies a uniform exterior cone condition, $B_1 \setminus \Omega$ contains a cone congruent to $V_h \cap B_1$. Poincaré's inequality implies that

$$\int_{B_1} w^2 dx \leqslant C \int_{B_1} |Dw|^2 dx.$$

The rest of the proof is similar to that of Lemma 1.3. □

Theorem 3.3. *Suppose that* Ω *satisfies a uniform exterior cone condition, and* (a^{ij}) *satisfies* (1.2). *Let* u *be a weak solution of* (1.1)′ *with* $[u]_{\varepsilon_1,\partial\Omega} < \infty$, *where* $\varepsilon_1 > 0$. *Then for any* $x_0 \in \partial\Omega$, $0 < R \leqslant h$, *there exist* $C > 0$, $0 < \gamma \leqslant \varepsilon_1$, *such that*

$$\operatorname*{osc}_{\Omega\cap B_R(x_0)} u \leqslant C\left(\frac{R}{h}\right)^\gamma\left[\operatorname*{osc}_{B_h(x_0)\cap\Omega} u + h^\gamma [u]_{\varepsilon_1;\partial\Omega}\right], \tag{3.5}$$

where C *and* γ *depend only on* $n, \Lambda/\lambda$ *and the solid angle of opening of the exterior cone.*

Proof. Set $\Omega_R = \Omega \cap B_R(x_0)$, $\partial\Omega_R = \partial\Omega \cap B_R(x_0)$ and

$$\begin{aligned} &M(R) = \sup_{\Omega_R} u, \quad m(R) = \inf_{\Omega_R} u, \quad \omega(R) = M(R) - m(R), \\ &M_0(R) = \sup_{\partial\Omega_R} u, \quad m_0(R) = \inf_{\partial\Omega_R} u. \end{aligned}$$

For any $m \geqslant 0$, we use the exterior cone condition for v_m^- defined in (3.2) to derive

$$\left(\fint_{B_R(x_0)} (v_m^-)^{p_0} dx\right)^{1/p_0} \geqslant m\left[\frac{|V_h \cap B_R(x_0)|}{|B_R(x_0)|}\right]^{1/p_0} = C^{-1} m \quad (R \leqslant h).$$

Applying Lemma 3.2 to the functions $v = M(R) - u$ and $v = u - m(R)$, we get

$$\begin{aligned} M(R) - M(\theta R) &\geqslant \frac{1}{C}[M(R) - M_0(R)], \\ m(\theta R) - m(R) &\geqslant \frac{1}{C}[m_0(R) - m(R)]. \end{aligned}$$

Adding these two inequalities, we obtain

$$\omega(R) - \omega(\theta R) \geqslant \frac{1}{C}[\omega(R) - \operatorname*{osc}_{\partial\Omega_R} u].$$

Therefore, for $0 < R \leqslant h$,

$$\omega(\theta R) \leqslant \left(1 - \frac{1}{C}\right)\omega(R) + \frac{1}{C}[u]_{\varepsilon_1;\partial\Omega_h} R^{\varepsilon_1}.$$

Applying Lemma 2.1, we now obtain the lemma. □

Next we consider the nonhomogeneous equation

$$-D_j(a^{ij}D_iu) = f + D_if^i. \tag{3.6}$$

Theorem 3.4. *Suppose that Ω satisfies a uniform exterior cone condition, the coefficients of (3.6) satisfy (1.2), and $f \in L^{q_*}(\Omega)$, $f^i \in L^q(\Omega)$ for some $q > n$, where $q_* = nq/(n+q)$. If u is a weak solution of (3.6) and $[u]_{\varepsilon_1;\partial\Omega} < \infty$ for some $\varepsilon_1 > 0$, then there exist $C > 0$ and $0 < \gamma < \varepsilon_1$ such that for $0 < R \leqslant h$, $x_0 \in \partial\Omega$,*

$$\operatorname*{osc}_{\Omega\cap B_R(x_0)} u \leqslant C\left(\frac{R}{h}\right)^\gamma \left[\operatorname*{osc}_{B_h(x_0)\cap\Omega} u + [u]_{\varepsilon_1;\partial\Omega} + \|f\|_{L^{q_*}} + \sum_i \|f^i\|_{L^q}\right)\Big], \tag{3.7}$$

where C and γ depend only on $n, \Lambda/\lambda, q$ and the solid angle of opening of the exterior cone.

The proof is exactly the same as that of Theorem 2.3.

Theorem 3.5. *Under the assumptions of Theorem 3.4, there exist $C > 0$ and $0 < \gamma < 1$ such that*

$$|u(x) - u(y)| \leqslant C|x-y|^\gamma\left(|u|_{0;\Omega} + [u]_{\varepsilon_1;\partial\Omega} + \|f\|_{L^{q_*}} + \sum_i \|f^i\|_{L^q}\right), \tag{3.8}$$

$x, y, \in \Omega$, where C and γ depend only on $n, \Lambda/\lambda, \varepsilon_1, q$ and Ω.

Proof. Assume without loss of generality that $d_{xy} = d_x$. For simplicity we let $\delta = d_x$. Theorem 3.4 implies the following interior estimate (similar to the corollary of Theorem 2.2):

$$|u(x) - u(y)| \leqslant C|x-y|^\gamma\left(\delta^{-\gamma}\operatorname*{osc}_{B_\delta(x)} u + F_0\right), \tag{3.9}$$

where $F_0 = \|f\|_{L^{q_*}} + \sum \|f^i\|_{L^q}$. The definition of d_x implies that there exists $x_0 \in \partial\Omega$ such that $|x - x_0| = \delta$. If $2\delta \geqslant h$ (the height of the exterior cone), then we can replace δ in (3.9) with $h/2$ to conclude the desired estimate. Now we assume that $2\delta < h$. Obviously,

$$\delta^{-\gamma}\operatorname*{osc}_{B_\delta(x)} u \leqslant \delta^{-\gamma}\operatorname*{osc}_{B_{2\delta}(x_0)\cap\Omega} u. \tag{3.10}$$

By Theorem 3.4,

$$\delta^{-\gamma} \operatorname*{osc}_{B_{2\delta}(x_0)\cap\Omega} u \leqslant Ch^{-\gamma}\Big(\operatorname*{osc}_{B_h(x_0)\cap\Omega} u + [u]_{\varepsilon_1;\partial\Omega} + F_0 \Big). \tag{3.11}$$

Combining the inequalities (3.9), (3.10) and (3.11), we immediately obtain the estimate (3.8). □

CHAPTER 5

Quasilinear Equations of Divergence Form

In this chapter, we study the following equation of divergence form:

$$-D_i[a_i(x,u,Du)] + b(x,u,Du) = 0.$$

This equation has a broad range of applications. For example, the minimization of the functional

$$F[u] = \int_\Omega f(x,u,Du)dx$$

gives its Euler equation, which is a divergence equation of this type.

We are mainly interested in the solvability of the Dirichlet problem for the above divergence equation. The method is very typical: we reduce the solvability problem to an a priori estimate by using the Leray-Schauder fixed point theorem. Therefore, the key is to obtain an a priori estimate under rather weak assumptions. In this chapter, we shall introduce some special techniques in a priori estimates which utilize the divergence structure. Comparing with equations of nondivergence form in Chapter 7, we find that the method here requires less regularity on the coefficients than the Bernstein estimate.

1. Boundedness of weak solutions

Consider the equation of divergence form

$$-D_i[a_i(x,u,Du)] + b(x,u,Du) = 0. \tag{1.1}$$

For the equation (1.1), we assume the following structure conditions: for any $(x,y,\eta) \in \Omega \times \mathbb{R} \times \mathbb{R}^n$,

$$a_i(x,y,\eta)\eta_i \geqslant |\eta|^2 - g^2(x), \tag{1.2}$$

$$|a_i(x,y,\eta)| \leqslant \Lambda[|\eta| + g(x)], \tag{1.3}$$

$$|b(x,y,\eta)| \leqslant \Lambda[|\eta|^2 + f(x)], \tag{1.4}$$

where $f, g \geqslant 0$ and $f, g \in L^1(\Omega)$. For simplicity, we assume that a_i, b are continuous with respect to the variables (x,y,η).

Definition 1.1. A function $u \in W^{1,2}_{loc}(\Omega)$ is said to be a *weak solution* of (1.1), if for any $\varphi \in C^1_0(\Omega)$,

$$Q[u,\varphi] \triangleq \int_\Omega a_i(x,u,Du)D_i\varphi dx + \int_\Omega b(x,u,Du)\varphi dx = 0. \tag{1.5}$$

A function $u \in W^{1,2}_{loc}(\Omega)$ is said to be a *weak subsolution* (*weak supersolution*), if

$$Q[u,\varphi] \leqslant 0 \quad (\geqslant 0) \quad \forall \varphi \in C^1_0(\Omega), \quad \varphi \geqslant 0. \tag{1.6}$$

The structure conditions (1.3) and (1.4) guarantee that the integrals in (1.5) are well defined.

To derive the L^∞ estimate, we also assume the following structure condition:

$$-b(x,z,\eta)\operatorname{sign} z \leqslant \Lambda[|\eta| + f(x)], \quad \forall (x,z,\eta) \in \Omega \times \mathbb{R} \times \mathbb{R}^n. \tag{1.7}$$

Theorem 1.1. *Suppose that the structure conditions* (1.2), (1.3), (1.4) *and* (1.7) *hold,* $F_0 \triangleq \|g\|_{L^q} + \|f\|_{L^{q_*}} < \infty$ *for some* $q > n$*, where* $q_* = nq/(n+q)$*. Then for a weak subsolution* $u \in W^{1,2}(\Omega)$ *of* (1.1),

$$\operatorname*{ess\,sup}_{\Omega} u \leqslant \sup_{\partial\Omega} u^+ + CF_0,$$

where C *depends only on* n, Λ *and* $\operatorname{diam}\Omega$.

Proof. Assume that $\sup_{\partial\Omega} u^+ < \infty$. For $k \geqslant \sup_{\partial\Omega} u^+$, we choose the test function $\varphi = (u-k)^+$ in (1.6), and we set $v = (u-k)^+$, $A(k) = \{x \in \Omega | u(x) > k\}$. Using the structure conditions (1.2) and (1.7) in (1.6), we obtain

$$\begin{aligned}
\int_\Omega |Dv|^2 dx &\leqslant \int_{A(k)} g^2 dx + \Lambda \int_{A(k)} (|Dv| + f) v dx \\
&\leqslant \frac{1}{4}\int_\Omega |Dv|^2 dx + C\int_\Omega |v|^2 dx \\
&\quad + \|g\|_{L^q}^2 |A(k)|^{1-2/q} + C\|v\|_{L^{2^*}} \|f\|_{L^{q_*}} |A(k)|^{(1/2)-(1/q)},
\end{aligned}$$

where $2^* = 2n/(n-2)$. We shall discuss only the case $n \geqslant 3$. By the Sobolev embedding theorem and Cauchy's inequality,

$$\int_\Omega |Dv|^2 dx \leqslant C\int_\Omega |v|^2 dx + C\Big[\|g\|_{L^q}^2 + \|f\|_{L^{q_*}}^2\Big] |A(k)|^{1-(2/q)}.$$

This estimate is similar to (4.14) of Chapter 1. Analogously to (4.18) of Chapter 1, we derive the bound

$$\operatorname*{ess\,sup}_{\Omega} u \leqslant \sup_{\partial\Omega} u^+ + C\|u\|_{L^2(\Omega)} + CF_0|\Omega|^{(1/n)-(1/q)}.$$

We can now follow the method of Step 2 in the proof of Theorem 4.2 in Chapter 1 to take care of the term $C\|u\|_{L^2(\Omega)}$ on the right-hand side. □

Remark 1.1. If we know a priori that the weak solution $u \in C(\overline{\Omega}) \cap C^1(\Omega)$, then the integrals in (1.5) are well defined without structure conditions (1.3) and (1.4); and therefore conditions (1.3) and (1.4) are no longer needed in Theorem 1.1. In this case, the above theorem is valid with only the structure conditions (1.2) and (1.7).

Remark 1.2. Let $u \in C(\overline{\Omega}) \cap C^1(\Omega)$. We can improve the structure conditions (1.2) and (1.7) as follows: for any $k > 0$,

$$a_i(x,z,\eta)\eta_i \geqslant |\eta|^2 - [\mu(z-k)^+]^2 - g^2(x), \tag{1.2$'$}$$

$$-b(x,z,\eta)\operatorname{sign} z \leqslant \Lambda[|\eta| + \mu(z-k)^+ + f(x)]. \tag{1.7$'$}$$

The theorem is still valid.

Remark 1.3. If we change the structure condition (1.2) to

$$a_i(x,z,\eta)\eta_i \geqslant |\eta|^\tau - g^\tau(x),$$

where $\tau > 1$, then we can modify the remaining structure conditions and the definition of a weak solution so that a similar theorem holds.

The proofs of the remarks are left as exercises for the reader.

2. Hölder estimates for bounded weak solutions

The methods in this section are similar to those in §1 of Chapter 4. However, quasilinear nonhomogeneous equations are discussed here. In the structure conditions (1.3)–(1.4), the growth order of b in η is one order higher than that of a_i. Such a growth order condition is called a *natural growth order condition* (or called a natural growth condition, for simplicity). Examples of nonexistence of a classical solution can be constructed for the Dirichlet problem if such a natural growth order condition is not satisfied.

Theorem 2.1. *Let the structure conditions* (1.2), (1.3) *and* (1.4) *be in force. Suppose that* $u \in W^{1,2}(B_R)$ *is a bounded weak subsolution of* (1.1) *on* B_R *and that*

$$F_0 = R^{1-n/q}\|g\|_{L^q} + R^{2-2n/q}\|f+g^2\|_{L^{q/2}} < \infty \tag{2.1}$$

for some $q > n$. *Then for any* $p > 0$ *and* $0 < \theta < 1$,

$$\operatorname*{ess\,sup}_{B_{\theta R}} \widetilde{u} \leqslant C\left[\fint_{B_R} \widetilde{u}^p dx\right]^{1/p}, \tag{2.2}$$

where $\widetilde{u} = u^+ + F_0$, *and* C *depends only on* $n, \Lambda, q, p, (1-\theta)^{-1}$ *and* $\|u\|_{L^\infty}$.

Proof. First, we assume that $0 < R \leqslant 1$, $p \geqslant 2$. We choose the test function

$$\varphi = \zeta^2 \widetilde{u}^{2p-1} e^{\Lambda u}, \tag{2.3}$$

where $\zeta(x)$ is a cutoff function on B_R. It follows from (1.6) and the structure condition (1.4) that

$$\begin{aligned}&\int_{B_R} a_i(x,u,Du) D_i(\zeta^2\widetilde{u}^{2p-1}e^{\Lambda u})dx\\ &\qquad \leqslant \Lambda \int_{B_R} (|Du|^2 + f)\zeta^2 \widetilde{u}^{2p-1} e^{\Lambda u} dx.\end{aligned} \tag{2.4}$$

Using the structure conditions (1.2)–(1.3), we derive

$$\begin{aligned}&(2p-1)\int_{B_R} \zeta^2\widetilde{u}^{2p-2}e^{\Lambda u}|Du|^2dx\\ &\leqslant \ (2p-1)\int_{B_R}\zeta^2\widetilde{u}^{2p-2}e^{\Lambda u}|g|^2dx + \Lambda\int_{B_R}(g^2+f)\zeta^2\widetilde{u}^{2p-1}e^{\Lambda u}dx\\ &\qquad +2\Lambda\int_{B_R}(|Du|+g)\zeta|D\zeta|\widetilde{u}^{2p-1}e^{\Lambda u}dx.\end{aligned}$$

Since $\widetilde{u} \geqslant F_0$ and $\|u\|_{L^\infty} < \infty$,

$$\begin{aligned}&(2p-1)\int_{B_R}\zeta^2\widetilde{u}^{2p-2}e^{\Lambda u}|Du|^2dx\\ &\leqslant \ Cp\int_{B_R}\left[\frac{g^2}{F_0^2}+\frac{g^2+f}{F_0}\right]\zeta^2\widetilde{u}^{2p}dx + (2p-1)\varepsilon\int_{B_R}\zeta^2\widetilde{u}^{2p-2}|Du|^2dx\\ &\qquad +\frac{C_\varepsilon}{2p-1}\int_{B_R}|D\zeta|^2\widetilde{u}^{2p}dx,\end{aligned}$$

where C depends only on n, Λ and $\|u\|_{L^\infty}$. Let $v = \widetilde{u}^p$ and $\varepsilon = 1/2$. Then

$$\frac{1}{p}\int_{B_R} \zeta^2 |Dv|^2 dx \leqslant Cp\int_{B_R} h(x)\zeta^2 \widetilde{u}^{2p} dx + \frac{C}{p}\int_{B_R} |D\zeta|^2 v^2 dx, \tag{2.5}$$

where

$$h(x) = \frac{g^2+f}{F_0} + \frac{g^2}{F_0^2}, \qquad \|h\|_{L^{q/2}} \leqslant 2.$$

By Hölder's inequality and the Sobolev embedding theorem,

$$\begin{aligned} \int_{B_R} h\zeta^2 \widetilde{u}^{2p} dx &\leqslant \|h\|_{L^{q/2}} \|\zeta^2 v^2\|_{L^{q/(q-2)}} \\ &\leqslant \varepsilon \|\zeta v\|_{L^{2^*}}^2 + C\varepsilon^{-n/(q-n)} \|\zeta v\|_{L^2}^2 \\ &\leqslant C\varepsilon \|D(\zeta v)\|_{L^2}^2 + C\varepsilon^{-n/(q-n)} \|\zeta v\|_{L^2}^2. \end{aligned} \tag{2.6}$$

We take $\varepsilon = \dfrac{1}{2C}p^{-2}$ in the above inequality and substitute (2.6) into (2.5). Then

$$\int_{B_R} |D(\zeta v)|^2 dx \leqslant C\Big(p^{2+2n/(q-n)} + \|\nabla\zeta\|_{L^\infty}^2\Big)\int_{B_R} v^2 dx.$$

We can now follow the standard Moser iteration to derive the desired estimate as in Lemma 1.2 of Chapter 4. □

Theorem 2.2. *Let the structure conditions* (1.2), (1.3) *and* (1.4) *be in force. Suppose that* $u \in W^{1,2}(B_{\sigma R})$ $(\sigma > 1)$ *is a bounded weak supersolution of* (1.1) *on* $B_{\sigma R}$ *and that* (2.1) *holds for some* $q > n$. *Then there exists* $p_0 > 0$ *such that for* $0 < \theta < 1$,

$$\operatorname*{ess\,inf}_{B_{\theta R}} \widetilde{u} \geqslant C^{-1}\left[\fint_{B_R} |\widetilde{u}|^{p_0} dx\right]^{1/p_0}, \tag{2.7}$$

where $\widetilde{u} = u + F_0$, *and* C *depends only on* n, Λ, q, $(1-\theta)^{-1}$, $(\sigma-1)^{-1}$ *and* $\|u\|_{L^\infty}$.

Proof. Using $F_0 + \varepsilon$ instead of F_0 if necessary, we may assume without loss of generality that $F_0 > 0$. We also assume that $R = 1$ and choose a cutoff function ζ on B_1. Choosing the test function $\varphi = \zeta^2\widetilde{u}^{-(2p+1)}e^{-\Lambda u}$ in the definition of a weak supersolution and using (1.4), we get

$$\begin{aligned} &\int_{B_1} a_i(x,u,Du) D_i[\zeta^2 \widetilde{u}^{-(2p+1)} e^{-\Lambda u}] dx \\ &\qquad \geqslant -\Lambda \int_{B_1} (|Du|^2 + f)\zeta^2 \widetilde{u}^{-(2p+1)} e^{-\Lambda u}] dx. \end{aligned}$$

Using a computation similar to that in Theorem 2.1, we get, for any $p > 0$,

$$\operatorname*{ess\,sup}_{B_\theta} \widetilde{u}^{-1} \leqslant C\left[\int_{B_1} \widetilde{u}^{-p} dx\right]^{1/p},$$

i.e.,

$$\begin{aligned} \operatorname*{ess\,inf}_{B_\theta} \widetilde{u} &\geqslant \frac{1}{C}\left[\int_{B_1} \widetilde{u}^{-p} dx\right]^{-1/p} \\ &\geqslant \frac{1}{C}\left[\int_{B_1} \widetilde{u}^{-p} dx \cdot \int_{B_1} \widetilde{u}^{p} dx\right]^{-1/p} \left[\int_{B_1} \widetilde{u}^{p} dx\right]^{1/p}. \end{aligned}$$

Now we argue the same way as in Lemma 1.3 of Chapter 4: it suffices to prove that, for some $p_0 > 0$,

$$\int e^{p_0|w|}dx \leqslant C,$$

where $w = \beta - \log \widetilde{u}$. We choose the test function $\varphi = \zeta^2\widetilde{u}^{-1}e^{-Au}$. Then

$$\begin{aligned}\int_{B_\sigma} \zeta^2|Dw|^2dx &\leqslant C\int_{B_\sigma}\left[g^2\widetilde{u}^{-2} + (g^2+f)\widetilde{u}^{-1}\right]dx + C\int_{B_\sigma}|D\zeta|^2dx \\ &\leqslant C\int_{B_\sigma} h(x)dx + C\int_{B_\sigma}|D\zeta|^2dx \\ &\leqslant C.\end{aligned}$$

We take $\beta = \fint_{B_{\overline{\sigma}}} \log\widetilde{u}dx$ ($\overline{\sigma} = \dfrac{1+\sigma}{2}$) and proceed as in Lemma 1.3 of Chapter 4 to obtain the theorem. □

Theorems 2.1 and 2.2 imply a Harnack's inequality for quasilinear equations with nonhomogeneous terms.

Theorem 2.3. *Under the assumptions of Theorem 2.1, a bounded weak solution u of* (1.1) *on B_R satisfies, for any $0 < \theta < 1$,*

$$\operatorname*{ess\,sup}_{B_{\theta R}} u \leqslant C\left[\operatorname*{ess\,inf}_{B_{\theta R}} u + F_0\right], \tag{2.8}$$

where F_0 is defined in (2.1), *and C depends only on n, Λ, q, $(1-\theta)^{-1}$ and $\|u\|_{L^\infty}$.*

Next, applying Lemma 2.1 in Chapter 4, we derive the interior Hölder continuity and local Hölder estimates for bounded weak solutions of (1.1).

Theorem 2.4. *Let the structure conditions* (1.2), (1.3) *and* (1.4) *be in force. Suppose that for some $q > n$,*

$$\widetilde{F}_0 = \|g\|_{L^q} + \|f\|_{L^{q/2}} < \infty, \tag{2.9}$$

and that u is a bounded weak solution of (1.1) *on Ω. Then there exist $C \geqslant 0$ and $0 < \gamma < 1$ such that for any $B_R(x) \subset \Omega$,*

$$\operatorname*{ess\,osc}_{B_R(x)} u \leqslant C\left(\frac{R}{d_x}\right)^\gamma\left[\operatorname*{ess\,osc}_{B_{d_x}(x)} u + \widetilde{F}_0\right], \tag{2.10}$$

where C and γ depend only on n, Λ, q, $\operatorname{diam}\Omega$ and $\|u\|_{L^\infty}$.

Similarly to Lemma 3.2 in Chapter 4, we can also derive a weak Harnack's inequality near the boundary.

Theorem 2.5. *Let Ω satisfy a uniform exterior cone condition, and let the structure conditions* (1.2), (1.3) *and* (1.4) *be in force. Suppose that u is a bounded weak supersolution of* (1.1) *on Ω, $x_0 \in \partial\Omega$ and $m = \inf_{B_R(x_0)\cap\partial\Omega} u$. Then there exists $p > 0$ such that, for any $0 < \theta < 1$ and $0 < R \leqslant h$ (the height of the cone),*

$$\inf_{B_{\theta R}(x_0)} u_m^- \geqslant \frac{1}{C}\left[\fint_{B_R(x_0)} (u_m^-)^p dx\right]^{1/p} - CF_0,$$

where C depends only on n, Λ, q, Ω, $(1-\theta)^{-1}$ and $\|u\|_{L^\infty}$.

Finally, we have the global Hölder estimate (cf. Theorem 3.5 of Chapter 4).

Theorem 2.6. *Assume, in addition to the assumptions of Theorem 2.4, that* Ω *satisfies a uniform exterior cone condition and that* $[u]_{\varepsilon_1,\partial\Omega} < \infty$ *for some* $\varepsilon_1 > 0$. *Then there exist* $C > 0$ *and* $0 < \gamma < 1$ *such that*

$$|u(x) - u(y)| \leqslant C|x - y|^{\gamma}\Big(|u|_{0;\Omega} + [u]_{\varepsilon_1;\partial\Omega} + \widetilde{F}_0\Big),$$

where C *and* γ *depend only on* n, Λ, q, Ω *and* $\|u\|_{L^\infty}$.

3. Gradient estimates

In order to obtain gradient estimates, we need stronger structure conditions. We shall impose a uniform ellipticity condition. For $(x, z, \eta) \in \Omega \times [-M, M] \times \mathbb{R}^n$, we assume

$$\frac{\partial a_i}{\partial \eta_j}(x, z, \eta)\xi_i\xi_j \geqslant |\xi|^2, \quad \forall \xi \in \mathbb{R}^n, \tag{3.1}$$

$$\left|\frac{\partial a_i}{\partial \eta_j}\right| \leqslant \Lambda, \tag{3.2}$$

$$\sum_i (1 + |\eta|)\left|\frac{\partial a_i}{\partial z}\right| + \sum_{i,j}\left|\frac{\partial a_i}{\partial x_j}\right| + |b| \leqslant \Lambda(1 + |\eta|^2). \tag{3.3}$$

In a uniform ellipticity condition, the ratio between the largest and smallest eigenvalues of the matrix $\left(\frac{\partial a_i}{\partial \eta_j}\right)$ is bounded by a constant. The structure conditions (3.1) and (3.2) are of this type.

For equations of divergence form, there are no distinctive methods on the boundary gradient estimates. One usually rewrites the equation in nondivergence form and uses a barrier function; such a technique belongs to equations of nondivergence form and will be introduced in Chapter 7. Here we state only the results.

Theorem 3.1. *Let the assumptions* (3.1), (3.2) *and* (3.3) *be in force. Suppose that* Ω *satisfies a uniform exterior sphere condition (see Definition 7.1 in Chapter 2). If* $u \in C^2(\Omega) \cap C^1(\overline{\Omega})$ *is a solution of* (1.1) *in* Ω *with* $\|u\|_{L^\infty} \leqslant M$, *then*

$$\sup_{\partial\Omega} |Du| \leqslant C,$$

where C *depends only on* n, Λ, M *and* Ω.

We are mainly interested in global estimates for Du in this section.

Theorem 3.2. *Let the assumptions* (3.1)–(3.3) *be in force. Suppose that* $u \in C^2(\Omega) \cap C^1(\overline{\Omega})$ *is a solution of* (1.1) *in* Ω *with* $\|u\|_{L^\infty} \leqslant M$, $[u]_{\alpha;\Omega} < \infty$ $(\alpha > 0)$. *Then*

$$\sup_{\Omega} |Du| \leqslant C\Big[1 + \sup_{\partial\Omega} |Du|\Big],$$

where C *depends only on* n, Λ, M, α *and* $[u]_{\alpha;\Omega}$.

Proof. u satisfies the equation

$$-D_i[a_i(x, u, Du)] + b(x, u, Du) = 0.$$

Differentiating with respect to x_k (the distribution derivative), we obtain

$$D_i\left[\frac{\partial a_i}{\partial \eta_j}D_{jk}u\right]+D_if_k^i=0 \qquad (\text{in the } \mathscr{D}'(\Omega) \text{ sense}), \tag{3.4}$$

where

$$f_k^i=\frac{\partial a_i}{\partial x_k}+\frac{\partial a_i}{\partial u}D_ku-b\delta_{ik}.$$

By the structure condition (3.3),

$$|f_k^i|\leqslant \Lambda(1+|Du|^2).$$

From (3.4),

$$\int_\Omega\left\{\frac{\partial a_i}{\partial \eta_j}D_{jk}uD_i\varphi+f_k^iD_i\varphi\right\}=0,\quad \forall\varphi\in W_0^{1,2}(\Omega). \tag{3.5}$$

The function $v\triangleq|Du|^2$ takes its maximum at a point $x_0\in\overline{\Omega}$. The theorem is clearly valid if $x_0\in\partial\Omega$. Now we assume that $x_0\in\Omega$. Assume without loss of generality that $N\triangleq\sqrt{v(x_0)}=\sup\limits_\Omega|Du|>1$. Take a cutoff function $\zeta(x)$ of $B_R(x_0)$ $(R=N^{-1})$ such that

$$\zeta\in C_0^\infty(B_R(x_0)),\quad \zeta(x_0)=1,\quad |D\zeta|\leqslant\frac{2}{R}=2N.$$

We choose the test function to be $\zeta^2(D_ku)\varphi$ in (3.5) and sum over k, where $\varphi\in C_0^\infty(\Omega)$, $\varphi\geqslant0$. Then

$$\begin{aligned}\int_\Omega\zeta^2\varphi\frac{\partial a_i}{\partial\eta_j}D_{ik}uD_{jk}udx+\int_\Omega f_k^iD_{ik}u\cdot\zeta^2\varphi dx&\\ +\int_\Omega\left[\frac{1}{2}\frac{\partial a_i}{\partial\eta_j}D_jv+f_k^iD_ku\right][2\zeta\varphi D_i\zeta+\zeta^2D_i\varphi]dx&=0.\end{aligned}$$

Let $w=\zeta^2v$. By the structure condition (3.1),

$$\begin{aligned}\int_\Omega\zeta^2\varphi|D^2u|^2dx+\int_\Omega\left[\zeta^2\varphi f_k^iD_{ik}u+\left(\frac{\partial a_i}{\partial\eta_j}D_jv+2f_k^iD_ku\right)\zeta\varphi D_i\zeta\right]dx&\\ +\int_\Omega\left[\frac{1}{2}\frac{\partial a_i}{\partial\eta_j}D_jw-\zeta\frac{\partial a_i}{\partial\eta_j}vD_j\zeta+\zeta^2f_k^iD_ku\right]D_i\varphi dx&\leqslant 0.\end{aligned}$$

Using Cauchy's inequality, we get

$$\begin{aligned}&\int_\Omega\left[\frac{\partial a_i}{\partial\eta_j}D_jw-2\zeta\frac{\partial a_i}{\partial\eta_j}vD_j\zeta+2\zeta^2f_k^iD_ku\right]D_i\varphi dx\\ &-C\int_\Omega\left(\zeta^2|f_k^i|^2+|D\zeta|^2(1+|Du|^2)+|f_k^i(D_ku)\zeta D_i\zeta|\right)\varphi dx\leqslant0,\\ &\qquad\qquad\forall\varphi\in C_0^\infty(\Omega),\ \varphi\geqslant0.\end{aligned}\tag{3.6}$$

This inequality implies that w is a weak subsolution of some linear equation on $B_R(x_0)\cap\Omega$. By the maximum principle (Theorem 4.2 of Chapter 1) and (3.2),

(3.3), for some $q > n$ (notice that $N = R^{-1} = \sup_\Omega |Du|$), we have

$$
\begin{aligned}
\sup_{\Omega\cap B_R(x_0)} w &\leqslant \sup_{\partial\Omega\cap B_R(x_0)} w + CR^{1-n/q}\Big\{\|N^2(1+|Du|^2)\|_{L^{q_*}} \\
&\qquad + \|N(1+|Du|^2)\|_{L^q}\Big\} \\
&\leqslant \sup_{\partial\Omega\cap B_R(x_0)} w + C\Big\{1 + N^{1+n/q}\|\,|Du|^2\|_{L^{q_*}} \\
&\qquad + N^{n/q}\|\,|Du|^2\|_{L^q}\Big\}.
\end{aligned} \tag{3.7}
$$

Next, we estimate $\|Du\|_{L^2(B_R(x_0))}$. Let $\zeta(x)$ be a cutoff function on $B_{2R}(x_0)$ such that $\zeta(x) = 1$ on $B_R(x_0)$. We multiply (1.1) by the test function $\zeta^2[u(x) - u(x_0)]$ and integrate by parts. Then

$$
\begin{aligned}
&\int_{\Omega_{2R}} a_i(x,u,Du)\zeta^2 D_i u dx + 2\int_{\Omega_{2R}} a_i \cdot (u - u(x_0))\zeta D_i\zeta dx \\
&\quad + \int_{\Omega_{2R}} b\zeta^2(u - u(x_0))dx \\
&\quad + \int_{\partial\Omega_{2R}} a_i(x,u,Du)\zeta^2(u - u(x_0))\cos(\nu, x_i)dS = 0,
\end{aligned}
$$

where $\Omega_{2R} = \Omega \cap B_{2R}(x_0)$. Utilizing the structure conditions (3.1)–(3.3) and noticing also that

$$
\begin{aligned}
&\eta_i a_i(x,z,\eta) \geqslant |\eta|^2 + \eta_i a_i(x,z,0), \\
&|u(x) - u(x_0)| \leqslant [u]_{\alpha;\Omega}\,(2R)^\alpha, \quad \forall x \in \Omega_{2R},
\end{aligned}
$$

we derive

$$
\int_{\Omega_{2R}} \zeta^2|Du|^2 dx \leqslant CR^{n-1+\alpha}N + CR^{n+\alpha}N^2 \leqslant CN^{2-\alpha-n}.
$$

In deriving the last inequality, we also used the fact that $R = N^{-1}$. Thus

$$
\begin{aligned}
&\|\,|Du|^2\|_{L^q(\Omega_R)} \leqslant N^{(2q-2)/q}\|Du\|^{2/q}_{L^2(\Omega_R)} \leqslant CN^{2-(n+\alpha)/q}, \\
&\|\,|Du|^2\|_{L^{q_*}(\Omega_R)} \leqslant N^{2-(n+\alpha)/q_*}.
\end{aligned}
$$

Substituting these estimates into (3.7), we obtain

$$
N^2 = w(x_0) \leqslant \sup_{\Omega_R} w \leqslant \sup_{\partial\Omega\cap B_R(x_0)} w + C\Big(N^{1+n/q} + N^{2-\alpha/q_*} + N^{2-\alpha/q}\Big).
$$

It follows immediately that

$$
\sup_\Omega |Du| \leqslant N \leqslant C\Big(\sup_{\partial\Omega} w\Big)^{1/2} + C,
$$

which is exactly what we want to prove. □

4. Gradient Hölder estimates

In the last section, we proved that if $u \in C^2(\Omega)$ (in fact, all we need is $u \in W^{2,2}(\Omega)$) is a solution, then $D_k u$ is a weak solution of (3.4). Since we have already established an L^∞ estimate for Du, the functions $\dfrac{\partial a_i}{\partial \eta_j}$ and f_k^i belong to $L^\infty(\Omega)$.

Applying the De Giorgi-Nash-Moser estimates for linear elliptic equations with bounded measurable coefficients, we immediately obtain

Theorem 4.1. *Let the structure conditions* (3.1), (3.2) *and* (3.3) *be in force. Suppose that* $u \in C^2(\Omega)$ *is a solution of* (1.1) *on* Ω, $\|u\|_{L^\infty} \leqslant M$, $\|Du\|_{L^\infty} \leqslant M_1$. *Then for any* $\Omega' \subset\subset \Omega$, *there exist* $C > 0$ *and* $0 < \gamma < 1$ *such that*

$$[Du]_{\alpha;\Omega'} \leqslant C, \tag{4.1}$$

where α *and* C *depend on* n, Λ, M *and* M_1, *and* C *depends also on* $\operatorname{dist}\{\Omega', \partial\Omega\}$.

For the global estimates, we need the Morrey theorem on Hölder continuity. The proof of this theorem is more compatible with the second part and therefore is postponed. Here we state only the result.

Theorem 4.2 (Morrey theorem). *Let* $u \in W^{1,p}(\Omega)$, $p > 1$. *Suppose that there exist* $K > 0$ *and* $0 < \alpha < 1$ *such that for any ball* B_R,

$$\int_{\Omega_R} |Du|^p dx \leqslant K R^{n-p+\alpha p}, \quad \Omega_R = \Omega \cap B_R. \tag{4.2}$$

Suppose also that $|\Omega_R| \geqslant AR^n$ *for some constant* $A > 0$. *Then* $u \in C^\alpha(\overline{\Omega})$ *and*

$$\operatorname*{osc}_{\Omega_R} u \leqslant CKR^\alpha,$$

where C *depends only on* n, α, p *and* Λ.

This theorem reduces the Hölder estimate to an integral estimate (4.2), which is more suitable for equations in divergence form.

Theorem 4.3. *Let the structure conditions* (3.1)–(3.3) *be in force. Suppose that* $\partial\Omega \in C^{1,1}$, $u \in C^2(\Omega) \cap C^1(\overline{\Omega})$ *is a solution of* (1.1), $\|u\|_{L^\infty} \leqslant M$, $\|Du\|_{L^\infty} \leqslant M_1$, *and* $u = g$ *on* $\partial\Omega$, *where* $g \in C^2(\overline{\Omega})$. *Then*

$$[Du]_{\alpha;\Omega} \leqslant C, \tag{4.3}$$

where $C > 0$ *and* $0 < \alpha < 1$ *depend only on* $n, \Lambda, M, M_1, \Omega$ *and* $|g|_{2;\Omega}$.

Proof. Assume without loss of generality that $g = 0$. Let $x_0 \in \partial\Omega$, and let ϕ and V be the map and neighborhood given in Definition 5.1 of Chapter 1. Under the change of variables $y = \phi(x)$, $v(y) = u \circ \phi^{-1}(y)$, the function v satisfies

$$-D_i[\overline{a}^i(y, v, Dv)] + \overline{b}(y, v, Dv) = 0, \tag{4.4}$$

where

$$\overline{a}^i = a_r \frac{\partial y_i}{\partial x_r}, \quad \overline{b} = a_r \frac{\partial}{\partial y_i}\left(\frac{\partial y_i}{\partial x_r}\right) + b.$$

Analogously to (3.3), the function $w = D_k v$ satisfies

$$\begin{gathered} -D_i\left(\frac{\partial \overline{a}^i}{\partial \eta_j} D_j w\right) + D_i \overline{f}^i_k = 0 \qquad (\mathscr{D}'(B_1^+)) \\ (k = 1, 2, \cdots, n-1), \end{gathered} \tag{4.5}$$

where $\left(\dfrac{\partial \overline{a}^i}{\partial \eta_j}\right)$ is positive definite, while $\dfrac{\partial \overline{a}^i}{\partial \eta_j}$ and $\overline{f}^i_k$ are bounded. Furthermore,

$$w = D_k v = 0 \quad (k = 1, 2, \cdots, n-1) \quad \text{on } y_n = 0. \tag{4.6}$$

By the boundary estimates for linear equations, there exist $C > 0$ and $\alpha \in (0,1)$ such that

$$[w]_{\alpha;B^+_{1/2}} = [D_k v]_{\alpha;B^+_{1/2}} \leqslant C \quad (k = 1, 2, \cdots, n-1), \tag{4.7}$$

where C, α depend only on the quantities stated in the theorem. Next we estimate $D_n v$. Let $y_0 \in B^+_{1/2}$. For $R \leqslant 1/6$, we take a cutoff function ζ on $B_{2R} = B_{2R}(y_0)$ and choose the test function $\varphi = \zeta^2(w(y) - l)$ $(w = D_k v,\ 1 \leqslant k \leqslant n-1)$, where $l = w(y_0)$ if $B_{2R} \subset B^+_1$ and $l = 0$ if $B_{2R} \cap \partial\mathbb{R}^n_+$ is nonempty. From (4.5),

$$\begin{aligned}\int_{B^+_1} \zeta^2 \frac{\partial \overline{a}^i}{\partial \eta_j} D_i w D_j w dy &+ \int_{B^+_1} 2\zeta(w-l) \frac{\partial \overline{a}^i}{\partial \eta_j} D_i \zeta D_j w dy \\ &= \int_{B^+_1} \overline{f}^i_k \left[\zeta^2 D_i w + 2\zeta(w-l) D_i \zeta\right] dy \quad (k = 1, 2, \cdots, n-1).\end{aligned}$$

By the ellipticity and Cauchy's inequality,

$$\int_{B^+_1} \zeta^2 |Dw|^2 dy \leqslant C\Big(R^n + \int_{B^+_1} |D\zeta|^2 (w-l)^2 dy\Big).$$

Choose ζ such that $\zeta = 1$ for $y \in B_R(y_0)$. Then by (4.7),

$$\int_{B^+_1 \cap B_R(y_0)} |Dw|^2 dy \leqslant C\Big(R^n + R^{n-2} \sup_{B_{2R}} |w - l|^2\Big) \leqslant C R^{n-2+2\alpha},$$

i.e.,

$$\sum_{i+j<2n} \int_{B^+_1 \cap B_R(y_0)} |D_{ij} v|^2 dy \leqslant C R^{n-2+2\alpha}. \tag{4.8}$$

Using (4.4), we get

$$D_{nn} v = \left(\frac{\partial \overline{a}^n}{\partial \eta_n}\right)^{-1} \Big[- \sum_{i+j<2n} \frac{\partial \overline{a}^i}{\partial \eta_j} D_{ij} v - \frac{\partial \overline{a}^i}{\partial y_i} - \frac{\partial \overline{a}^i}{\partial z} D_i v + \overline{b}\Big]. \tag{4.9}$$

From (4.8) and (4.9), we deduce that

$$\int_{B^+_1 \cap B_R(y_0)} |D_{nn} v|^2 dy \leqslant C R^{n-2+2\alpha}. \tag{4.10}$$

By the Morrey theorem, (4.8) and (4.10) imply that

$$[Dv]_{\alpha;B^+_{1/2}} \leqslant C.$$

After changing the variables back to x and using a finite covering argument, we derive the global $[Du]_{\alpha;\Omega}$ estimate. □

5. Solvability of the Dirichlet problem

In this century people realized that the solvability problem, by means of topological fixed point theorems, can be reduced to establishing a priori estimates for solutions. The Leray-Schauder fixed point theorem is one of the most frequently used fixed point theorems. We shall only state the theorem and refer the readers to [GT, Chapter 11] for a proof.

Leray-Schauder Theorem. *Let X be a Banach space and $T(x, \sigma)$ a map from $X \times [0,1]$ to X. Suppose that T satisfies:*

(1) *T is compact (i.e., T is continuous and maps any bounded set to a precompact set);*

(2) $T(x,0) = 0, \quad \forall x \in X$;

(3) *there exists $M > 0$ such that*

$$\|x_0\|_X \leqslant M, \quad \forall x_0 \in \{x \in X \mid \exists \sigma \in [0,1],\ x = T(x,\sigma)\}.$$

Then $T(\cdot, 1)$ has a fixed point; i.e., there exists $x_0 \in X$ such that $T(x_0, 1) = x_0$.

Now we consider the Dirichlet problem

$$-D_i[a_i(x,u,Du)] + b(x,u,Du) = 0 \quad \text{on } \Omega, \tag{5.1}$$

$$u = \varphi \quad \text{in } \partial\Omega. \tag{5.2}$$

For (5.1), we shall assume the structure conditions: for any $(x,z,\eta) \in \Omega \times \mathbb{R} \times \mathbb{R}^n$,

$$\frac{\partial a_i}{\partial \eta_j}\xi_i\xi_j \geqslant |\xi|^2, \quad \forall \xi \in \mathbb{R}^n, \tag{5.3}$$

$$|a_i(x,z,0)| \leqslant g(x) \qquad (i = 1,2,\cdots,n), \tag{5.4}$$

$$\begin{aligned} &(1+|\eta|^2)\left|\frac{\partial a_i}{\partial \eta_j}\right| + (1+|\eta|)\left(\left|\frac{\partial a_i}{\partial z}\right| + |a_i|\right) + \left|\frac{\partial a_i}{\partial x_j}\right| + |b| \\ &\leqslant \mu(|z|)(1+|\eta|^2) \qquad (i,j = 1,2,\cdots,n), \end{aligned} \tag{5.5}$$

where $g \in L^q(\Omega)$ $(q > n)$ and $\mu(s)$ is monotonically increasing on $[0,\infty)$. The above structure conditions are still natural; there are two aspects: (1) the growth order of b in $|\eta|$ is one order higher than that of a_i (or two orders higher than that of $\frac{\partial a_i}{\partial \eta_j}$); (2) the growth orders of a_i and b in $|\eta|$ behave like polynomials.

Theorem 5.1. *Let the assumptions (5.3)–(5.5), (1.7) be in force. Suppose that $\partial\Omega \in C^{2,\alpha}$, $a_i \in C^{1,\alpha}(\overline{\Omega} \times \mathbb{R} \times \mathbb{R}^n)$, $b \in C^{0,\alpha}(\overline{\Omega} \times \mathbb{R} \times \mathbb{R}^n)$ and $\varphi \in C^{2,\alpha}(\overline{\Omega})$, where $0 < \alpha < 1$. Then there exists a solution $u \in C^{2,\alpha}(\overline{\Omega})$ for the Dirichlet problem (5.1), (5.2).*

Proof. We shall apply the Leray-Schauder fixed point theorem. Take $C^{1,\alpha}(\overline{\Omega})$ as the Banach space X in the theorem. For any $v \in C^{1,\alpha}(\overline{\Omega})$ and $0 \leqslant \sigma \leqslant 1$, consider the following Dirichlet problem:

$$\begin{cases} -\left[\sigma \dfrac{\partial a_i}{\partial \eta_j}(x,v,Dv)D_{ij}u + (1-\sigma)\Delta u\right] \\ \quad +\sigma\left[\dfrac{\partial a_i}{\partial x_i} + \dfrac{\partial a_i}{\partial z}\eta_i + b\right]_{(x,z,\eta)=(x,v,Dv)} = 0 \quad \text{in } \Omega, \\ u = \sigma\varphi \quad \text{on } \partial\Omega. \end{cases} \tag{5.6}$$

Notice that the coefficients of the linear equation for u are in $C^{\alpha^2}(\overline{\Omega})$. Therefore the Schauder theory implies that (5.6) admits a unique solution $u \in C^{2,\alpha^2}(\overline{\Omega}) \subset C^{1,\alpha}(\overline{\Omega})$. Hence $u = T(v,\sigma)$ defines a map from $C^{1,\alpha}(\overline{\Omega}) \times [0,1]$ to $C^{1,\alpha}(\overline{\Omega})$. We shall verify all three conditions in the Leray-Schauder theorem.

(1) Let K be a bounded set in $C^{1,\alpha}(\overline{\Omega})$. The Schauder theory implies that TK is a bounded set in $C^{2,\alpha^2}(\overline{\Omega})$, which is precompact in $C^{1,\alpha}(\overline{\Omega})$ by the Ascoli-Arzelá theorem. Next, we show that T is continuous. Suppose that $v_m \to v$ $(C^{1,\alpha}(\overline{\Omega}))$,

$\sigma_m \to \sigma$ as $m \to \infty$. Denote $u_m = T(v_m, \sigma_m)$; we want to show that $u_m \to u = T(v, \sigma)$. In fact, from the above arguments we deduce that u_m is precompact in $C^2(\overline{\Omega})$. For any convergent subsequence u_{m_k} in $C^2(\overline{\Omega})$, if u_{m_k} converges to some $\widetilde{u} \in C^2(\overline{\Omega})$, then it is obvious that $\widetilde{u}$ is a solution of (5.6) and therefore $\widetilde{u} \equiv u$, by the maximum principle for linear equations. It follows that $\{u_m\}$ itself must converge in $C^2(\overline{\Omega})$ to $u = T(v, \sigma)$. Thus T is compact.

(2) Obviously, $T(v, 0) = 0$.

(3) If $u = T(u, \sigma)$ for some $\sigma \in [0, 1]$, then u satisfies

$$\begin{cases} -D_i[\sigma a_i(x, u, Du) + (1-\sigma) D_i u] + \sigma b(x, u, Du) = 0 & \text{in } \Omega, \\ u = \sigma\varphi \quad \text{on } \partial\Omega. \end{cases} \tag{5.7}$$

The structure conditions are assumed in this theorem. By a priori estimates in the previous sections, there exist $M > 0$ and $0 < \gamma < 1$ such that

$$|u|_{1,\gamma;\Omega} \leqslant M. \tag{5.8}$$

Now we rewrite (5.7) in nondivergence form:

$$-\left[\sigma \frac{\partial a_i}{\partial \eta_j} D_{ij} u + (1-\sigma)\Delta u\right] + \sigma\left[\frac{\partial a_i}{\partial x_i} + \frac{\partial a_i}{\partial z} D_i u + b\right] = 0, \tag{5.9}$$

where the coefficients belong to $C^{\alpha\gamma}(\overline{\Omega})$, by (5.8). The Schauder theory implies that $u \in C^{2,\alpha\gamma}(\overline{\Omega})$ and that there exists $C > 0$, independent of σ and u, such that

$$|u|_{2,\alpha\gamma} \leqslant C; \tag{5.10}$$

this implies that

$$|u|_{1,\alpha;\Omega} \leqslant M.$$

The third condition of the Leray-Schauder theorem is satisfied.

The Leray-Schauder theorem implies that $T(\cdot, 1)$ has a fixed point, which is exactly a solution of (5.1), (5.2). The previous argument shows that $u \in C^{2,\alpha\gamma}(\overline{\Omega})$ and therefore the coefficients in (5.9) ($\sigma = 1$) belong to $C^{\alpha}(\overline{\Omega})$. Applying the Schauder theory again, we conclude that $u \in C^{2,\alpha}(\overline{\Omega})$. The proof is complete. □

We summarize the above proof as follows: if we define the map T appropriately, then condition (2) is obviously satisfied. By the Schauder theory, it is not difficult to verify condition (1). Thus the solvability problem is reduced to a priori estimates of solutions (condition (3)).

There is another technique in the proof that we should pay special attention to. After the estimate (5.8) is derived, we use the Schauder theory repeatedly to get higher regularities. Such an iteration technique is frequently used in studying partial differential equations and should be skillfully mastered.

CHAPTER 6

Krylov-Safonov Estimates

As we saw in the last chapter, the De Giorgi-Nash-Moser estimates played a key role in studying quasilinear equations in divergence form. For general equations in nondivergence form, people hoped that a similar estimate might be valid, which would be indispensable in studying fully nonlinear elliptic equations. After more than 20 years of effort, Krylov and Safonov obtained such an estimate in 1980. Another proof was given by Trudinger. Both proofs rely on the Alexandroff-Bakelman-Pucci maximum principle.

1. The Alexandroff-Bakelman-Pucci maximum principle

First, we introduce the concept of a normal mapping. It is a generalization of a gradient mapping.

Definition 1.1. Let $u \in C(\Omega)$, where Ω is a bounded open domain in $\mathbb{R}^n$. For $y \in \Omega$, we set

$$\chi(y) = \{p \in \mathbb{R}^n \mid u(x) \leqslant u(y) + p \cdot (x - y),\ \forall x \in \Omega\}. \tag{1.1}$$

χ defines a map from Ω to a class consisting of subsets of $\mathbb{R}^n$. We say that χ is a *normal mapping* defined by u.

A normal mapping has its clear geometric meaning. The *lower space* of the graph $z = u(x)$ is the set (in $\mathbb{R}^{n+1}$)

$$\{(x, z) \in \mathbb{R}^n \times \mathbb{R} \mid x \in \Omega, -\infty < z < u(x)\}.$$

If $p \in \chi(y)$, then the hyperplane $z = u(y) + p \cdot (x - y)$ is a supporting plane for the lower space of the graph $z = u(x)$ at $(y, u(y))$. $\chi(y)$ is the set of all p's corresponding to supporting planes for the lower space of the graph $z = u(x)$ at $(y, u(y))$ such that $(-p, 1)$ is a normal vector of the supporting plane.

Definition 1.2. Let $u \in C(\Omega)$. The set

$$\begin{aligned} \Gamma_u &= \{y \in \Omega \mid \chi(y) \neq \emptyset\} \\ &= \{y \in \Omega \mid \exists p \in \mathbb{R}^n \text{ such that } u(x) \leqslant u(y) + p \cdot (x - y),\ \forall x \in \Omega\} \end{aligned} \tag{1.2}$$

is said to be the *contact set* of u.

Next, we consider the *convex hull* of the lower space of the graph $z = u(x)$; this convex hull is the lower space of some graph $z = \widehat{u}(x)$. It is clear that $\widehat{u}(x) \geqslant u(x)$

and that it is the smallest concave function. Γ_u consists of the projections to the plane $z = 0$ of those points where the hypersurfaces $z = \widehat{u}(x)$ and $z = u(x)$ meet. This is where the name *contact set* comes from.

If $u \in C^1(\Omega)$, $y \in \Gamma_u$, then $\chi(y) = \{Du(y)\}$; if furthermore $u \in C^2(\Omega)$ and $\chi(y)$ is nonempty, then $-D^2u(y) \geqslant 0$ (i.e., the Hessian matrix is negative semidefinite). In fact, the function associated with the normal mapping

$$w(x) = u(y) + p \cdot (x - y) - u(x), \quad \forall x \in \Omega, \tag{1.3}$$

attains its minimum at y, and therefore $Dw(y) = 0$ and $D^2w(y) \geqslant 0$. This implies the above result. More generally, we have

Lemma 1.1. *Let $u \in W^{2,1}_{loc}(\Omega) \cap C(\Omega)$. Then*

$$\chi(y) = \{Du(y)\}, \quad -D^2u(y) \geqslant 0, \quad a.e.\ y \in \Gamma_u. \tag{1.4}$$

Proof. Let $w(x)$ be defined by (1.3). For each fixed direction $\xi \in \mathbb{R}^n$, $|\xi| = 1$, we have

$$\frac{w(y + h\xi) - w(y)}{h} \to \frac{\partial w}{\partial \xi}, \tag{1.5}$$

$$\frac{w(y + h\xi) + w(y - h\xi) - 2w(y)}{h^2} \to \frac{\partial^2 w}{\partial \xi^2}, \tag{1.6}$$

where the convergence is in the space $L^1_{loc}(\Omega)$. If we take subsequences, then the above limits are valid for almost every $y \in \Omega$. If $y \in \Gamma_u$, then $w(x)$ takes its minimum at y. Letting $h \to 0^+$ and $h \to 0^-$ in (1.5), we deduce that

$$\frac{\partial w}{\partial \xi} = 0, \quad \text{a.e. } y \in \Gamma_u.$$

If we take ξ to be the direction of coordinate axes, then

$$\chi(y) = \{Du(y)\}, \quad \text{a.e. } y \in \Gamma_u.$$

Similarly, by (1.6),

$$-\frac{\partial^2 u}{\partial \xi^2} = \frac{\partial^2 w}{\partial \xi^2} \geqslant 0, \quad \text{a.e. } y \in \Gamma_u.$$

It follows that for ξ in a dense countable subset of the unit sphere,

$$-\frac{\partial^2 u}{\partial \xi^2} \geqslant 0, \quad \text{a.e. } y \in \Gamma_u.$$

From this inequality we deduce that, for any ξ, $|\xi| = 1$,

$$-\frac{\partial^2 u}{\partial \xi^2} \geqslant 0, \quad \text{a.e. } y \in \Gamma_u.$$

This implies that $-D^2u(y) \geqslant 0$ for almost every $y \in \Gamma_u$. The proof is complete. □

Definition 1.3. The set

$$\chi(\Omega) = \chi(\Gamma_u) = \bigcup_{y \in \Omega} \chi(y)$$

is said to be the *image set* of the normal mapping determined by u.

Example. Let $\Omega = B_d(x_0)$. Consider the function

$$u(x) = \frac{\lambda}{d}(d - |x - x_0|). \tag{1.7}$$

Its graph is a cone surface with vertex at (x_0, λ), base $B_d(x_0)$, and height λ. Clearly, $\Gamma_u = \Omega$ and

$$\chi(y) = \begin{cases} B_{\lambda/d}(0) & \text{if } y = x_0, \\ -\dfrac{\lambda}{d}\dfrac{y - x_0}{|y - x_0|} & \text{if } y \neq x_0. \end{cases}$$

The image set of the normal mapping is given by

$$\chi(\Omega) = B_{\lambda/d}(0). \tag{1.8}$$

Definition 1.4. Let $\Omega \subset \mathbb{R}^n$, $x_0 \in \Omega$. Let w be the function such that its graph is a cone surface with vertex at (x_0, λ) and base Ω (see Fig. 1). We denote its image set of the normal mapping by

$$\Omega[x_0, \lambda] = \chi_w(\Omega). \tag{1.9}$$

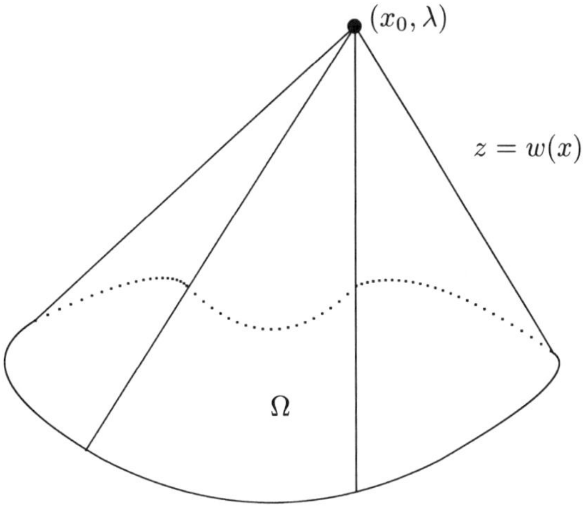

Fig. 1

Lemma 1.2. *Let $u \in C(\Omega)$. Then*

(1) *for any $y \in \Gamma_u$,*

$$|p| \leqslant \frac{2 \sup |u|}{\operatorname{dist}\{y, \partial\Omega\}}, \quad \forall p \in \chi(y); \tag{1.10}$$

(2) *the normal mapping maps any compact subset of Ω to a closed set in $\mathbb{R}^n$.*

Proof. For $y \in \Gamma_u$,

$$u(y) + p \cdot (x - y) \geqslant u(x), \quad \forall x \in \Omega. \tag{1.11}$$

The ray starting at y with direction $-p$ intersects $\partial\Omega$ at x_0, i.e.,

$$x_0 = y - \frac{1}{|p|}|x_0 - y|p. \tag{1.12}$$

Using compact subsets of Ω to approximate Ω if necessary, we may assume without loss of generality that u is continuous on $\overline{\Omega}$. Choosing x to be x_0 in (1.11), we obtain

$$u(y) - |x_0 - y|\,|p| \geqslant u(x_0),$$

and therefore

$$|p| \leqslant \frac{2\sup|u|}{|x_0 - y|} \leqslant \frac{2\sup|u|}{\operatorname{dist}\{y, \partial\Omega\}}.$$

Now we prove (2). Let F be a compact subset of Ω. Suppose that $\{p_n\} \subset \chi(F)$ and $p_n \to p_0$ $(n \to \infty)$. We want to show that $p_0 \in \chi(F)$. Since $p_n \in \chi(F)$, there exists $y_n \in F$ such that $p_n \in \chi(y_n)$. From the definition of a normal mapping,

$$u(y_n) + p_n \cdot (x - y_n) \geqslant u(x), \quad \forall x \in \Omega.$$

Since F is compact, a subsequence $\{y_{n_k}\}$ converges to some $y_0 \in F$ as $k \to \infty$. Letting $n = n_k \to \infty$ in the above inequality, we easily see that $p_0 \in \chi(y_0)$. □

Lemma 1.3. *Suppose that Ω, A are open domains in $\mathbb{R}^n$.*

(1) *If $\Omega \subset A$, then for $x_0 \in \Omega$,*

$$\Omega[x_0, \lambda] \supset A[x_0, \lambda].$$

(2) *If the diameter of Ω is d, then*

$$|\Omega[x_0, \lambda]| \geqslant \left(\frac{\lambda}{d}\right)^n \omega_n, \tag{1.13}$$

where $|\cdot|$ denotes the measure of the set and ω_n is the volume of an n-dimensional unit ball.

Proof. (1) is obvious. We prove (2). Clearly, $B_d(x_0) \supset \Omega$. Let $A = B_d(x_0)$. By (1) and (1.8),

$$|\Omega[x_0, \lambda]| \geqslant |A[x_0, \lambda]| = |B_{\lambda/d}(0)| = \omega_n \left(\frac{\lambda}{d}\right)^n,$$

where we use Lemma 1.2 (2) to deduce that $\Omega[x_0, \lambda]$ and $A[x_0, \lambda]$ are measurable sets. □

Lemma 1.4. *Suppose that $u \in C^2(\Omega)$, $g \in C(\overline{\Omega})$, $g \geqslant 0$, and E is a measurable subset of Γ_u. Then*

$$\int_{Du(E)} g(\xi(p))dp \leqslant \int_E g(x)\det(-D^2u)dx, \tag{1.14}$$

where $\xi(p) = (Du)^{-1}(p)$ is well defined and is continuous on $Du(E)$ except on a zero measure set.

Proof. Let $J(x) = \det(-D^2u)$, $S = \{x \in \Omega \mid J(x) = 0\}$. By Sard's theorem (cf. Appendix 2), $|Du(S)| = 0$.

First we assume that E is open. Then $E \setminus S$ is an open set. Thus there exist cubes $\{C_l\}_{l=1}^{\infty}$ with disjoint interior and sides parallel to the coordinate axes such that $E \setminus S = \bigcup_{l=1}^{\infty} C_l$. We assume without loss of generality that the C_l are small enough so that $Du : C_l \to Du(C_l)$ is a diffeomorphism. Then

$$\int_{Du(C_l)} g(\xi(p))dp = \int_{C_l} g(x)J(x)dx.$$

It follows that

$$\begin{aligned}\int_{Du(E\setminus S)} g(\xi(p))dp &\leqslant \sum_l \int_{Du(C_l)} g(\xi(p))dp \\ &= \sum_l \int_{C_l} g(x)J(x)dx = \int_{E\setminus S} g(x)J(x)dx.\end{aligned}$$

Using the definition of S and the fact that $|Du(S)| = 0$, we obtain (1.14).

Now we assume that E is a measurable subset of Γ_u. Then there exists an open set $G \supset E \setminus S$ such that $J(x) > 0$ on G. Since $E \setminus S$ is a measurable set, there exist open sets $\{O_l\}_{l=1}^{\infty}$ such that $E \setminus S \subset O_l$ and $|O_l \setminus (E \setminus S)| \to 0$ $(l \to \infty)$. Using the above results, we derive

$$\int_{Du(E)} g(\xi(p))dp \leqslant \int_{Du(G\cap O_l)} g(\xi(p))dp \leqslant \int_{G\cap O_l} g(x)J(x)dx.$$

Letting $l \to \infty$ in the above inequality, we get (1.14). □

Lemma 1.5. *Suppose that $u \in C(\overline{\Omega})$, $u \leqslant 0$ on $\partial\Omega$, $x_0 \in \Omega$, and $u(x_0) > 0$. Then*

$$\Omega[x_0, u(x_0)] \subset \chi(\Gamma_u^+), \tag{1.15}$$

where $\Gamma_u^+ = \Gamma_u \cap \{u \geqslant 0\}$.

Proof. This lemma is obvious from the geometric picture. However, here we shall give a rigorous analytical proof. Let $p \in \Omega[x_0, u(x_0)]$. From Definition 1.4,

$$u(x_0) + p \cdot (x - x_0) \geqslant 0, \quad \forall x \in \Omega. \tag{1.16}$$

Define

$$\lambda_0 = \inf\{\lambda \mid \lambda + p \cdot (x - x_0) \geqslant u(x), \quad \forall x \in \Omega\}.$$

Since u is continuous,

$$\lambda_0 + p \cdot (x - x_0) \geqslant u(x), \quad \forall x \in \Omega; \tag{1.17}$$

furthermore, there exists $\xi \in \overline{\Omega}$ such that

$$\lambda_0 + p \cdot (\xi - x_0) = u(\xi). \tag{1.18}$$

We consider two cases:

Case (i): $\lambda_0 = u(x_0)$. Then (1.17) implies that $x_0 \in \Gamma_u^+$, $p \in \chi(\Gamma_u^+)$.

Case (ii): $\lambda_0 > u(x_0)$. We claim that $\xi \notin \partial\Omega$. In fact, (1.18) and (1.16) imply

$$u(\xi) > u(x_0) + p \cdot (\xi - x_0) \geqslant 0.$$

Since $u\big|_{\partial\Omega} \leqslant 0$, we must have $\xi \notin \partial\Omega$. We subtract (1.18) from (1.17) to obtain

$$u(\xi) + p \cdot (x - \xi) \geqslant u(x), \quad \forall x \in \Omega.$$

It follows that $\xi \in \Gamma_u^+$ and $p \in \chi(\Gamma_u^+)$. □

Now we are ready to prove an Alexandroff-Bakelman-Pucci type of maximum principle.

Lemma 1.6. *Suppose that $u \in C^2(\overline{\Omega})$ and $u \leqslant 0$ on $\partial\Omega$. Then*

$$\sup_{\Omega} u \leqslant \frac{d}{\sqrt[n]{\omega_n}} \left[\int_{\Gamma_u^+} \det(-D^2 u) dx \right]^{1/n}, \tag{1.19}$$

where $d = \operatorname{diam} \Omega$.

Proof. For $x_0 \in \Omega$, $u(x_0) > 0$, we can derive from (1.15) and (1.13) that

$$|\chi(\Gamma_u^+)| \geqslant |\Omega[x_0, u(x_0)]| \geqslant \omega_n \left[\frac{u(x_0)}{d} \right]^n,$$

where $\chi(\Gamma_u^+)$ is a measurable set, by Lemma 1.2 (2). It follows that

$$u(x_0) \leqslant \frac{d}{\sqrt[n]{\omega_n}} |\chi(\Gamma_u^+)|^{1/n}.$$

By Lemma 1.4,

$$|\chi(\Gamma_u^+)| = \int_{\chi(\Gamma_u^+)} dp \leqslant \int_{\Gamma_u^+} \det(-D^2 u) dx.$$

Combining the above two inequalities, we get (1.19). □

We now relax the smoothness assumption on u in the above lemma.

Theorem 1.7. *If $u \in C(\overline{\Omega}) \cap W^{2,n}_{loc}(\Omega)$, then*

$$\sup_{\Omega} u \leqslant \sup_{\partial\Omega} u + \frac{d}{\sqrt[n]{\omega_n}} \left[\int_{\Gamma_v^+} \det(-D^2 u) dx \right]^{1/n}, \tag{1.20}$$

where $v = u - \sup_{\partial\Omega} u$.

Proof. Under our assumptions, there exists a sequence $\{u_m\} \subset C^2(\overline{\Omega})$ such that u_m converges to u in $W^{2,n}_{loc}(\Omega)$; i.e., for any $\Omega' \subset\subset \Omega$, u_m converges to u in $W^{2,n}(\Omega')$ as $m \to \infty$. For any $\varepsilon > 0$, set $\Omega_\varepsilon = \{x \in \Omega \mid \operatorname{dist}\{x, \partial\Omega\} > \varepsilon\}$,

$$v_\varepsilon = u - \sup_{\Omega \setminus \Omega_\varepsilon} u - \varepsilon, \tag{1.21}$$

and

$$v_{m,\varepsilon} = u_m - \sup_{\Omega \setminus \Omega_\varepsilon} u_m - \varepsilon. \tag{1.22}$$

Applying Lemma 1.6 to the function $v_{m,\varepsilon}$, we get

$$\sup_{\Omega} v_{m,\varepsilon} \leqslant \frac{d}{\sqrt[n]{\omega_n}}\left[\int_{\Gamma^+_{v_{m,\varepsilon}}} \det(-D^2 u_m)dx\right]^{1/n}. \tag{1.23}$$

Notice that $\Gamma^+_{v_{m,\varepsilon}} \subset \Omega_\varepsilon$, and

$$\begin{aligned}\int_{\Gamma^+_{v_{m,\varepsilon}}} \det(-D^2u_m)dx &= \int_{\Gamma^+_{v_{m,\varepsilon}}} \left[\det(-D^2u_m) - \det(-D^2u)\right]dx \\ &\quad + \int_{\Gamma^+_{v_{m,\varepsilon}}} \det(-D^2u)dx \\ &\triangleq I_{1m} + I_{2m}.\end{aligned} \tag{1.24}$$

Clearly, $I_{1m} \to 0$ as $m \to \infty$. Next we claim that

$$\limsup_{m\to\infty} \Gamma^+_{v_{m,\varepsilon}} \subset \Gamma^+_{v_\varepsilon}. \tag{1.25}$$

In fact, if $x_0 \in \limsup_{m\to\infty} \Gamma^+_{v_{m,\varepsilon}}$, then there exists a subsequence m_k such that $x_0 \in \Gamma^+_{v_{m_k,\varepsilon}}$. By Lemma 1.2 (1),

$$|p_k| \leqslant \frac{4\sup|u_m|}{\operatorname{dist}\{x_0, \partial\Omega\}} \leqslant \frac{4}{\varepsilon}\sup|u_m|, \quad \forall p_k \in \mathcal{X}_{v_{m_k,\varepsilon}}(x_0).$$

Thus there is a convergent subsequence of p_k. We assume without loss of generality that $p_k \to p_0$. It is easy to verify that $p_0 \in \mathcal{X}_{v_\varepsilon}(x_0)$, i.e., $x_0 \in \Gamma^+_{v_\varepsilon}$. (1.25) is proved. Hence

$$\limsup_{m\to\infty} I_{2m} \leqslant \int_{\Gamma^+_{v_\varepsilon}} \det(-D^2u)dx.$$

Letting $m \to \infty$ in (1.24), we get

$$\sup_{\Omega} v_\varepsilon \leqslant \frac{d}{\sqrt[n]{\omega_n}}\left[\int_{\Gamma^+_{v_\varepsilon}} \det(-D^2u)dx\right]^{1/n}. \tag{1.26}$$

In the process of taking the limit, we made use of the fact that $v_{m,\varepsilon}$ converges to v_ε in $C(\overline{\Omega}_\varepsilon)$, which is a corollary of the Sobolev embedding theorem. Since $\Gamma^+_{v_\varepsilon} \subset \Gamma^+_v$, we can let $\varepsilon \to 0$ in (1.26) and obtain (1.20). □

Now we discuss the following elliptic equation:

$$Lu = -a^{ij}D_{ij}u + b^iD_iu + cu = f, \tag{1.27}$$

where we assume that

$$(a^{ij}) \geqslant 0 \quad \text{in } \Omega, \tag{1.28}$$

$$\sum_i \|b^i/\mathcal{D}^*\|_{L^n(\Omega)} \leqslant B, \tag{1.29}$$

$$c \geqslant 0 \quad \text{on } \Omega, \tag{1.30}$$

where $\mathcal{D}^* = [\det(a^{ij})]^{1/n}$.

First we consider a special case:

$$L_0u = -a^{ij}D_{ij}u = f. \tag{1.31}$$

Theorem 1.8. *Suppose that* $u \in C(\overline{\Omega}) \cap W^{2,n}_{loc}(\Omega)$ *satisfies* $L_0 u \leqslant f$ *almost everywhere in* Ω *and the coefficients* a^{ij} *satisfy* (1.28). *Then*

$$\sup_{\Omega} u(x) \leqslant \sup_{\partial\Omega} u(x) + \frac{d}{n\sqrt[n]{\omega_n}} \left\| \frac{f^+}{\mathcal{D}^*} \right\|_{L^n(\Gamma_v^+)}, \tag{1.32}$$

where $v = u - \sup_{\partial\Omega} u(x)$.

Proof. We assume that the right-hand side of (1.32) is finite. Set $A = (a^{ij})$, $U = (-D^2 u)$. By Lemma 1.1, $U \geqslant 0$ almost everywhere in Γ_v. Using the inequality that the arithmetic mean is greater than the geometric mean, we get

$$\begin{aligned} -a^{ij} D_{ij} u &= \operatorname{Trace}(AU) \geqslant n[\det(AU)]^{1/n} \\ &= n\mathcal{D}^*[\det(-D^2u)]^{1/n}, \quad \text{a.e. } x \in \Gamma_v^+. \end{aligned}$$

It follows that

$$[\det(-D^2u)]^{1/n} \leqslant \frac{1}{n\mathcal{D}^*}[-a^{ij}D_{ij}u] \leqslant \frac{f^+}{n\mathcal{D}^*}$$

almost everywhere in Γ_v^+. Applying Theorem 1.7, we immediately obtain the theorem. □

Finally, we have the following Alexandroff-Bakelman-Pucci maximum principle for (1.27):

Theorem 1.9. *Suppose that* $u \in C(\overline{\Omega}) \cap W^{2,n}_{loc}(\Omega)$ *satisfies the inequality* $Lu \leqslant f$ *almost everywhere in* Ω *and the coefficients satisfy* (1.28), (1.29) *and* (1.30). *Then*

$$\sup_{\Omega} u(x) \leqslant \sup_{\partial\Omega} u(x) + C \left\| \frac{f^+}{\mathcal{D}^*} \right\|_{L^n(\Gamma_u)}, \tag{1.33}$$

where C *depends only on* n, B *and* $\operatorname{diam}\Omega$.

Proof. Set $v = u - \sup_{\partial\Omega} u(x)$. Since $Lu \leqslant f$, we have

$$-a^{ij}D_{ij}v \leqslant -b^i D_i v + f \leqslant |Du|\,|\boldsymbol{b}| + f^+$$

a.e. on Γ_v^+, where $\boldsymbol{b}$ is the vector $(b^1, b^2, \cdots, b^n)$. By Young's inequality,

$$-a^{ij}D_{ij}v \leqslant \Big[|\boldsymbol{b}|^n + \mu^{-n}(f^+)^n\Big]^{1/n} \Big[|Du|^{n/(n-1)} + \mu^{n/(n-1)}\Big]^{(n-1)/n},$$

where μ is a constant to be determined. Set

$$g(p) = \Big[|p|^{n/(n-1)} + \mu^{n/(n-1)}\Big]^{1-n}.$$

Then we can rewrite the inequality as

$$-a^{ij}D_{ij}v \cdot [g(Du)]^{1/n} \leqslant \Big[|\boldsymbol{b}|^n + \mu^{-n}(f^+)^n\Big]^{1/n}. \tag{1.34}$$

First, we assume that $u \in C^2(\overline{\Omega})$. Then Lemma 1.4 implies that

$$\int_{\chi(\Gamma_v^+)} g(p)dp \leqslant \int_{\Gamma_v^+} g(Du)\det(-D^2u)dx. \tag{1.35}$$

Since v is continuous, there exists $x_0 \in \overline{\Omega}$ such that $v(x_0) = \sup_\Omega v(x) \triangleq M$. If $x_0 \in \partial\Omega$, then $M = 0$. Now we assume that $x_0 \in \Omega$. Then by Lemma 1.3 (2) and Lemma 1.5,

$$B_{M/d}(0) \subset \Omega[x_0, v(x_0)] \subset \chi(\Gamma_v^+).$$

Notice that

$$g(p) \geqslant 2^{2-n}(|p|^n + \mu^n)^{-1};$$

therefore,

$$\begin{aligned}\int_{\chi(\Gamma_v^+)} g(p)dp &\geqslant \int_{B_{M/d}(0)} 2^{2-n}(|p|^n + \mu^n)^{-1}dp \\ &= 2^{2-n}\omega_n \log\left[1 + \left(\frac{M}{\mu d}\right)^n\right].\end{aligned}$$

Combining this inequality with (1.35), we derive

$$\log\left[1 + \left(\frac{M}{\mu d}\right)^n\right] \leqslant \frac{2^{2-n}}{\omega_n}\int_{\Gamma_v^+} g(Du)\det(-D^2u)dx. \tag{1.36}$$

Using a proof similar to that of Theorem 1.7 and noticing also that $g(p)$ is bounded, we can extend the above inequality to functions $u \in C(\overline{\Omega}) \cap W^{2,n}_{loc}(\Omega)$. Using (1.36) and (1.34), we now proceed as in the proof of Theorem 1.8 to deduce that

$$\begin{aligned}\log\left[1 + \left(\frac{M}{\mu d}\right)^n\right] &\leqslant \frac{2^{2-n}}{\omega_n}\int_{\Gamma_v^+} g(Du)\left[\frac{-a^{ij}D_{ij}u}{n\mathcal{D}^*}\right]^n dx \\ &\leqslant \frac{2^{2-n}}{\omega_n n^n}\int_{\Gamma_v^+}\left[\left|\frac{\boldsymbol{b}}{\mathcal{D}^*}\right|^n + \frac{1}{\mu^n}\left(\frac{f^+}{\mathcal{D}^*}\right)^n\right]dx.\end{aligned}$$

If $f^+ \not\equiv 0$, we set $\mu = \|f^+/\mathcal{D}^*\|_{L^n(\Gamma_v^+)}$. Then

$$\left(\frac{M}{\mu d}\right)^n \leqslant \exp\left\{\frac{2^{2-n}}{\omega_n n^n}\int_{\Gamma_v^+}\left[\left|\frac{\boldsymbol{b}}{\mathcal{D}^*}\right|^n + 1\right]dx\right\} - 1,$$

and

$$M = \sup_\Omega v(x) \leqslant Cd\|f^+/\mathcal{D}^*\|_{L^n(\Gamma_v^+)}.$$

Since $\Gamma_v^+ \subset \Gamma_v = \Gamma_u$, we get the theorem in this case.

On the other hand, if $f^+ \equiv 0$, we choose $\mu > 0$ and follow the above calculations. Then we let $\mu \to 0$ to obtain the theorem. □

2. Harnack inequalities and interior Hölder estimates

We shall follow the procedure for establishing the De Giorgi-Nash-Moser estimates. We shall first establish a local maximum principle, then the weak Harnack inequality, and finally the Harnack inequality.

For simplicity, we consider only the following equation:

$$Lu = -a^{ij}D_{ij}u = f \quad \text{in } \Omega. \tag{2.1}$$

Assume that the coefficient matrix (a^{ij}) satisfies a uniform ellipticity condition:

$$\lambda > 0, \quad \Lambda/\lambda \leqslant \gamma, \quad \forall x \in \Omega, \tag{2.2}$$

where $\lambda = \lambda(x)$ and $\Lambda = \Lambda(x)$ are the smallest and the largest eigenvalue of the matrix $(a^{ij}(x))$, respectively, and γ is a constant.

Theorem 2.1 (Local maximum principle). *Let the assumption* (2.2) *be in force. Suppose that* $u \in W^{2,n}(\Omega)$ *satisfies* $Lu \leqslant f$ *almost everywhere in* Ω *and* $f/\lambda \in L^n(\Omega)$. *Then for any* $p > 0$, $B_{2R}(y) \subset \Omega$, *we have*

$$\sup_{B_R(y)} u(x) \leqslant C\left\{\left[\fint_{B_{2R}(y)} (u^+)^p dx\right]^{1/p} + R\|f/\lambda\|_{L^n(B_{2R}(y))}\right\}, \tag{2.3}$$

where C *depends only on* n, γ, p.

Proof. Using the change of variables $x \longleftrightarrow \frac{x-y}{2R}$ if necessary, we may assume without loss of generality that y is the origin and $R = 1/2$. Take the cutoff function

$$\eta(x) = (1 - |x|^2)^\beta,$$

where $\beta \geqslant 2$ is to be determined. A simple calculation shows that

$$\begin{aligned} D_i\eta &= -2\beta x_i(1 - |x|^2)^{\beta-1}, \\ D_{ij}\eta &= -2\beta\delta_{ij}(1 - |x|^2)^{\beta-1} + 4\beta(\beta - 1)x_i x_j(1 - |x|^2)^{\beta-2}, \\ L\eta &= [2\beta\mathcal{T}(1 - |x|^2) - 4\beta(\beta - 1)a^{ij}x_i x_j](1 - |x|^2)^{\beta-2}, \end{aligned}$$

where $\mathcal{T}$ =Trace(a^{ij}). Set $v = \eta u$. By the product rule for derivatives,

$$Lv = uL\eta + \eta Lu - 2a^{ij}D_i\eta D_j u. \tag{2.4}$$

Since $v = 0$ on ∂B_1, the proof of Lemma 1.2 (1) implies that

$$|Dv| \leqslant \frac{v(x)}{\operatorname{dist}\{x, \partial B_1\}} \leqslant \frac{v(x)}{1 - |x|}, \quad \text{a.e. } x \in \Gamma_v.$$

Therefore

$$\begin{aligned} |Du| &= \frac{1}{\eta}|Dv - uD\eta| \leqslant \frac{1}{\eta}\left[\frac{v(x)}{1 - |x|} + u|D\eta|\right] \\ &\leqslant 2(1 + \beta)\eta^{-1/\beta}u, \quad \text{a.e. } x \in \Gamma_v. \end{aligned}$$

Using these inequalities in (2.4), we obtain

$$\begin{aligned} Lv &\leqslant \eta f + (16\beta^2 + 2n\beta)\Lambda\eta^{-2/\beta}v \\ &\leqslant C\lambda\eta^{-2/\beta}v + f, \quad \text{a.e. } x \in \Gamma_v, \end{aligned}$$

where C depends only on n, β and γ. By the Alexandroff-Bakelman-Pucci maximum principle,

$$\begin{aligned} \sup_{B_1} v &\leqslant C\Big(\|\eta^{-2/\beta}v^+\|_{L^n(B_1)} + \|f/\lambda\|_{L^n(B_1)}\Big) \\ &\leqslant C\Big\{(\sup v^+)^{1-2/\beta}\|(u^+)^{2/\beta}\|_{L^n(B_1)} + \|f/\lambda\|_{L^n(B_1)}\Big\}. \end{aligned}$$

Using Young's inequality, we get

$$\sup_{B_1} v \leqslant C\|f/\lambda\|_{L^n(B_1)} + \varepsilon \sup_{B_1} v^+ + C_\varepsilon \|(u^+)^{2/\beta}\|^{\beta/2}_{L^n(B_1)}.$$

We let $\varepsilon = 1/2$ in the above inequality. If $p < n$, we choose $\beta = 2n/p > 2$. Then

$$\sup_{B_{1/2}} u \leqslant C\Big[\|u^+\|_{L^p(B_1)} + \|f/\lambda\|_{L^n(B_1)}\Big]. \tag{2.5}$$

If $p \geqslant n$, we can use the result for $p < n$ and Hölder's inequality to obtain the theorem. □

The proof of the weak Harnack inequality is the most difficult part of this section. We shall use K_0 and $K_R(y)$ to denote cubes with their sides parallel to the coordinate axes. We employ the Krylov-Safonov measure theory method; it is based on the following lemma:

Lemma 2.2. *Let K_0 be a cube in $\mathbb{R}^n$. Suppose that Γ is a measurable subset of K_0. For $0 < \delta < 1$, we set*

$$\Gamma_\delta = \bigcup \{K_{3R}(y) \cap K_0 \mid K_R(y) \subset K_0,\ |\Gamma \cap K_R(y)| \geqslant \delta |K_R(y)|\}. \tag{2.6}$$

If $\Gamma_\delta \neq K_0$, then $|\Gamma| \leqslant \delta |\Gamma_\delta|$.

Proof. If $|K_0 \cap \Gamma| > \delta |K_0|$, then obviously $\Gamma_\delta = K_0$. Therefore, if $\Gamma_\delta \neq K_0$, then $|K_0 \cap \Gamma| \leqslant \delta |K_0|$. We divide K_0 into 2^n congruent smaller cubes and denote them by $\{K(i_1)\}_{i_1=1}^{2^n}$. For each $K(i_1)$ there are two possibilities:

(1) $|\Gamma \cap K(i_1)| \leqslant \delta |K(i_1)|$,

(2) $|\Gamma \cap K(i_1)| > \delta |K(i_1)|$.

We denote by $\mathscr{F}_1$ the collection of all cubes $K(i_1)$ belonging to case (2). For a cube $K(i_1)$ belonging to case (1), we further divide it into 2^n smaller cubes and denote them by $K(i_1, i_2)$ ($i_2 = 1, 2, \cdots, 2^n$). There are also two possibilities for $K(i_1, i_2)$. We denote by $\mathscr{F}_2$ the collection of all cubes $K(i_1, i_2)$ belonging to case (2). Continuing this process, we get a sequence $\mathscr{F}_1, \mathscr{F}_2, \cdots, \mathscr{F}_m, \cdots$ of collections of cubes belonging to case (2). Now we set

$$\mathscr{F} = \{K(i_1, \cdots, i_{m-1}) \mid K(i_1, \cdots, i_{m-1}, i_m) \in \mathscr{F}_m\}. \tag{2.7}$$

For $K(i_1, \cdots, i_m) \in \mathscr{F}_m$, it is clear that

$$|K(i_1, \cdots, i_m) \cap \Gamma| > \delta |K(i_1, \cdots, i_m)|, \tag{2.8}$$

$$|K(i_1, \cdots, i_{m-1}) \cap \Gamma| \leqslant \delta |K(i_1, \cdots, i_{m-1})|. \tag{2.9}$$

The definition of Γ_δ implies that $K(i_1, \cdots, i_{m-1}) \subset \Gamma_\delta$. It follows that

$$\widetilde{\Gamma}_\delta \triangleq \bigcup_{K \in \mathscr{F}} K \subset \Gamma_\delta.$$

On the other hand, (2.9) implies that

$$\begin{aligned} |\widetilde{\Gamma}_\delta \cap \Gamma| &= \Big|\bigcup_{K\in\mathscr{F}} (K\cap\Gamma)\Big| = \sum_{K\in\mathscr{F}} |K\cap\Gamma| \\ &\leqslant \delta \sum_{K\in\mathscr{F}} |K| = \delta|\widetilde{\Gamma}_\delta| \leqslant \delta|\Gamma_\delta|. \end{aligned}$$

It is clear from the division procedure that any density point of Γ also belongs to $\widetilde{\Gamma}_\delta$. Since Γ is a measurable set, measure theory implies that almost every point of Γ is a density point. It follows that

$$|\Gamma| = |\widetilde{\Gamma}_\delta \cap \Gamma| \leqslant \delta|\Gamma_\delta|.$$

The proof is complete. □

Theorem 2.3 (Weak Harnack inequality). *Let the coefficients of L satisfy the assumption* (2.2). *Suppose that $u \in W^{2,n}(\Omega)$ satisfies $Lu \geqslant f$ almost everywhere in Ω and $f/\lambda \in L^n(\Omega)$, and $u \geqslant 0$ on $B_{2R}(y) \subset \Omega$. Then there exist $p > 0$ and $C > 1$ such that*

$$\left[\fint_{B_R(y)} (u^+)^p dx\right]^{1/p} \leqslant C\left[\inf_{B_R(y)} u + R\|f/\lambda\|_{L^n(B_{2R}(y))}\right], \tag{2.10}$$

where C and p depend only on n and γ.

Proof. As the proof is long, we divide it into five steps.

Step 1. Using the change of variables $x \longleftrightarrow \frac{x-y}{2R}$ if necessary, we may assume without loss of generality that y is the origin and $R = 1/2$. Set

$$\widetilde{u} = u + \|f/\lambda\|_{L^n(B_1)}, \quad \Gamma = \{x \in B_1 \mid \widetilde{u}(x) \geqslant 1\}. \tag{2.11}$$

We shall prove that there exist $C > 0$ and $0 < \delta < 1$ such that if $|\Gamma \cap K_\alpha| \geqslant \delta|K_\alpha|$, then

$$\inf_{K_{3\alpha}} \widetilde{u} \geqslant C^{-1}, \tag{2.12}$$

where K_α denotes the cube centered at the origin with side length α and sides parallel to the coordinate axes, $\alpha = 1/(6\sqrt{n})$, and C and δ depend only on n and γ.

For any $\varepsilon > 0$, we set $w = \log \dfrac{1}{\widetilde{u}+\varepsilon}$, $g = \dfrac{f}{\widetilde{u}+\varepsilon}$. A simple calculation shows that

$$\begin{aligned} -a^{ij}D_{ij}w &= \frac{1}{\widetilde{u}+\varepsilon} a^{ij}D_{ij}u - a^{ij}D_iwD_jw \\ &\leqslant -g - a^{ij}D_iwD_jw. \end{aligned}$$

We choose the cutoff function $\eta = (1-|x|^2)^\beta$ and set $v = \eta w$. Then

$$\begin{aligned} Lv &= \eta Lw + wL\eta - 2a^{ij}D_iwD_j\eta \\ &\leqslant -\eta g - \eta a^{ij}D_iwD_jw - 2a^{ij}D_iwD_j\eta + \frac{vL\eta}{\eta} \\ &\leqslant -\eta g + \frac{1}{\eta}a^{ij}D_i\eta D_j\eta + \frac{vL\eta}{\eta}. \end{aligned} \tag{2.13}$$

We have already computed in Theorem 2.1 that

$$
\begin{aligned}
\frac{L\eta}{\eta} &= (1-|x|^2)^{-2}\cdot 2\beta[\mathcal{T}(1-|x|^2)-2(\beta-1)a^{ij}x_ix_j] \\
&= 2\beta(1-|x|^2)^{-2}\Big[\mathcal{T}-\Big(\mathcal{T}|x|^2+2(\beta-1)a^{ij}x_ix_j\Big)\Big].
\end{aligned}
\tag{2.14}
$$

If

$$2(\beta-1)\lambda|x|^2 \geqslant n\Lambda,$$

then (2.14) implies that $L\eta\leqslant 0$. Therefore for $\alpha\in(0,1)$,

$$L\eta\leqslant 0 \quad \text{for } |x|>\alpha,$$

provided we take β such that

$$\beta-1\geqslant\frac{n\gamma}{2\alpha^2}.$$

In the above derivation we also made use of the uniform ellipticity condition (2.1): $\Lambda/\lambda\leqslant\gamma$. It follows from (2.13) that

$$Lv\leqslant|g|+4\beta^2\Lambda+\sup_{B_\alpha}\left(\frac{L\eta}{\eta}\right)\chi(B_\alpha)v \quad \text{on } B^+=\{x\in B_1\mid w(x)>0\},$$

where $\chi(B_\alpha)$ is the characteristic function on B_α. Since

$$\|g/\lambda\|_{L^n(B_1)}\leqslant 1,$$

the Alexandroff-Bakelman-Pucci maximum principle implies that

$$\sup_{B_1} v\leqslant C\Big[1+\|v^+\|_{L^n(B_\alpha)}\Big],\tag{2.15}$$

where C depends only on n,γ,α.

In order to take advantage of Lemma 2.2, we work in cubes instead of balls. From (2.15),

$$\sup_{B_1} v\leqslant C\Big[1+\|v^+\|_{L^n(K_\alpha)}\Big]\leqslant C\Big[1+|K_\alpha^+|^{1/n}\sup_{B_1} v\Big],$$

where $K_\alpha^+=\{x\in K_\alpha\mid v>0\}=\{x\in K_\alpha\mid \widetilde{u}+\varepsilon<1\}$. If

$$\frac{|K_\alpha^+|}{|K_\alpha|}\leqslant\theta\triangleq\frac{1}{(2C)^n|K_\alpha|}=\frac{1}{(4C\alpha)^n},$$

then

$$\sup_{B_1} v\leqslant 2C,$$

i.e.,

$$\inf_{B_{1/2}}(\widetilde{u}+\varepsilon)\geqslant\frac{1}{C}.\tag{2.16}$$

Here we use the same C as long as it depends on the same quantities; its value may change from one line to another. Letting $\varepsilon\to 0$ in (2.16), we deduce that if

$$\frac{|K_\alpha^+|}{|K_\alpha|}\leqslant\theta,$$

where $K_\alpha^+=\{x\in K_\alpha\mid\widetilde{u}\leqslant 1\}$, then

$$\inf_{B_{1/2}}\widetilde{u}\geqslant C^{-1}.$$

We take $\delta = 1 - \theta$. Recall that $\alpha = 1/(6\sqrt{n})$; it is obvious that $K_{3\alpha} \subset B_{1/2}$. Therefore, if $|\Gamma \cap K_\alpha| \geqslant \delta |K_\alpha|$, then

$$|K_\alpha^+| = |K_\alpha \setminus (\Gamma \cap K_\alpha)| \leqslant \theta |K_\alpha|.$$

It follows that

$$\inf_{K_{3\alpha}} \widetilde{u} \geqslant \inf_{B_{1/2}} \widetilde{u} \geqslant C^{-1},$$

which is exactly what we want to prove.

Step 2. We claim that for any positive integer m, if

$$|\Gamma \cap K_\alpha| \geqslant \delta^m |K_\alpha|, \tag{2.17}$$

then

$$\inf_{K_\alpha} \widetilde{u} \geqslant C^{-m}, \tag{2.18}$$

where C is the constant determined in Step 1.

(2.18) is clearly valid when $m = 1$. We now use induction on m. Suppose that it is valid for m; we want to show that it is also valid for $m+1$. Suppose that

$$|\Gamma \cap K_\alpha| \geqslant \delta^{m+1} |K_\alpha|. \tag{2.19}$$

Define $\widetilde{K}_0 = K_\alpha$ and

$$\Gamma_\delta = \bigcup \{K_{3r}(x) \cap \widetilde{K}_0 \mid K_r(x) \subset \widetilde{K}_0, \;\; |\Gamma \cap K_r(x)| \geqslant \delta |K_r(x)|\}.$$

By Lemma 2.2,

$$\text{either } \Gamma_\delta = \widetilde{K}_0, \quad \text{or} \quad |\Gamma \cap \widetilde{K}_0| \leqslant \delta |\Gamma_\delta|.$$

By the definition of Γ_δ and the conclusion in Step 1,

$$\inf_{\Gamma_\delta} \widetilde{u} \geqslant C^{-1}. \tag{2.20}$$

If $\Gamma_\delta = \widetilde{K}_0 = K_\alpha$, then (2.20) implies (2.18). Now we assume that $|\Gamma \cap \widetilde{K}_0| \leqslant \delta |\Gamma_\delta|$. The function $v = Cu$ satisfies

$$-a^{ij} D_{ij} v \leqslant Cf.$$

Using the notation from Step 1, we obtain

$$\widetilde{v} = v + \|Cf/\lambda\|_{L^n(B_1)} = C\widetilde{u}.$$

It follows from (2.20) that

$$\Gamma_\delta \subset \widetilde{\Gamma} \triangleq \{x \in B_1 \mid \widetilde{v} \geqslant 1\}.$$

Combining these facts and (2.19), we derive

$$\begin{aligned} |\widetilde{\Gamma} \cap \widetilde{K}_0| &\geqslant |\Gamma_\delta| \geqslant \frac{1}{\delta} |\Gamma \cap \widetilde{K}_0| \\ &= \frac{1}{\delta} |\Gamma \cap K_\alpha| \geqslant \delta^m |K_\alpha| = \delta^m |\widetilde{K}_0|. \end{aligned}$$

By the induction assumption

$$\inf_{K_\alpha} \widetilde{v} \geqslant C^{-m},$$

i.e.,

$$\inf_{K_\alpha} \widetilde{u} \geqslant C^{-(m+1)}.$$

Step 3. Set

$$\Gamma_t = \{x \in B_1 \mid \widetilde{u}(x) > t\}. \tag{2.21}$$

We claim that there exist $C > 1$, $\mu > 0$ such that for all $t > 0$,

$$|B_\alpha \cap \Gamma_t| \leqslant C|B_\alpha| \Big[\inf_{B_\alpha} \widetilde{u}/t \Big]^\mu, \tag{2.22}$$

where C, μ depend only on n, γ.

Set $v = u/t$. Using the notation from Step 1, $\widetilde{v} = \widetilde{u}/t$. Set

$$\widetilde{\Gamma} \triangleq \{x \in B_1 \mid \widetilde{v}(x) > 1\} = \Gamma_t.$$

If $|B_\alpha \cap \Gamma_t| = 0$, then (2.22) is clearly valid. Now we assume that $|B_\alpha \cap \Gamma_t| \neq 0$; then there exists a positive integer m such that

$$\delta^m |K_\alpha| \leqslant |\widetilde{\Gamma} \cap K_\alpha| \leqslant \delta^{m-1} |K_\alpha|,$$

i.e.,

$$\log \frac{|\widetilde{\Gamma} \cap K_\alpha|}{|K_\alpha|} \cdot (\log \delta)^{-1} \leqslant m \leqslant 1 + \log \frac{|\widetilde{\Gamma} \cap K_\alpha|}{|K_\alpha|} \cdot (\log \delta)^{-1}.$$

From (2.18) of Step 2, we deduce that

$$\inf_{K_\alpha} \widetilde{v} \geqslant C^{-m} \geqslant C^{-1} \left[\frac{|\widetilde{\Gamma} \cap K_\alpha|}{|K_\alpha|} \right]^{\log C/\log(\delta^{-1})}.$$

Take $\mu = \log(\delta^{-1})/\log C$. Then

$$|\widetilde{\Gamma} \cap K_\alpha| \leqslant \Big(C \inf_{K_\alpha} \widetilde{v} \Big)^\mu |K_\alpha|.$$

From this inequality, it is not difficult to derive (2.22).

Step 4. We now prove that there exists $p > 0$ such that

$$\left[⨍_{B_\alpha} |u|^p dx \right]^{1/p} \leqslant C \Big[\inf_{B_\alpha} u + \|f/\lambda\|_{L^n(B_1)} \Big]. \tag{2.23}$$

By Lemma 1.1 in Chapter 3,

$$\begin{aligned} \int_{B_\alpha} |u|^p dx &= p \int_0^\infty t^{p-1} |B_\alpha \cap \Gamma_t| dt \\ &= p \int_0^b t^{p-1} |B_\alpha \cap \Gamma_t| dt + p \int_b^\infty t^{p-1} |B_\alpha \cap \Gamma_t| dt, \end{aligned}$$

where b is to be determined. Applying the estimate in Step 3 to the second term in the right-hand side of the above equation, we obtain

$$\int_{B_\alpha} |u|^p dx \leqslant p \int_0^b t^{p-1} |B_\alpha| dt + p \int_b^\infty C m_0^\mu |B_\alpha| t^{p-\mu-1} dt,$$

where $m_0 = \inf_{B_\alpha} \widetilde{u}$. We take $p = \mu/2$. Then

$$\int_{B_\alpha} |u|^p dx \leqslant b^p |B_\alpha| + C m_0^{2p}\, b^{-p} |B_\alpha|.$$

If we choose $b = C^{1/(2p)}m_0$, then

$$\int_{B_\alpha} |u|^p dx \leqslant 2C^{1/2}m_0^p \leqslant 2C^{1/2}\Big(\inf_{B_\alpha} u + \|f/\lambda\|_{L^n(B_1)}\Big)^p.$$

Step 5. Using a finite covering argument, we prove that

$$\left(\fint_{B_{1/2}} |u|^p dx\right)^{1/p} \leqslant C\Big(\inf_{B_{1/2}} u + \|f/\lambda\|_{L^n(B_1)}\Big). \tag{2.24}$$

Since $u \in W^{2,n}(B_{1/2})$, there exists $x_0 \in \overline{B}_{1/2}$ such that

$$u(x_0) = \inf_{B_{1/2}} u. \tag{2.25}$$

Clearly,

$$\begin{aligned}\int_{B_{(2+\alpha)/4}} dy \int_{B_{\alpha/4}(y)} |u|^p dx &= \int_{B_{\alpha/4}} d\xi \int_{B_{(2+\alpha)/4}} |u(y+\xi)|^p dy \\ &\geqslant |B_{\alpha/4}| \int_{B_{1/2}} |u(y)|^p dy.\end{aligned}$$

By the mean value theorem, there exists $y_0 \in \overline{B}_{(2+\alpha)/4}$ such that

$$\int_{B_{\alpha/4}(y_0)} |u|^p dx \geqslant \left(\frac{\alpha}{2+\alpha}\right)^n \int_{B_{1/2}} |u(x)|^p dx. \tag{2.26}$$

We want to cover the line segment connecting x_0 (defined in (2.25)) to y_0 by a sequence of balls of radius $\alpha/4$. Take a sequence of balls $\{B_{\alpha/4}(x_k)\}_{k=1}^N$ such that

$$\begin{aligned}&x_k \in \overline{B}_{(2+\alpha)/4}, \quad x_{k+1} \in B_{\alpha/4}(x_k) \quad (k = 0, 1, 2, \cdots, N-1), \\ &x_N = y_0, \quad N \leqslant \frac{8}{\alpha}.\end{aligned}$$

On each $B_{\alpha/4}(x_k)$ we can use (2.23) from Step 4 to conclude that

$$\begin{aligned}\inf_{B_{\alpha/4}(x_k)} u &\geqslant \frac{1}{C}\left[\fint_{B_{\alpha/4}(x_k)} |u|^p dx\right]^{1/p} - \|f/\lambda\|_{L^n(B_{1/4}(x_k))} \\ &\geqslant \frac{1}{C}\left[\frac{1}{|B_{\alpha/4}|}\int_{B_{\alpha/4}(x_k)\cap B_{\alpha/4}(x_{k+1})} |u|^p dx\right]^{1/p} - \|f/\lambda\|_{L^n(B_1)} \\ &\geqslant \frac{1}{\widetilde{C}} \inf_{B_{\alpha/4}(x_{k+1})} u - \|f/\lambda\|_{L^n(B_1)}.\end{aligned}$$

We assume that $\widetilde{C} \geqslant 2$. By iterating this inequality, we derive

$$\inf_{B_{\alpha/4}(x_0)} u \geqslant \frac{1}{\widetilde{C}^N} \inf_{B_{\alpha/4}(y_0)} u - 2\|f/\lambda\|_{L^n(B_1)}. \tag{2.27}$$

Using (2.23) again on $B_{\alpha/4}(y_0)$ and recalling also (2.26), we obtain

$$\begin{aligned}\inf_{B_{\alpha/4}(y_0)} u &\geqslant \frac{1}{C}\left[\fint_{B_{\alpha/4}(y_0)} |u|^p dx\right]^{1/p} - 3\|f/\lambda\|_{L^n(B_1)} \\ &\geqslant \frac{1}{C'}\left[\fint_{B_{1/2}} |u|^p dx\right]^{1/p} - 3\|f/\lambda\|_{L^n(B_1)}.\end{aligned} \tag{2.28}$$

Combining (2.27), (2.28) and (2.25), we obtain (2.24). The proof is now complete. □

Combining the local maximum principle and weak Harnack inequality, we can establish interior Hölder estimates.

Theorem 2.4 (Harnack inequality). *Let the coefficients of L satisfy the uniform ellipticity assumption (2.2). Suppose that $u \in W^{2,n}(\Omega)$ satisfies $Lu = f$ almost everywhere in Ω, $f/\lambda \in L^n(\Omega)$, and $u \geqslant 0$ on $B_{2R}(y) \subset \Omega$. Then*

$$\sup_{B_{R/2}(y)} u \leqslant C\Big[\inf_{B_{R/2}(y)} u + R\|f/\lambda\|_{L^n(B_{2R}(y))}\Big], \tag{2.29}$$

where C depends only on n and γ.

Proof. For p determined in Theorem 2.3, we apply Theorems 2.1 and 2.3 to conclude that

$$\begin{aligned}
\sup_{B_{R/2}(y)} u &\leqslant C\Big[\Big(\fint_{B_R(y)} u^p dx\Big)^{1/p} + R\|f/\lambda\|_{L^n(B_{2R}(y))}\Big] \\
&\leqslant C\Big[\inf_{B_R(y)} u + R\|f/\lambda\|_{L^n(B_{2R}(y))}\Big] \\
&\leqslant C\Big[\inf_{B_{R/2}(y)} u + R\|f/\lambda\|_{L^n(B_{2R}(y))}\Big].
\end{aligned}$$

□

Theorem 2.5. *Let the coefficients of L satisfy the uniform ellipticity assumption (2.2). Suppose that $u \in W^{2,n}_{loc}(\Omega)$ satisfies $Lu = f$ almost everywhere in Ω and $f/\lambda \in L^n(\Omega)$. Then for any $B_{R_0}(y) \subset \Omega$ and $0 < R \leqslant R_0$,*

$$\operatorname*{osc}_{B_R(y)} u \leqslant C\Big(\frac{R}{R_0}\Big)^{\alpha}\Big[\operatorname*{osc}_{B_{R_0}(y)} u + R_0\|f/\lambda\|_{L^n(\Omega)}\Big],$$

where C depends only on n and γ.

Proof. Set

$$\begin{aligned}
m(R) &= \inf_{B_R(y)} u, \qquad & M(R) &= \sup_{B_R(y)} u, \\
\omega(R) &= M(R) - m(R), \qquad & F_0 &= \|f/\lambda\|_{L^n(\Omega)}.
\end{aligned}$$

It is clear that $\widetilde{u} = u - m(R)$ is a nonnegative solution of $Lu = f$ on $B_R(y)$. By the Harnack inequality (Theorem 2.4),

$$\begin{aligned}
\omega\Big(\frac{R}{4}\Big) &= \sup_{B_{R/4}(y)} \widetilde{u} - \inf_{B_{R/4}(y)} \widetilde{u} \\
&\leqslant \sup_{B_{R/4}(y)} \widetilde{u} - \frac{1}{C}\sup_{B_{R/4}(y)} \widetilde{u} + RF_0 \\
&\leqslant \Big(1 - \frac{1}{C}\Big)\omega(R) + RF_0.
\end{aligned}$$

Applying Lemma 2.1 in Chapter 4, we immediately obtain the theorem. □

3. Global Hölder estimates

We can use a method similar to that for equations of divergence form to derive estimates near the boundary: namely, first consider u_M^+ and u_m^- and then deduce the weak Harnack inequality. We shall not go into the details. For equations of non-divergence form, the more general barrier function method can be used for Hölder estimates near the boundary, although a slightly stronger assumption, a *uniform exterior sphere condition* (i.e., the exterior sphere in Definition 7.1 of Chapter 2 has a uniform radius ρ with respect to $\partial\Omega$), is necessary.

Lemma 3.1. *Suppose that Ω satisfies a uniform exterior sphere condition, the coefficients of L satisfy (2.2), $u \in W^{2,n}_{loc}(\Omega) \cap C(\overline{\Omega})$ satisfies $Lu \leqslant f$, $f/\lambda \in L^\infty(\Omega)$, and $[u]_{\alpha;\partial\Omega} < \infty$ for some $\alpha \in (0,1)$. Then for any $x_0 \in \partial\Omega$, $x \in \Omega$,*

$$u(x) - u(x_0) \leqslant C|x - x_0|^{\alpha/(1+\alpha)}\Big([u]_{\alpha;\partial\Omega} + |u|_0 + \|f/\lambda\|_{L^\infty(\Omega)}\Big), \tag{3.1}$$

where C depends only on n, γ, α and the radius ρ of the uniform exterior sphere.

Proof. For any $\overline{x} \in \Omega$, there exists $x_0 \in \partial\Omega$ such that $|x_0 - \overline{x}| = \text{dist}\{\overline{x}, \partial\Omega\}$. Assume without loss of generality that $|x_0 - \overline{x}| \leqslant \rho$, where ρ is the radius of the uniform exterior sphere. We assume that $\rho \leqslant 1$. Consider the exterior sphere $B_r(y)$ at x_0:

$$B_r(y) \subset \mathbb{R}^n \setminus \overline{\Omega}, \quad \partial B_r(y) \cap \partial\Omega = \{x_0\},$$

where r satisfies $|x_0 - \overline{x}| \leqslant r \leqslant \rho$ and is to be determined later on. We construct the barrier function

$$w(x) = \frac{1}{r^p} - \frac{1}{|x-y|^p},$$

where p is large and is to be determined. A direct computation shows that

$$\begin{aligned} Lw &= a^{ij}\left[p(p+2)\frac{(x_i - y_i)(x_j - y_j)}{|x-y|^{p+4}} - p\frac{\delta_{ij}}{|x-y|^{p+2}}\right] \\ &\geqslant \frac{p\lambda}{|x-y|^{p+2}}(p + 2 - \gamma n), \end{aligned}$$

where the ellipticity condition (2.2) is used. Take a neighborhood of x_0,

$$\mathcal{N}_r = \{x \in \Omega \mid r < |x-y| < 3r\}.$$

If $p > \gamma n$, then

$$Lw \geqslant \frac{2p\lambda}{(3r)^{p+2}} \quad \text{on } \mathcal{N}_r. \tag{3.2}$$

We choose the following auxiliary function on $\mathcal{N}_r$:

$$v(x) = Kw(x) - [u(x) - u(x_0)] + (6r)^\alpha[u]_{\alpha;\partial\mathcal{N}_r \cap \partial\Omega}, \tag{3.3}$$

where K is a positive constant to be determined. By (3.2),

$$Lv \geqslant \lambda\left(\frac{2pK}{(3r)^{p+2}} - \frac{f(x)}{\lambda}\right) > 0,$$

provided $K \geqslant 3^{p+2}\|f/\lambda\|_{L^\infty}\ r^{p+2}$. Clearly,

$$v(x) \geqslant 0 \quad \text{on } \partial\mathcal{N}_r \cap \partial\Omega.$$

If we choose $K \geqslant 4r^p|u|_0$, then

$$v(x) \geqslant K\frac{1}{r^p}\left(1 - \frac{1}{3^p}\right) - 2|u|_0 \geqslant 0 \quad \text{on } \partial\mathcal{N}_r \cap \Omega.$$

It follows from the Alexandroff-Bakelman-Pucci maximum principle that if

$$K \geqslant r^p \max\left\{4|u|_0,\ 3^{p+2}\|f/\lambda\|_{L^\infty}\right\},$$

then $v(x) \geqslant 0$ on $\mathcal{N}_r$; i.e., for $x \in \mathcal{N}_r$,

$$\begin{aligned} u(x) - u(x_0) &\leqslant Kw(x) + (6r)^\alpha[u]_{\alpha;\partial\mathcal{N}_r\cap\partial\Omega} \\ &\leqslant \frac{pK}{r^{p+1}}(|x-y| - r) + (6r)^\alpha[u]_{\alpha;\partial\Omega}. \end{aligned}$$

In particular,

$$u(\overline{x}) - u(x_0) \leqslant \frac{pK|\overline{x} - x_0|}{r^{p+1}} + (6r)^\alpha[u]_{\alpha;\partial\Omega}.$$

If we choose $r = |\overline{x} - x_0|^{1/(1+\alpha)} \cdot \rho^{\alpha/(1+\alpha)}$, then

$$u(\overline{x}) - u(x_0) \leqslant C|\overline{x} - x_0|^{\alpha/(1+\alpha)}\left\{[u]_{\alpha;\partial\Omega} + |u|_0 + \|f/\lambda\|_{L^\infty}\right\},$$

where C depends only on n, γ, ρ and α. From this estimate, we easily derive the desired conclusion. □

Theorem 3.2. *Let the assumptions of Lemma 3.1 be in force. If u is a solution of (2.1), then there exist $C > 0$ and $0 < \beta < 1$ such that*

$$|u(x) - u(y)| \leqslant C|x-y|^\beta\left(|u|_0 + [u]_{\alpha;\partial\Omega} + \|f/\lambda\|_{L^\infty}\right), \tag{3.4}$$

where C depends only on n, γ, α, ρ, and diam Ω, *while β depends only on n, γ and α.*

Proof. If we apply Lemma 3.1 to u and $-u$, then $\forall x \in \Omega, x_0 \in \partial\Omega$,

$$|u(x) - u(x_0)| \leqslant C|x - x_0|^{\alpha/(1+\alpha)} \cdot \left(|u|_0 + [u]_{\alpha;\partial\Omega} + \|f/\lambda\|_{L^\infty}\right);$$

i.e., for any $x_0 \in \partial\Omega$,

$$\operatorname*{osc}_{B_R(x_0)} u \leqslant CR^{\alpha/(1+\alpha)}\left(|u|_0 + [u]_{\alpha;\partial\Omega} + \|f/\lambda\|_{L^\infty}\right).$$

Just as in the proof of Theorem 3.5 in Chapter 4, the above estimate, together with the interior Hölder estimate, implies the global Hölder estimate (3.4). □

CHAPTER 7

Fully Nonlinear Elliptic Equations

In this chapter, we study the fully nonlinear elliptic equation

$$F(x, u, Du, D^2u) = 0,$$

where $F(x, z, p, r)$ is defined on $\Gamma = \Omega \times \mathbb{R} \times \mathbb{R}^n \times \mathcal{S}^n$, $\mathcal{S}^n$ being the space of all n by n symmetric matrices.

There are two typical examples of fully nonlinear elliptic equations. The first one is from geometry — the Monge-Ampère equation:

$$\det(D^2u) = f(x) > 0,$$

or more generally, the Gaussian curvature equation:

$$\det(D^2u) = K(x)(1 + |Du|^2)^{(n+2)/2};$$

another example comes from control theory — the Bellman equation:

$$\sup_k (L^k u - f^k) = 0,$$

where L^k is a linear elliptic operator. The first equation is a nonuniformly elliptic equation, while the last equation is a uniformly elliptic equation. This being a basic textbook, we shall discuss only uniformly elliptic equations.

For the nonlinear function F, we shall impose the following structure conditions:

(F1) There exist $\lambda = \lambda(x, z, p)$, $\Lambda = \Lambda(x, z, p)$ such that on Γ,

$$\lambda I \leqslant \left(\frac{\partial F}{\partial r_{ij}}\right) \leqslant \Lambda I, \qquad \lambda > 0,\ \Lambda/\lambda \leqslant \mu_1(|z|);$$

(F2) $|F(x, z, p, 0)| \leqslant \lambda \mu_2(|z|)(1 + |p|^2)$;

(F3) $(1 + |p|)^{-1}|F_x| + |F_z| + (1 + |p|)|F_p| \leqslant \lambda \mu_3(|z|)(1 + |p|^2 + |r|)$;

(F4) on Γ, $F(x, z, p, r)$ is concave with respect to r;

(F5) $F(x, z, p, 0) \operatorname{sign} z \leqslant \lambda \overline{\mu}(1 + |p|)$,

where $\overline{\mu}$ is a constant and the $\mu_i(t)$ $(i = 1, 2, 3)$ are nonnegative nondecreasing functions defined on $[0, \infty)$.

(F1) is a uniform ellipticity condition. (F2) and (F3) are referred to as natural growth conditions. (F4) is a concavity condition (i.e., $\left(\dfrac{\partial^2 F}{\partial r_{ij}\partial r_{kl}}\right) \leqslant 0$ in the distribution sense). (F5) will be used for the maximum norm estimate.

1. Maximum norm and Hölder estimates for solutions

Theorem 1.1. *Under the structure conditions* (F1) *and* (F5), *a solution* $u \in C(\overline{\Omega}) \cap W^{2,n}_{loc}(\Omega)$ *of the Dirichlet problem*

$$F(x, u, Du, D^2u) = 0 \quad \text{in } \Omega, \tag{1.1}$$

$$u = \varphi \quad \text{on } \Omega \tag{1.2}$$

satisfies the estimate

$$\sup_{\Omega} |u| \leqslant |\varphi|_{0;\partial\Omega} + C \triangleq M_0,$$

where C depends only on $n, \overline{\mu}$ and $\operatorname{diam}\Omega$.

Proof. We rewrite the equation (1.1) as

$$-a^{ij}D_{ij}u - F(x, u, Du, 0) = 0,$$

where

$$a^{ij}(x, z, p, r) = \int_0^1 \frac{\partial F}{\partial r_{ij}}(x, z, p, \tau r)d\tau.$$

Using (F5) on the subdomain $\Omega^+ = \{x \in \Omega \mid u(x) > 0\}$, we get

$$\begin{aligned} 0 &= -a^{ij}D_{ij}u - F(x, u, Du, 0)\operatorname{sign} u \\ &\geqslant -a^{ij}D_{ij}u - \lambda\overline{\mu}\operatorname{sign}(D_iu)D_iu - \lambda\overline{\mu}. \end{aligned}$$

Applying the Alexandroff-Bakelman-Pucci maximum principle on Ω^+, we obtain

$$\sup_{\Omega} u \leqslant |\varphi|_{0;\partial\Omega} + C,$$

where C depends only on $n, \overline{\mu}$ and $\operatorname{diam}\Omega$. Similarly, we can derive the lower bound for u. □

For simplicity, we denote $\mu_i = \mu_i(M_0)$ $(i = 1, 2, 3)$.

Lemma 1.2. *Suppose that $u \in W^{2,n}_{loc}(\Omega)$ satisfies the inequality*

$$Lu = -a^{ij}(x)D_{ij}u \geqslant -\lambda(\mu_0|Du|^2 + g(x)), \tag{1.3}$$

where $\mu_0 > 0$, $g \in L^n(\Omega)$, and (a^{ij}) satisfies

$$\lambda I \leqslant (a^{ij}) \leqslant \Lambda I, \quad \frac{\Lambda}{\lambda} \leqslant \mu_1 \ (\textit{constant}), \quad \lambda(x) > 0 \quad \text{in } \Omega.$$

If u is nonnegative on $B_{2R} \subset \Omega$, then there exist $\kappa > 0$ and $C \geqslant 1$ such that

$$\left(⨍_{B_R} u^\kappa dx\right)^{1/\kappa} \leqslant C\Big(\inf_{B_R} u + R\|g\|_{L^n(B_{2R})}\Big), \tag{1.4}$$

where κ depends only on n and μ_1, while C depends only on n, μ_1 and $\mu_0|u|_{0,B_{2R}}$.

Proof. Set

$$v = 1 - e^{-\mu_0 u}, \quad M = |u|_{0,B_{2R}}.$$

From (1.3), we deduce that

$$\begin{aligned} Lv &= -a^{ij}D_{ij}v = \mu_0 e^{-\mu_0 u}[Lu + \mu_0 a^{ij}D_iuD_ju] \\ &\geqslant -\lambda\mu_0|g(x)|. \end{aligned}$$

By Theorem 2.3 in Chapter 6, there exist $\kappa > 0$ and $C \geqslant 1$, depending only on n, μ_1, such that

$$\left(\fint_{B_R} v^\kappa dx\right)^{1/\kappa} \leqslant C\left(\inf_{B_R} v + \mu_0 R\|g\|_{L^n(B_{2R})}\right).$$

It follows that

$$\left(\fint_{B_R} u^\kappa dx\right)^{1/\kappa} \leqslant Ce^{\mu_0 M}\left(\inf_{B_R} u + R\|g\|_{L^n(B_{2R})}\right).$$

The proof is complete. □

Theorem 1.3. *Suppose that $u \in W^{2,n}_{loc}(\Omega)$ satisfies the inequality*

$$|Lu| \leqslant \lambda(\mu_0|Du|^2 + g(x)) \quad \text{in } \Omega, \tag{1.5}$$

where we assume the same conditions on L, μ_0 and g as in Lemma 1.2. Then for any $B_R \subset \Omega$ and $0 < \sigma < 1$, we have

$$\operatorname*{osc}_{B_{\sigma R}} u \leqslant C\sigma^\alpha\left[\operatorname*{osc}_{B_R} u + R\|g\|_{L^n(B_R)}\right], \tag{1.6}$$

where $C \geqslant 1$ and $0 < \alpha < 1$ depend only on $n, \Lambda/\lambda$ and $\mu_0|u|_{0;\Omega}$.

Proof. Set

$$M(R) = \sup_{B_R} u, \quad m(R) = \inf_{B_R} u,$$
$$\omega(R) = M(R) - m(R), \quad F_0 = \|g\|_{L^n(B_R)}.$$

Applying the weak Harnack inequality (Lemma 1.2) to $M(R) - u$ and $u - m(R)$, we obtain

$$\left[\fint_{B_{R/2}} \left(M(R) - u\right)^\kappa dx\right]^{1/\kappa} \leqslant C\left[M(R) - M\left(\frac{R}{2}\right) + RF_0\right],$$
$$\left[\fint_{B_{R/2}} \left(u - m(R)\right)^\kappa dx\right]^{1/\kappa} \leqslant C\left[m\left(\frac{R}{2}\right) - m(R) + RF_0\right].$$

It follows that

$$\begin{aligned}
\omega(R) &= \left(\fint_{B_{R/2}} [M(R) - m(R)]^\kappa dx\right)^{1/\kappa} \\
&\leqslant C\left[\left(\fint_{B_{R/2}} \left(M(R) - u\right)^\kappa dx\right)^{1/\kappa} + \left(\fint_{B_{R/2}} \left(u - m(R)\right)^\kappa dx\right)^{1/\kappa}\right] \\
&\leqslant C\left[\omega(R) - \omega\left(\frac{R}{2}\right) + RF_0\right].
\end{aligned}$$

Therefore,

$$\omega\left(\frac{R}{2}\right) \leqslant \frac{C-1}{C}\omega(R) + RF_0.$$

Applying Lemma 2.1 in Chapter 4, we immediately get (1.6). □

Theorem 1.4. *Suppose that* $u \in W^{2,n}_{loc}(\Omega)$ *is a solution of* (1.1) *satisfying* $|u|_{0;\Omega} \leqslant M_0$ *and that* $F(x,z,r,p)$ *satisfies the structure conditions* (F1) *and* (F2). *Then for any* $B_R \subset \Omega$ *and* $0 < \sigma \leqslant 1$, *we have*

$$\underset{B_{\sigma R}}{\operatorname{osc}}\, u \leqslant C\sigma^\alpha \Big\{ \underset{B_R}{\operatorname{osc}}\, u + \mu_2 R \Big\}, \tag{1.7}$$

where $C \geqslant 1$ *and* $0 < \alpha < 1$ *depend only on* n, μ_1 *and* $\mu_2 M_0$.

Proof. The structure condition (F2) implies that

$$|a^{ij} D_{ij} u| = |F(x,u,Du,0)| \leqslant \lambda\mu_2(1 + |Du|^2),$$

where

$$a^{ij} = \int_0^1 \frac{\partial F}{\partial r_{ij}}(x,u,Du,\tau D^2 u)d\tau.$$

Now (1.7) follows immediately from Theorem 1.3. □

In order to establish the global Hölder estimates, we use the barrier function method as in §3 of the last chapter.

Lemma 1.5. *Suppose that* $u \in C^2(\Omega) \cap C(\overline{\Omega})$ *satisfies*

$$Lu \leqslant \lambda\mu_0(1 + |Du|^2), \tag{1.8}$$

where μ_0, L *satisfy the conditions in Lemma 1.2. We further assume that* Ω *satisfies a uniform exterior sphere condition and that* $[u]_{\alpha;\partial\Omega} < \infty$ *for some* $0 < \alpha \leqslant 1$. *Then for any* $x_0 \in \partial\Omega$ *and* $x \in \Omega$,

$$u(x) - u(x_0) \leqslant C|x - x_0|^{\alpha/(1+\alpha)} \Big([u]_{\alpha;\partial\Omega} + |u|_{0;\Omega} + 1 \Big), \tag{1.9}$$

where C *depends only on* $n, \Lambda/\lambda, \mu_0, \mu_0 |u|_{0;\alpha}$ *and the radius* ρ *of the exterior sphere.*

Proof. For any $\overline{x} \in \Omega$, there exists $x_0 \in \partial\Omega$ such that $|\overline{x} - x_0| = \operatorname{dist}\{\overline{x}, \partial\Omega\}$. Assume without loss of generality that the uniform exterior sphere radius $\rho \leqslant 1$ and $|\overline{x} - x_0| \leqslant \rho$. Suppose that $B_r(y)$ is the exterior sphere at x_0, where r satisfies $|\overline{x} - x_0| \leqslant r \leqslant \rho$ and is to be determined. We consider the barrier function:

$$w(x) = \varphi(d), \quad d = |x - y| - r, \tag{1.10}$$

where φ is a function to be determined, satisfying $\varphi' > 0$, $\varphi'' < 0$. In the x_0-neighborhood

$$\mathcal{N} = \Omega \cap \{0 < d < \delta\} \quad (\delta \text{ to be determined}),$$

the barrier function w satisfies

$$\begin{aligned} \widetilde{L}w &\triangleq -a^{ij} D_{ij} w - \lambda\mu_0(|Dw|^2 + 1) \\ &= -a^{ij}\Big[\varphi'' \cdot \frac{(x_i - y_i)(x_j - y_j)}{|x-y|^2} \\ &\qquad + \varphi' \cdot \Big(\frac{\delta_{ij}}{|x-y|} - \frac{(x_i - y_i)(x_j - y_j)}{|x-y|^3} \Big) \Big] - \lambda\mu_0 \Big[(\varphi')^2 + 1 \Big] \\ &\geqslant \lambda(-\varphi'') - \frac{\Lambda(n-1)}{r}\varphi' - \lambda\mu_0 \Big[(\varphi')^2 + 1 \Big]. \end{aligned} \tag{1.11}$$

If we assume (recall that $\Lambda/\lambda \leqslant \mu_1,\ \mu_0 > 0$)

$$\varphi' \geqslant \max\Big\{1, \frac{\mu_1(n-1)}{r\mu_0}\Big\}, \tag{1.12}$$

then (1.11) implies that

$$\widetilde{L}w \geqslant \lambda(\varphi')^2\Big[-\frac{\varphi''}{(\varphi')^2} - 4\mu_0\Big] \quad \text{in } \mathcal{N}.$$

If we further assume

$$-\frac{\varphi''}{(\varphi')^2} = 8\mu_0, \tag{1.13}$$

then

$$\widetilde{L}w > 0 \geqslant \widetilde{L}u \quad \text{in } \mathcal{N}. \tag{1.14}$$

We can solve (1.13) to obtain

$$\varphi(d) = \frac{1}{8\mu_0}\log(1+Kd), \tag{1.15}$$

where $K > 0$ is to be determined in such a way that

$$w = \varphi(d) \geqslant 2M_0 \quad \text{on } \partial\mathcal{N}\cap\Omega, \tag{1.16}$$

where $M_0 \triangleq |u|_{0;\Omega}$. In order to satisfy (1.12) and (1.16), i.e.,

$$\begin{aligned} \varphi'(d) &= \frac{1}{8\mu_0}\frac{K}{(1+Kd)} \geqslant \frac{1}{8\mu_0}\frac{K\delta}{(1+K\delta)}\frac{1}{\delta} \\ &\geqslant \max\Big\{1, \frac{\mu_1(n-1)}{r\mu_0}\Big\}, \\ \varphi(\delta) &= \frac{1}{8\mu_0}\log(1+K\delta) \geqslant 2M_0, \end{aligned}$$

it suffices to choose

$$\delta \leqslant \min\Big\{\frac{1}{16\mu_0}, \frac{r}{16\mu_1(n-1)}\Big\}, \tag{1.17}$$

$$K\delta = \exp(16\mu_0 M_0) - 1. \tag{1.18}$$

It is clear that the auxiliary function

$$v(x) = w(x) - [u(x) - u(x_0)] + (6r)^\alpha[u]_{\alpha;\partial\Omega}$$

satisfies

$$v(x) \geqslant 0 \quad \text{on } \partial\mathcal{N}\cap\partial\Omega.$$

Since we also have (1.14) and (1.16), the maximum principle implies that

$$v(x) \geqslant 0 \quad \text{in } \mathcal{N}.$$

In particular,

$$u(\overline{x}) - u(x_0) \leqslant \frac{1}{8\mu_0}\log(1+K|\overline{x}-x_0|) + (6r)^\alpha[u]_{\alpha;\partial\Omega}.$$

We choose $r \leqslant \min\left\{\rho, \frac{\mu_1}{\mu_0}(n-1)\right\}$, $\delta = \frac{r}{16\mu_1(n-1)}$. Then we can rewrite the above inequality as (recalling also (1.18))

$$u(\overline{x}) - u(x_0) \leqslant \frac{1}{8\mu_0}\log\left(1 + \frac{C|\overline{x} - x_0|}{r}\right) + (6r)^\alpha [u]_{\alpha;\partial\Omega}, \tag{1.19}$$

where C depends only on n, μ_1 and $\mu_0 M_0$. If

$$|\overline{x} - x_0| \leqslant \min\left\{\rho, \left(\frac{\mu_1}{\mu_0}(n-1)\right)^{1+\alpha}\rho^{-\alpha}\right\},$$

then we can take $r = \rho^{\alpha/(1+\alpha)}|\overline{x} - x_0|^{1/(1+\alpha)}$. From (1.19), we deduce that

$$u(\overline{x}) - u(x_0) \leqslant C|\overline{x} - x_0|^{\alpha/(1+\alpha)}\left([u]_{\alpha;\partial\Omega} + 1\right).$$

The proof is complete. □

Theorem 1.6. *Let the structure conditions* (F1), (F2) *be in force. Suppose that $u \in C^2(\Omega) \cap C(\overline{\Omega})$ is a solution of* (1.1), *Ω satisfies a uniform exterior sphere condition, $|u|_{0;\Omega} \leqslant M_0$, and $[u]_{\alpha;\partial\Omega} < \infty$. Then there exist $C \geqslant 1$ and $0 < \beta < 1$ such that*

$$[u]_{\beta;\Omega} \leqslant C([u]_{\alpha;\partial\Omega} + |u|_{0;\Omega} + 1),$$

where C, β depend only on $n, \mu_1, \mu_2, \mu_2 M_0$ and the radius ρ of the uniform exterior sphere.

The proof is similar to that of Theorem 3.2 in Chapter 6.

2. Gradient estimates

A slight modification of the proof of Lemma 1.5 yields the following boundary gradient estimate.

Theorem 2.1. *Let the structure conditions* (F1), (F2) *be in force. Suppose that $u \in C^2(\Omega) \cap C(\overline{\Omega})$ is a solution of* (1.1), *Ω satisfies a uniform exterior sphere condition, $\partial\Omega \in C^1$, and $u = 0$ on $\partial\Omega$, $|u|_{0;\Omega} \leqslant M_0$. Then*

$$\sup_{x\in\partial\Omega} |Du| \leqslant C, \tag{2.1}$$

where C depends only on $n, \mu_1, \mu_2, \mu_2 M_0$ and the radius ρ of the uniform exterior sphere.

Proof. Take $r = \rho$ in Lemma 1.5, where x_0 is an arbitrary point on $\partial\Omega$. Since $u\Big|_{\partial\Omega} = 0$,

$$|u(x)| \leqslant |\varphi(d)|, \quad 0 < d < \delta.$$

It follows that

$$\left|\frac{\partial u}{\partial n}\right|_{x=x_0} \leqslant \varphi'(0) \leqslant \frac{1}{8\mu_2}K,$$

which is exactly the conclusion of the theorem. □

Theorem 2.2. *Suppose that* $u \in C^3(\Omega) \cap C^1(\overline{\Omega})$ *is a solution of* (1.1), $|u|_{0;\Omega} \leqslant M_0$. $[u]_{\beta;\Omega} \leqslant M_\beta$ $(0 < \beta \leqslant 1)$, $[Du]_{0;\partial\Omega} \leqslant \overline{M}_1$. *We also assume the structure conditions* (F1) *and* (F3). *Then*

$$[Du]_{0;\Omega} \leqslant M_1, \tag{2.2}$$

where M_1 *depends only on* $n, \mu_1, \mu_3, M_0, M_\beta, \overline{M}_1$ *and* β^{-1}.

Proof. Bernstein's method is frequently used in gradient estimates. Under the natural structure conditions, a nonlinear transform is necessary, and therefore tedious calculations are involved. In what follows, we shall give a modified Bernstein's method which avoids the nonlinear transform and is valid under the natural structure conditions.

Since $u \in C^1(\overline{\Omega})$, there exists $x_0 \in \overline{\Omega}$ such that $|Du(x_0)| = \sup_\Omega |Du| \triangleq M$. Assume without loss of generality that $x_0 \in \Omega$ and $M \geqslant 1$. We choose a cutoff function $\zeta(x)$ on the ball $B_{1/M}(x_0)$ such that

$$\begin{aligned} &\zeta(x_0) = 1, \quad 0 \leqslant \zeta(x) \leqslant 1, \\ &|D\zeta| \leqslant CM, \quad |D^2\zeta| \leqslant CM^2, \end{aligned} \tag{2.3}$$

where C depends only on n. Define the auxiliary function

$$w(x) = \zeta^2(x)|Du|^2 + KM^2[u(x) - u(x_0)]^2 \qquad \text{for } x \in B_{1/M}(x_0) \cap \Omega, \tag{2.4}$$

where K is a positive constant to be determined. Denote by

$$Lw = -\frac{\partial F}{\partial r_{ij}}(x, u, Du, D^2u)D_{ij}w - \frac{\partial F}{\partial p_i}(x, u, Du, D^2u)D_iw.$$

By the product rule for derivatives,

$$\begin{aligned} Lw &= \zeta^2 L(|Du|^2) + |Du|^2 L(\zeta^2) - 2\frac{\partial F}{\partial r_{ij}} D_i\zeta^2 D_j(|Du|^2) \\ &\quad + 2KM^2[u(x) - u(x_0)]Lu - 2KM^2\frac{\partial F}{\partial r_{ij}} D_iuD_ju. \end{aligned} \tag{2.5}$$

Since $|Du|^2 = \sum_k (D_ku)^2$,

$$L(|Du|^2) = 2D_ku \cdot L(D_ku) - 2\frac{\partial F}{\partial r_{ij}} D_{ki}uD_{kj}u.$$

Differentiating (1.1) with respect to x_k, we obtain

$$L(D_ku) = \frac{\partial F}{\partial z} D_ku + \frac{\partial F}{\partial x_k}.$$

By the structure condition (F3),

$$|L(D_ku)| \leqslant \lambda\mu_3(M^3 + M|D^2u|),$$

where $|D^2u|^2 = \sum_{i,j} (D_{ij}u)^2$. Thus

$$L(|Du|^2) \leqslant -2\lambda|D^2u|^2 + C\lambda M^2(M^2 + |D^2u|). \tag{2.6}$$

Clearly,

$$|u(x) - u(x_0)| \leqslant M_\beta M^{-\beta} \quad \text{for } x \in B_{1/M}(x_0) \cap \Omega, \tag{2.7}$$

$$|Lu| \leqslant C\lambda(M^2 + |D^2u|). \tag{2.8}$$

Substituting (2.6), (2.7) and (2.8) into (2.5) and also using (2.3), we derive, for $x \in B_{1/M}(x_0) \cap \Omega$,

$$\begin{aligned} Lw \leqslant\ & -2\lambda\zeta^2|D^2u|^2 - 2\lambda KM^2|Du|^2 \\ & + C\lambda M^2(M^2 + |D^2u|) + C\lambda KM^{2-\beta}(M^2 + |D^2u|). \end{aligned}$$

Using Cauchy's inequality, we get, for $x \in B_{1/M}(x_0) \cap \Omega$,

$$Lw \leqslant -2\lambda KM^2|Du|^2 + \frac{C\lambda M^4}{\zeta^2} + C\lambda\Big(KM^{4-\beta} + \frac{K^2M^{4-2\beta}}{\zeta^2}\Big). \tag{2.9}$$

If $w(x)$ attains its maximum at some point $y \in \partial B_{1/M}(x_0) \cap \Omega$, then

$$M^2 = w(x_0) \leqslant w(y) = KM^2[u(y) - u(x_0)]^2 \leqslant CKM^{2-2\beta};$$

it follows that

$$M \leqslant (CK)^{1/(2\beta)}.$$

If $w(x)$ attains its maximum at some point $y \in B_{1/M}(x_0) \cap \partial\Omega$, then

$$M^2 = w(x_0) \leqslant w(y) \leqslant \overline{M}_1^2 + KM^2[u(y) - u(x_0)]^2 \leqslant \overline{M}_1^2 + CKM^{2-2\beta};$$

Young's inequality implies that

$$M^2 \leqslant 2\overline{M}_1^2 + CK^{2/\beta}.$$

Finally, if $w(x)$ attains its maximum at some interior point $y \in B_{1/M}(x_0) \cap \Omega$, then at the point $x = y$,

$$0 \leqslant Lw \leqslant -2\lambda KM^2|Du|^2 + \frac{C\lambda M^4}{\zeta^2} + C\lambda\Big(KM^{4-\beta} + \frac{K^2M^{4-2\beta}}{\zeta^2}\Big),$$

i.e.,

$$\zeta^2|Du|^2\Big|_{x=y} \leqslant \frac{C}{K}M^2 + C(M^{2-\beta} + KM^{2-2\beta}). \tag{2.10}$$

Using the formula for $w(x)$, we find that

$$M^2 = w(x_0) \leqslant w(y) \leqslant \Big[\zeta^2|Du|^2\Big]_{x=y} + CKM^{2-2\beta}.$$

Substituting (2.10) into the above inequality, we obtain

$$M^2 \leqslant \frac{C}{K}M^2 + C(M^{2-\beta} + KM^{2-2\beta}).$$

We now take $K = 2C$. Then

$$M^2 \leqslant C(M^{2-\beta} + KM^{2-2\beta});$$

it follows that $M \leqslant C$. The proof is complete. □

3. Gradient Hölder estimates

A gradient is a vector. For vectors, we need an iteration result similar to Lemma 2.1 in Chapter 4.

Lemma 3.1. *Suppose that $\omega_i(R)$ $(i = 1, 2, \cdots, N)$ are nonnegative, nondecreasing functions defined on $(0, R_0]$. For each $R \in (0, R_0/2]$, we assume that there exists an index set $A(R) \subset \{1, 2, \cdots, N\}$ (the index set can be empty) such that*

$$\sum_{i\in A(R)} \omega_i(R) \leqslant \gamma \sum_{i\in A(R)} \omega_i(2R) + C_0\Big(\frac{R}{R_0}\Big)^\alpha, \quad \forall 0 < R \leqslant \frac{R_0}{2}, \tag{3.1}$$

$$\sum_{i\notin A(R)} \omega_i(R) \leqslant C_1\Big[\sum_{i\in A(R)} \omega_i(2R) + \Big(\frac{R}{R_0}\Big)^\alpha\Big], \quad \forall 0 < R \leqslant \frac{R_0}{2}, \tag{3.2}$$

where C_0, $C_1 > 0$, $\gamma \in (0, 1)$, $\alpha \in (0, 1]$ are constants. Then there exist $\alpha_1 \in (0, \alpha]$ and $C_2 > 0$ such that

$$\sum_{i=1}^N \omega_i(R) \leqslant C_2\Big(\frac{R}{R_0}\Big)^{\alpha_1}\Big(\sum_{i=1}^N \omega_i(R_0) + 1\Big), \quad \forall 0 < R \leqslant R_0, \tag{3.3}$$

where α_1, C_2 depend only on N, C_0, C_1, α and γ.

Proof. For $0 < \beta < 1$, we write

$$\sum_{i=1}^N \omega_i(R) = \sum_{i\in A(R)} \omega_i(R) + \beta \sum_{i\notin A(R)} \omega_i(R) + (1-\beta) \sum_{i\notin A(R)} \omega_i(R).$$

Using (3.1), (3.2), we immediately get

$$\begin{aligned}\sum_{i=1}^N \omega_i(R) \quad &\leqslant \quad (\gamma + C_1\beta) \sum_{i\in A(R)} \omega_i(2R) \\ &\qquad + (1-\beta) \sum_{i\notin A(R)} \omega_i(2R) + (C_0 + \beta C_1)\Big(\frac{R}{R_0}\Big)^\alpha.\end{aligned}$$

We choose β such that $\gamma + C_1\beta = 1 - \beta$, i.e., $\beta = (1-\gamma)/(C_1 + 1)$. Then

$$\sum_{i=1}^N \omega_i(R) \leqslant (1-\beta)\sum_{i=1}^N \omega_i(2R) + (C_0 + \beta C_1)\Big(\frac{R}{R_0}\Big)^\alpha.$$

By applying Lemma 2.1 in Chapter 4, we immediately obtain the lemma. □

Theorem 3.2. *Suppose that $u \in C^3(\Omega)$ is a solution of (1.1), $|u|_{0;\Omega} \leqslant M_0$, $[u]_{1;\Omega} \leqslant M_1$. We also assume the structure conditions (F1) and (F3). Then for any $B_R \subset \Omega$, we have*

$$\sum_{i=1}^n \operatorname*{osc}_{B_{\sigma R}} D_i u \leqslant C\sigma^\alpha\Big\{\sum_{i=1}^n \operatorname*{osc}_{B_R} D_i u + R\Big\}, \quad \forall 0 < \sigma < 1, \tag{3.4}$$

where $C \geqslant 1$ and $0 < \alpha < 1$ depend only on n, μ_1, μ_3, M_0 and M_1.

Proof. Set

$$w_l^\theta = (-1)^\theta D_l u + \varepsilon |Du|^2 \quad (l = 1, 2, \cdots, n; \theta = 1, 2), \tag{3.5}$$

where ε is a small positive constant to be determined (we shall take $\varepsilon = (16M_1 n)^{-1}$). Denote

$$\omega_l(R) = \operatorname*{osc}_{B_R} D_l u, \quad \omega_{\theta n+l}(R) = \operatorname*{osc}_{B_R} w_l^\theta \quad (l = 1, 2, \cdots, n;\ \theta = 1, 2). \tag{3.6}$$

Let L be the elliptic operator

$$L = -\frac{\partial F}{\partial r_{ij}}(x, u, Du, D^2u) D_{ij} - \frac{\partial F}{\partial p_i}(x, u, Du, D^2u) D_i.$$

Using the product rule for derivatives, we obtain

$$Lw_l^\theta = (-1)^\theta L(D_l u) + 2\varepsilon D_k u L(D_k u) - 2\varepsilon \frac{\partial F}{\partial r_{ij}} D_{ki} u D_{kj} u.$$

Differentiating (1.1) in x_l, we get

$$L(D_l u) = \frac{\partial F}{\partial z} D_l u + \frac{\partial F}{\partial x_l}.$$

By the structure condition (F3) and the fact that $[Du]_{0;\Omega} \leqslant M_1$,

$$Lw_l^\theta \leqslant C\lambda(1 + |D^2u|) - 2\varepsilon\lambda |D^2u|^2.$$

Using Cauchy's inequality, we derive

$$\begin{aligned} -\frac{\partial F}{\partial r_{ij}} D_{ij} w_l^\theta &= Lw_l^\theta + \frac{\partial F}{\partial p_i} D_i w_l^\theta \\ &\leqslant -2\varepsilon\lambda |D^2u|^2 + C\lambda(1 + |D^2u|)(1 + |Dw_l^\theta|) \\ &\leqslant C\lambda(1 + |Dw_l^\theta|^2), \end{aligned}$$

where the constant C in the last inequality depends on ε. Since we shall take $\varepsilon = (16nM_1)^{-1}$, we do not specify the dependence on ε. Set

$$W_l^\theta = \sup_{B_R} w_l^\theta. \tag{3.7}$$

Applying the weak Harnack inequality (Lemma 1.2) to the function $W_l^\theta - w_l^\theta$, we obtain

$$\begin{aligned} \Phi_l^\theta\Big(\frac{R}{2}\Big) &\triangleq \Big(⨍_{B_{R/2}} (W_l^\theta - w_l^\theta)^\kappa dx\Big)^{1/\kappa} \\ &\leqslant C\Big(W_l^\theta - \sup_{B_{R/2}} w_l^\theta + R^2\Big) \qquad (l = 1, 2, \cdots, n;\ \theta = 1, 2). \end{aligned} \tag{3.8}$$

For each fixed R, set

$$\begin{aligned} A(R) &= \Big\{ s = \theta n + l \;\Big|\; \omega_l(R) \geqslant \frac{1}{2n} \sum_{i=1}^n \omega_i(R), \\ &\qquad \Phi_l^\theta\Big(\frac{R}{2}\Big) = \max\Big[\Phi_l^1\Big(\frac{R}{2}\Big),\ \Phi_l^2\Big(\frac{R}{2}\Big)\Big] \Big\}. \end{aligned} \tag{3.9}$$

We shall apply Lemma 3.1 to $\omega_s(R)$ $(s = 1, 2, \cdots, 3n - 1, 3n)$.

For $s = \theta n + l \in A(R)$, (3.6), (3.5) and (3.9) imply

$$\begin{aligned} \omega_s(R) &\geqslant \omega_l(R) - 2\varepsilon M_1 \sum_{i=1}^n \omega_i(R) \\ &\geqslant \Big(\frac{1}{2n} - 2\varepsilon M_1\Big) \sum_{i=1}^n \omega_i(R) \\ &\geqslant \frac{1}{4n} \sum_{i=1}^n \omega_i(R), \end{aligned} \tag{3.10}$$

where we also use our choice of $\varepsilon = (16nM_1)^{-1}$. Similarly, for $s = \theta n + l \in A(R)$,

$$\begin{aligned} \sum_{\sigma=1}^2 \Big(W_l^\sigma - w_l^\sigma(x)\Big) &\geqslant W_l^1 + W_l^2 - 4\varepsilon M_1 \sum_{i=1}^n \omega_i(R) \\ &\geqslant \omega_l(R) - 4\varepsilon M_1 \sum_{i=1}^n \omega_i(R) \\ &\geqslant \frac{1}{4n} \sum_{i=1}^n \omega_i(R) \quad \text{for } x \in B_R. \end{aligned} \tag{3.11}$$

From (3.5), we deduce that

$$\omega_s(R) \leqslant \Big(1 + \frac{1}{4n}\Big) \sum_{i=1}^n \omega_i(R) \qquad (s = 1, 2, \cdots, 3n). \tag{3.12}$$

By the definition of Φ_l^θ in (3.8),

$$\bigg[\fint_{B_{R/2}} \sum_{\sigma=1}^2 (W_l^\sigma - w_l^\sigma)^\kappa dx \bigg]^{1/\kappa} \leqslant C \sum_{\sigma=1}^2 \Phi_l^\sigma\Big(\frac{R}{2}\Big).$$

Thus for $s = \theta n + l \in A(R)$, the definition of $A(R)$ and (3.11) imply that

$$\begin{aligned} \Phi_l^\theta\Big(\frac{R}{2}\Big) &\geqslant \frac{1}{2} \sum_{\sigma=1}^2 \Phi_l^\sigma\Big(\frac{R}{2}\Big) \\ &\geqslant \frac{1}{2C} \bigg[\fint_{B_{R/2}} \sum_{\sigma=1}^2 (W_l^\sigma - w_l^\sigma)^\kappa dx \bigg]^{1/\kappa} \\ &\geqslant \frac{1}{8Cn} \sum_{i=1}^n \omega_i(R). \end{aligned}$$

Using also (3.12), we now obtain, for $s = \theta n + l \in A(R)$,

$$\Phi_l^\theta\Big(\frac{R}{2}\Big) \geqslant \frac{1}{8C(4n+1)} \omega_s(R).$$

Combining this inequality with (3.8), we see that, for $s = \theta n + l \in A(R)$,

$$\begin{aligned} \omega_s(R) &\leqslant C\Big(W_l^\theta - \sup_{B_{R/2}} w_l^\theta + R^2\Big) \\ &\leqslant C\Big\{\omega_s(R) - \omega_s\Big(\frac{R}{2}\Big) + R^2\Big\}, \end{aligned}$$

i.e.,

$$\omega_s\Big(\frac{R}{2}\Big) \leqslant \frac{C-1}{C}\omega_s(R) + R^2 \quad \text{for } s \in A(R). \tag{3.13}$$

On the other hand, by (3.12) and (3.10),

$$\begin{aligned} \sum_{p=1}^{3n} \omega_p(R) &\leqslant 3n\Big(1+\frac{1}{4n}\Big)\sum_{i=1}^{n}\omega_i(R) \\ &\leqslant 12n^2\Big(1+\frac{1}{4n}\Big)\sum_{s\in A(R)}\omega_s(R). \end{aligned} \tag{3.14}$$

From the estimates (3.13) and (3.14), we find that $\omega_s(R)$ $(s = 1, 2, \cdots, 3n)$ satisfies the conditions of Lemma 3.1. Therefore we can apply Lemma 3.1 to conclude the proof. □

The above theorem implies the interior gradient Hölder estimates. In order to establish the global gradient Hölder estimate, we still need estimates near the boundary. Here, we shall employ a lemma on gradient Hölder estimates near the boundary, established by Krylov in 1983. This lemma plays an important role in global $C^{2,\alpha}$ estimates for fully nonlinear equations. The proof was later simplified by Caffarelli and Trudinger, whom we follow here.

On $B_{R_0}^+$, we consider the following linear elliptic equation:

$$Lu = -a^{ij}(x)D_{ij}u = f(x). \tag{3.15}$$

Let λ, Λ be the smallest and largest eigenvalues of the matrix $(a^{ij}(x))$, respectively. We assume that

$$\lambda > 0,\ \ \Lambda/\lambda \leqslant \mu_1 \quad \text{for } x \in B_{R_0}^+, \tag{3.16}$$

where μ_1 is a constant.

For $\delta > 0$, we denote the cylindrical domains by

$$\begin{aligned} B_{R,\delta} &= \{x \mid |x'| < R,\ 0 < x_n < \delta R\}, \\ B_{R/2,\delta}^* &= \Big\{x \,\Big|\, |x'| < R,\ \frac{\delta R}{2} < x_n < \frac{3}{2}\delta R\Big\}, \end{aligned} \tag{3.17}$$

where $x' = (x_1, \cdots, x_{n-1})$.

Lemma 3.3. *There exists $\delta = \delta(n, \mu_1) > 0$ such that if (a^{ij}) satisfies (3.16) on $B_{2R,\delta}$, $u \in C(\overline{B}_{2R,\delta}) \cap W^{2,n}_{loc}(B_{2R,\delta})$ satisfies $Lu \geqslant f$, $u \geqslant 0$ on $B_{2R,\delta}$, $u = 0$ on $x_n = 0$ and $f/\lambda \in L^\infty(B_{2R,\delta})$, then there exist $\kappa > 0$ and $C \geqslant 1$ such that*

$$\Big[\fint_{B_{R/2,\delta}^*}\Big(\frac{u}{x_n}\Big)^\kappa dx\Big]^{1/\kappa} \leqslant C\Big[\inf_{B_{R/2,\delta}}\Big(\frac{u}{x_n}\Big) + R\|f/\lambda\|_{L^\infty(B_{2R,\delta})}\Big], \tag{3.18}$$

where κ and C depend only on n and μ_1.

Proof. Set $v = u/x_n$. Using the barrier function method as in Lemmas 1.5 and 2.1, we can establish that v is bounded. We first prove that there exists $\delta = \delta(n, \mu_1) > 0$ such that

$$\inf_{|x'|<R,\ x_n=\delta R} v \leqslant 2\Big(\inf_{B_{R/2,\delta}} v + R\|f/\lambda\|_{L^\infty(B_R^+)}\Big). \tag{3.19}$$

To prove (3.19), we assume for simplicity that $R=1$, $\lambda=1$ and $\inf_{|x'|<R} v(x',\delta R)=1$ (rescale if necessary). Construct the barrier function on $B_{1,\delta}$:

$$w(x)=\left[1-|x'|^2+(1+\sup|f|)\frac{x_n-\delta}{\sqrt{\delta}}\right]x_n. \tag{3.20}$$

A simple computation shows that

$$\begin{aligned} Lw &= 2\sum_{i=1}^{n-1} a^{ii}x_n+4\sum_{i=1}^{n-1} a^{in}x_i-\frac{2}{\sqrt{\delta}}a^{nn}(1+\sup|f|) \\ &\leqslant -\frac{2}{\sqrt{\delta}}(1+\sup|f|)+6(n-1)\mu_1. \end{aligned}$$

We choose $\delta=\min\left\{\frac{1}{16},\frac{1}{9(n-1)^2\mu_1^2}\right\}$. Then

$$Lw\leqslant f\leqslant Lu \quad \text{for } x\in B_{1,\delta},$$

and for $x\in\partial B_{1,\delta}$,

$$\begin{aligned} w\Big|_{x_n=0} &= u\Big|_{x_n=0}=0, \\ w\Big|_{x_n=\delta} &\leqslant \delta\leqslant u\Big|_{x_n=\delta}, \\ w\Big|_{|x'|=1} &\leqslant 0\leqslant u\Big|_{|x'|=1}. \end{aligned}$$

By the Alexandroff-Bakelman-Pucci maximum principle, $u\geqslant w$ on $B_{1,\delta}$. In particular,

$$v\geqslant\frac{3}{4}-(1+\sup|f|)\sqrt{\delta}\geqslant\frac{1}{2}-\sup|f| \quad \text{on } B_{1/2,\delta}.$$

This inequality implies (3.19).

Applying the weak Harnack inequality (Theorem 3.3 in Chapter 6) to the nonnegative supersolution u on $B_{2R,\delta}$, we get

$$\left[\fint_{B^*_{R/2,\delta}} u^\kappa dx\right]^{1/\kappa}\leqslant C\left[\inf_{B^*_{R/2,\delta}} u+R^2\|f/\lambda\|_{L^\infty}\right].$$

We notice that

$$\frac{2u}{3\delta R}\leqslant v\leqslant\frac{2u}{\delta R} \quad \text{on } B^*_{R/2,\delta},$$

and

$$\begin{aligned} \left[\fint_{B^*_{R/2,\delta}} v^\kappa dx\right]^{1/\kappa} &\leqslant C\left[\inf_{B^*_{R/2,\delta}} v+R\|f/\lambda\|_{L^\infty}\right] \\ &\leqslant C\left[\inf_{|x'|<R,\ x_n=\delta R} v+R\|f/\lambda\|_{L^\infty}\right]. \end{aligned}$$

Substituting (3.19) into the above inequality, we immediately obtain (3.18). □

Lemma 3.4 (Krylov). *Suppose that the coefficient matrix* (a^{ij}) *satisfies* (3.16) *on* B_1^+, $f/\lambda \in L^\infty(B_1^+)$, *and* $u \in C(\overline{B_1^+}) \cap W^{2,n}_{loc}(B_1^+)$ *is a solution of* (3.15). *Then there exist* $C \geqslant 1$ *and* $0 < \alpha < 1$ *such that, for any* $0 < R \leqslant 1$,

$$\operatorname*{osc}_{B_R^+} \frac{u}{x_n} \leqslant CR^\alpha \Big(\operatorname*{osc}_{B_1^+} \frac{u}{x_n} + \|f/\lambda\|_{L^\infty(B_1^+)} \Big), \tag{3.21}$$

where α, C *depend only on* n, μ_1. *This estimate implies that, for* $0 < \theta < 1$,

$$[Du]_{\alpha;\partial B_\theta^+ \cap \{x_n=0\}} \leqslant C(1-\theta)^{-\alpha}\Big(|u|_{1;B_1^+} + \|f/\lambda\|_{L^\infty(B_1^+)}\Big). \tag{3.22}$$

Proof. Set

$$M(R) = \sup_{B_{R,\delta}} v, \quad m(R) = \inf_{B_{R,\delta}} v, \quad \omega(R) = M(R) - m(R),$$

where $v = u/x_n$ and δ is a constant determined in Lemma 3.3. We now apply Lemma 3.3 to the functions $x_n M(2R) - u$ and $u - x_n m(2R)$ to derive

$$\Big[⨍_{B^*_{R/2,\delta}} [M(2R) - v]^\kappa dx \Big]^{1/\kappa} \leqslant C\Big[M(2R) - M\Big(\frac{R}{2}\Big) + R\|f/\lambda\|_{L^\infty}\Big],$$

$$\Big[⨍_{B^*_{R/2,\delta}} [v - m(2R)]^\kappa dx \Big]^{1/\kappa} \leqslant C\Big[m\Big(\frac{R}{2}\Big) - m(2R) + R\|f/\lambda\|_{L^\infty}\Big].$$

It follows that

$$\begin{aligned} \omega(2R) &\leqslant C\Big\{\Big[⨍_{B^*_{R/2,\delta}} [M(2R) - v]^\kappa dx \Big]^{1/\kappa} \\ &\qquad + \Big[⨍_{B^*_{R/2,\delta}} [v - m(2R)]^\kappa dx \Big]^{1/\kappa}\Big\} \\ &\leqslant C\Big[\omega(2R) - \omega\Big(\frac{R}{2}\Big) + R\|f/\lambda\|_{L^\infty}\Big], \end{aligned}$$

i.e.,

$$\omega\Big(\frac{R}{2}\Big) \leqslant \frac{C-1}{C}\omega(2R) + R\|f/\lambda\|_{L^\infty}.$$

By Lemma 2.1 in Chapter 4, we get (3.21). □

Theorem 3.5. *Let the structure conditions* (F1) *and* (F3) *be in force. Suppose that* $\partial\Omega \in C^2$, $\varphi \in C^2(\overline{\Omega})$, *and that* $u \in C^3(\Omega) \cap C^1(\overline{\Omega})$ *is a solution of* (1.1)–(1.2), *with* $|u|_{0;\Omega} \leqslant M_0$, $[u]_{1;\Omega} \leqslant M_1$. *Then there exist* $C > 0$ *and* $0 < \beta < 1$ *such that*

$$\sum_{i=1}^n \operatorname*{osc}_{B_R(x_0)} D_i u \leqslant CR^\beta, \quad \forall R > 0, \ \forall x_0 \in \partial\Omega, \tag{3.23}$$

where β *and* C *depend only on* $n, \mu_1, \mu_3, M_0, M_1$, $|\varphi|_2$ *and* $\partial\Omega$.

Proof. The function $v = u - \varphi$ satisfies the equation

$$-a^{ij} D_{ij} v = -a^{ij} D_{ij}\varphi + F(x, u, Du, 0),$$

where

$$a^{ij} = \int_0^1 \frac{\partial F}{\partial r_{ij}}(x, u, Du, \tau D^2 u) d\tau.$$

Let $x_0 \in \partial\Omega$. Since $\partial\Omega \in C^2$, there is a C^2 diffeomorphism which maps Ω in a neighborhood of x_0 to B_1^+ and maps $\partial\Omega$ in a neighborhood of x_0 to a subset of $\{x_n = 0\}$. By Krylov's Lemma 3.4,

$$[Dv]_{\alpha;\partial B^+_{1/2}\cap\{x_n=0\}} \leqslant C.$$

Mapping back to the original domain Ω and using a finite covering argument, we obtain

$$[Du]_{\alpha;\partial\Omega} \leqslant C. \tag{3.24}$$

Analogously to the interior estimate, we consider the function

$$w_l^\theta = (-1)^\theta D_l u + \varepsilon \sum_{i=1}^n (D_i u)^2 \quad (l = 1, 2, \cdots, n;\ \theta = 1, 2). \tag{3.25}$$

As proved in Lemma 3.2,

$$-\frac{\partial F}{\partial r_{ij}} D_{ij} w_l^\theta \leqslant C_\varepsilon \lambda (1 + |Dw_l^\theta|^2).$$

By Lemma 1.5,

$$w_l^\theta(x) - w_l^\theta(x_0) \leqslant C_\varepsilon |x - x_0|^{\alpha/(1+\alpha)} \Big(1 + \sum_l [w_l^\theta]_{\alpha;\partial\Omega}\Big), \quad \forall x_0 \in \partial\Omega,\ x \in \Omega.$$

Using the definition of w_l^θ and (3.24), we deduce from the above inequality that

$$|D_l u(x) - D_l u(x_0)| - 2\varepsilon M_1 \sum_{i=1}^n |D_i u(x) - D_i u(x_0)| \leqslant C_\varepsilon |x - x_0|^{\alpha/(1+\alpha)}.$$

Choosing $\varepsilon = 1/(4nM_1)$ and summing over $l = 1, 2, \cdots, n$, we conclude that

$$\sum_{i=1}^n |D_i u(x) - D_i u(x_0)| \leqslant C|x - x_0|^{\alpha/(1+\alpha)}, \quad x_0 \in \partial\Omega,\ x \in \Omega.$$

(3.23) follows immediately. □

Combining the interior estimate (Theorem 3.2) and the estimate near the boundary (Theorem 3.5), we can establish the global gradient Hölder estimate: namely, there exist $C \geqslant 1$, $0 < \beta < 1$ such that

$$\sum_{i=1}^n [D_i u]_{\beta;\Omega} \leqslant C,$$

where β, C depend only on n, μ_1, μ_3, $|u|_0, [u]_1, |\varphi|_2$ and Ω.

4. Solvability for quasilinear equations of nondivergence form

Now we consider the following quasilinear equation:

$$-a^{ij}(x, u, Du) D_{ij} u + b(x, u, Du) = 0; \tag{4.1}$$

in other words, in (1.1) we choose

$$F(x, z, p, r) = a^{ij}(x, z, p) r_{ij} - b(x, z, p).$$

The corresponding structure conditions are as follows:

(F1)′ On $\Omega \times \mathbb{R} \times \mathbb{R}^n$, $\lambda > 0$, and $\Lambda/\lambda \leqslant \mu_1(|z|)$, where λ and Λ are the smallest and largest eigenvalues of the matrix (a^{ij});

(F2)′ $\quad |b(x,z,p)| \leqslant \lambda\mu_2(|z|)(1+|p|^2)$;

(F3)′
$$\begin{cases} \sum_{i,j}\Big[(1+|p|)^{-1}|D_x a^{ij}| + |D_z a^{ij}| + (1+|p|)|D_p a^{ij}|\Big] \leqslant \lambda\mu_3(|z|), \\ (1+|p|)^{-1}|D_x b| + |D_z b| + (1+|p|)|D_p b| \leqslant \lambda\mu_3(|z|)(1+|p|^2); \end{cases}$$

(F5)′ $\quad -b(x,z,p)\,\mathrm{sign}\, z \leqslant \lambda\overline{\mu}(1+|p|)$.

Under the structure conditions (F1)′, (F2)′, (F3)′ and (F5)′, we can apply the results from §§1–3. If $\partial\Omega \in C^2$ and $\varphi \in C^2(\overline{\Omega})$, then a solution $u \in C^3(\Omega)\cap C^1(\overline{\Omega})$ of the Dirichlet problem (4.1), (1.2) satisfies an a priori estimate in $C^{1,\beta}(\overline{\Omega})$, for some $\beta \in (0,1)$.

To study the solvability of the Dirichlet problem (4.1), (1.2), we shall use a special case of the Leray-Schauder fixed point theorem:

Let T be a compact map (i.e., T is continuous and maps any bounded set in X to a precompact set) from a Banach space X to itself. Suppose that there exists M such that

$$\|x_0\|_X \leqslant M, \quad \forall x_0 \in \{x \in X \mid \exists\sigma \in [0,1],\ x = \sigma T x\}.$$

Then T has a fixed point.

Theorem 4.1. *Let the structure conditions* (F1)′, (F2)′, (F3)′ *and* (F5)′ *be in force. Suppose that* $0<\alpha<1$, $\partial\Omega \in C^{2,\alpha}$, $a^{ij}, b \in C^1(\overline{\Omega}\times\mathbb{R}\times\mathbb{R}^n)$, $\varphi \in C^{2,\alpha}(\overline{\Omega})$. *Then the Dirichlet problem* (4.1), (1.2) *admits a solution* $u \in C^{2,\alpha}(\overline{\Omega})$.

Proof. First we assume that $a^{ij}, b \in C^{1,\delta}(\overline{\Omega}\times\mathbb{R}\times\mathbb{R}^n)$ for some $0<\delta<1$. Take the Banach space $X = C^{1,\alpha}(\overline{\Omega})$. For any $v \in X$, we solve the Dirichlet problem

$$\begin{cases} -a^{ij}(x,v,Dv)D_{ij}u + b(x,v,Dv) = 0 & \text{in } \Omega, \\ u = \varphi & \text{on } \partial\Omega. \end{cases} \tag{4.2}$$

Under our assumptions, this problem has a unique solution $u \in C^{2,\alpha}(\overline{\Omega})$, by the Schauder theory. We define the map T by $u = Tv$. Using a proof similar to that of Theorem 6.1 in Chapter 5, we can show that T is compact. For $u \in C^{1,\alpha}(\overline{\Omega})$, the solution of the equation $u = \sigma T u$ solves the following Dirichlet problem:

$$\begin{cases} -a^{ij}(x,u,Du)D_{ij}u + \sigma b(x,u,Du) = 0 & \text{in } \Omega, \\ u = \sigma\varphi & \text{on } \partial\Omega. \end{cases} \tag{4.3}$$

According to the definition of T, we already know that $u \in C^{2,\alpha}(\overline{\Omega})$. Since $a^{ij}, b \in C^{1,\delta}(\Omega\times\mathbb{R}\times\mathbb{R}^n)$, the Schauder theory implies that $u \in C^{3,\delta}(\Omega)\cap C^{2,\alpha}(\overline{\Omega})$. Thus we can apply the a priori estimate from §§1–3; namely, there exist $C>0$ and $0<\beta<1$ such that any solution of (4.3) satisfies

$$|u|_{1,\beta;\Omega} \leqslant C.$$

The Schauder theory now implies that

$$|u|_{2,\beta;\Omega} \leqslant C.$$

Hence, any solution of the equation $u = \sigma T u$ satisfies

$$|u|_{1,\alpha;\Omega} \leqslant C.$$

By the Leray-Schauder fixed point theorem, there is a fixed point u for T; i.e., the Dirichlet problem (4.1), (1.2) admits a solution u. It is clear that $u \in C^{2,\alpha}(\overline{\Omega})$.

In the general case, we can approximate a^{ij}, b by $a^{ij}_N, b_N \in C^{1,\delta}(\overline{\Omega} \times \mathbb{R} \times \mathbb{R}^n)$. The details are left as exercises for the reader. □

It is clear from the proof that further regularity of the coefficients is not important, while the structure conditions are essential.

5. Solvability for fully nonlinear equations

For the Dirichlet problem for fully nonlinear equations, the Leray-Schauder fixed point theorem is not applicable, because maps similar to the one in §4 are not compact. With the help of the implicit function theorem in nonlinear functional analysis, we shall use the continuation method.

We shall first state some concepts in nonlinear functional analysis. The implicit function theorem will be stated without proof.

Let X_1, X_2 be two real Banach spaces and F be a map from X_1 to X_2. F is said to have a *Fréchet derivative* at $u \in X_1$ if there exists a bounded linear operator $L : X_1 \to X_2$ such that

$$\|F[u+h] - F[u] - Lh\|_{X_2} = o(\|h\|_{X_1}), \quad \forall h \in X_1; \tag{5.1}$$

the operator L is said to be the *Fréchet derivative* of F at u, denoted by F_u.

Example 1. If $F : X_1 \to X_2$ is a bounded linear operator, then $F_u = F$.

Example 2. If $F[u] = F(D^2u)$, $X_1 = C^{2,\alpha}(\overline{\Omega})$, $X_2 = C^{\alpha}(\overline{\Omega})$ $(0 < \alpha < 1)$, and the function $F \in C^1(\mathcal{S}^n)$, then

$$F_u = \frac{\partial F}{\partial r_{ij}}(D^2u)D_{ij}. \tag{5.2}$$

In fact,

$$\begin{aligned}
&\left\|F[u+h] - F[u] - \frac{\partial F}{\partial r_{ij}}(D^2u)D_{ij}h\right\|_{C^\alpha} \\
=\ & \left\|F(D^2u + D^2h) - F(D^2u) - \frac{\partial F}{\partial r_{ij}}(D^2u)D_{ij}h\right\|_{C^\alpha} \\
=\ & \left\|\int_0^1 \frac{\partial F(D^2u + \tau D^2h)}{\partial r_{ij}}D_{ij}h\,d\tau - \frac{\partial F}{\partial r_{ij}}(D^2u)D_{ij}h\right\|_{C^\alpha} \\
\leqslant\ & \|h\|_{C^{2,\alpha}(\overline{\Omega})}\int_0^1 \left\|\frac{\partial F(D^2u + \tau D^2h)}{\partial r_{ij}} - \frac{\partial F}{\partial r_{ij}}(D^2u)\right\|_{C^\alpha} d\tau \\
=\ & o(\|h\|_{C^{2,\alpha}(\overline{\Omega})}).
\end{aligned}$$

Now suppose that X_1, Σ, X_2 are real Banach spaces. We denote the Fréchet derivative of the map $G : X_1 \times \Sigma \to X_2$ at the point (u, σ) by $G_{(u,\sigma)}$. For any $h \in X_1$ and $\lambda \in \Sigma$,

$$G_{(u,\sigma)}(h, \lambda) = G^1_{(u,\sigma)}h + G^2_{(u,\sigma)}\lambda.$$

The linear operators $G^1_{(u,\sigma)} : X_1 \to X_2$, $G^2_{(u,\sigma)} : \Sigma \to X_2$ are called the *partial Fréchet derivatives* of G at (u, σ).

Implicit function theorem. *Suppose that X_1, Σ, X_2 are real Banach spaces and that G is a map from an open subset of $X_1 \times \Sigma$ to an open subset of X_2. Let $(u_0, \sigma_0) \in X_1 \times \Sigma$. We assume that*

(1) $G[u_0, \sigma_0] = 0$,

(2) *G is Fréchet differentiable in a neighborhood of (u_0, σ_0) and its Fréchet derivative is continuous at (u_0, σ_0),*

(3) *the partial Fréchet derivative $G^1_{(u_0,\sigma_0)}$ is invertible.*

Then there exists a neighborhood $\mathcal{N}$ of σ_0 in Σ such that the equation $G[u, \sigma] = 0$ is solvable for every $\sigma \in \mathcal{N}$; i.e., for every $\sigma \in \mathcal{N}$, there exists u_σ such that $G[u_\sigma, \sigma] = 0$.

Set

$$F[u] = F(x, u, Du, D^2u).$$

Theorem 5.1. *Suppose that $F \in C^1(\Gamma)$ and satisfies the structure condition (F1), $\partial\Omega \in C^{2,\beta}$ for some $0 < \beta < 1$, and $\varphi \in C^{2,\beta}(\overline{\Omega})$. We assume, under some additional structure conditions (may depend on β), that any possible solution $u \in C^{2,\beta}(\overline{\Omega})$ of the equation $F[u] - \sigma F[\varphi] = 0$ ($\forall \sigma \in [0, 1]$) satisfying boundary condition (1.2) has the a priori estimate*

$$|u|_{2,\beta;\Omega} \leqslant M, \tag{5.3}$$

where $M \geqslant 1$ depends only on n, β, the structure conditions, φ and $\partial\Omega$. In addition, we assume

(F6) $F_z(x, z, p, r) \leqslant 0, \quad \forall (x, z, p, r) \in \Gamma.$

Then the Dirichlet problem (1.1), (1.2) admits a solution $u \in C^{2,\beta}(\overline{\Omega})$.

Proof. Take $X_1 = \{v \in C^{2,\beta}(\overline{\Omega}) \mid v = 0 \quad \text{on } \partial\Omega\}$, $X_2 = C^\beta(\overline{\Omega})$, $\Sigma = \mathbb{R}$. Define the map $G : X_1 \times [0, 1] \to X_2$:

$$G[v, \sigma] = F[v + \varphi] - \sigma F[\varphi]. \tag{5.4}$$

We want to solve the equation $G[v, \sigma] = 0$. It is obvious that for $\sigma = 0$, if $v_0 \in X_1$ is a solution of $G[v, 0] = 0$, then $u = v_0 + \varphi$ is a solution of the Dirichlet problem (1.1), (1.2). Now we set

$$S = \{\sigma \in [0, 1] \mid \exists v \in X_1,\ G[v, \sigma] = 0\}. \tag{5.5}$$

First, S is nonempty. In fact, when $\sigma = 1$, the function $v \equiv 0$ is a solution of the equation $G[v, \sigma] = 0$. Therefore $1 \in S$. We shall show that, relative to $[0, 1]$, S is both open and closed. It follows that $S = [0, 1]$, and the theorem follows.

We use the implicit function theorem to prove that S is open relative to $[0, 1]$. Suppose that $\sigma_0 \in S$. By definition, there exists $v_0 \in X_1$ such that $G[v_0, \sigma_0] = 0$. It is easy to compute the partial Fréchet derivative:

$$G^1_{(v_0,\sigma_0)} = \left[\frac{\partial F}{\partial r_{ij}} D_{ij} + \frac{\partial F}{\partial p_i} D_i + \frac{\partial F}{\partial z}\right]_{(z,p,r)=(v_0+\varphi, D(v_0+\varphi), D^2(v_0+\varphi))}.$$

This is a linear elliptic operator from X_1 to X_2. By the assumption (F6), the maximum principle and the Schauder theory, $G^1_{(v_0,\sigma_0)}$ is invertible. By the implicit function theorem, there exists a neighborhood $\mathcal{N}\cap[0,1]$ of σ_0 such that the equation $G[v,\sigma]=0$ is solvable for $\sigma\in\mathcal{N}\cap[0,1]$. It follows that $\mathcal{N}\cap[0,1]\subset S$.

Now we use the a priori estimate (5.3) to show that S is closed. If $\sigma_n\in S$ and $\sigma_n\to\sigma_0$ as $n\to\infty$, we want to prove that $\sigma_0\in S$. By the definition of S, there exists $v_n\in X_1$ such that $G[v_n,\sigma_n]=0$, i.e.,

$$\begin{gathered} v_n\in C^{2,\beta}(\overline{\Omega}),\qquad v_n\Big|_{\partial\Omega}=0,\\ F[v_n+\varphi]-\sigma_nF[\varphi]=0. \end{gathered}\tag{5.6}$$

By (5.3), $|v_n+\varphi|_{2,\beta;\Omega}\leqslant M$. Using the Ascoli-Arzelá theorem, a subsequence of v_n converges in $C^2(\overline{\Omega})$ to some function v_0, and $v_0\in C^{2,\beta}(\overline{\Omega})$. Taking the limit in (5.7) along this subsequence, we conclude that

$$F[v_0+\varphi]-\sigma_0F[\varphi]=0,\tag{5.7}$$

i.e., $\sigma_0\in S$. The proof is complete. □

Remark. Under the assumptions of Theorem 5.1, if we further assume $F\in C^{k,\alpha}(\Gamma)$ $(k\geqslant 1)$, then the solution u of the Dirichlet problem belongs to $C^{k+2,\alpha}(\Omega)\cap C^{2,\beta}(\overline{\Omega})$. If we assume $F\in C^{k,\alpha}(\overline{\Gamma})$ $(k\geqslant 1)$ and $\partial\Omega\in C^{k+2,\alpha}$, then $u\in C^{k+2,\alpha}(\overline{\Omega})$.

For the Dirichlet problem for fully nonlinear equations, Theorem 5.1 reduces the solvability problem to the a priori estimate (5.3), i.e., the global $C^{2,\beta}$ estimate.

6. A special class of equations

In this section, we consider the following form of fully nonlinear equation:

$$F(D^2u)=0,\tag{6.1}$$

where we only assume the structure conditions (F1) and (F4).

Theorem 6.1. *Suppose that F satisfies the structure conditions* (F1) *and* (F4), *$F\in C^2(\Gamma)$, $\partial\Omega\in C^3$, $\varphi\in C^3(\overline{\Omega})$. Then there exist $C\geqslant 1$ and $0<\beta<1$ such that a solution $u\in C^{2,\beta}(\overline{\Omega})$ of the Dirichlet problem* (6.1), (1.2) *satisfies*

$$[D^2u]_{\beta;\Omega}\leqslant C(|u|_{0;\Omega}+|\varphi|_{3;\Omega}),\tag{6.2}$$

where β and C depend only on $n,\mu_1(|u|_0)$ and $\partial\Omega$.

First, we notice that for any $1<p<\infty$, $u\in W^{4,p}_{loc}(\Omega)\cap C^{2,\beta}(\overline{\Omega})$. In fact, using finite differences, we can prove that $Du\in C^{2,\beta}(\Omega)$, and

$$\frac{\partial F}{\partial r_{ij}}(D^2u)D_{ij}(D_ku)=0\qquad(k=1,2,\cdots,n).$$

Thus L^p theory implies that $D_ku\in W^{2,p}_{loc}(\Omega)$. Similarly, we can get $D^2u\in W^{2,p/2}_{loc}(\Omega)$, where p is any real number in $(2,\infty)$.

In order to prove Theorem 6.1, we need the following algebraic lemma.

Lemma 6.2. *For $0 < \theta \leqslant \Theta < \infty$, we set*

$$S[\theta, \Theta] = \{A \in \mathcal{S}^n \mid \theta I \leqslant A \leqslant \Theta I\}, \tag{6.3}$$

where $\mathcal{S}^n$ is the set of all symmetric matrices. Then there exist N unit vectors $\gamma_1, \gamma_2, \cdots, \gamma_N$ such that any $A \in S[\theta, \Theta]$ can be represented as

$$A = \sum_{i=1}^{N} \beta_i(A)\gamma_i \otimes \gamma_i, \tag{6.4}$$

$$0 < \theta_1 \leqslant \beta_i(A) \leqslant \Theta_1 < \infty \qquad (i = 1, 2, \cdots, N), \tag{6.5}$$

where N, γ_k $(k = 1, 2 \cdots, N)$, θ_1, Θ_1 depend only on n, θ, Θ.

Proof. Let e_i be the unit vector in the x_i coordinate, and set $e_{ij} = e_i + e_j$. Then

$$\begin{pmatrix} n & 1 & \cdots & 1 \\ 1 & n & \cdots & 1 \\ \vdots & \vdots & \ddots & \vdots \\ 1 & 1 & \cdots & n \end{pmatrix} = \sum_{i=1}^{n} e_i \otimes e_i + \sum_{1 \leqslant i < j \leqslant n} e_{ij} \otimes e_{ij}.$$

For any $A \in S[\theta/2, \Theta]$, there exists a nondegenerate matrix C such that

$$A = C^T \begin{pmatrix} n & 1 & \cdots & 1 \\ 1 & n & \cdots & 1 \\ \vdots & \vdots & \ddots & \vdots \\ 1 & 1 & \cdots & n \end{pmatrix} C.$$

Therefore

$$A = \sum_{i=1}^{n} Ce_i \otimes Ce_i + \sum_{1 \leqslant i < j \leqslant n} Ce_{ij} \otimes Ce_{ij};$$

i.e., for this matrix A, there exist $n(n+1)/2$ unit vectors $\widetilde{\gamma}_1, \widetilde{\gamma}_2, \cdots, \widetilde{\gamma}_{n(n+1)/2}$ such that $\widetilde{\gamma}_i \otimes \widetilde{\gamma}_i$ $(i = 1, 2, \cdots, n(n+1)/2)$ are linearly independent vectors in $\mathcal{S}^n$, and

$$A = \sum_{i=1}^{n(n+1)/2} \beta_i \widetilde{\gamma}_i \otimes \widetilde{\gamma}_i \qquad (\beta_i > 0, 1 \leqslant i \leqslant \frac{n(n+1)}{2}).$$

The set

$$U(\widetilde{\gamma}_1, \widetilde{\gamma}_2, \cdots, \widetilde{\gamma}_{N_1}) = \Big\{ \sum_{i=1}^{N_1} \beta_i \widetilde{\gamma}_i \otimes \widetilde{\gamma}_i \;\Big|\; \beta_i > 0 \Big\} \qquad (N_1 = \frac{n(n+1)}{2})$$

is obviously an open set in $\mathcal{S}^n$. For each $A \in S[\theta/2, \Theta]$, we have the corresponding $U(\widetilde{\gamma}_1, \widetilde{\gamma}_2, \cdots, \widetilde{\gamma}_{N_1})$ $(A \in U(\widetilde{\gamma}_1, \widetilde{\gamma}_2, \cdots, \widetilde{\gamma}_{N_1}))$, which forms an open cover for $S[\theta/2, \Theta]$. Since $S[\theta/2, \Theta]$ is a closed set in $\mathcal{S}^n$, there is a finite subcover $U^{(1)}, U^{(2)}, \cdots, U^{(N_2)}$ for $S[\theta/2, \Theta]$. If we collect all the associated unit vectors $\gamma_1, \gamma_2, \cdots, \gamma_N$ $(N = N_1 N_2)$, then for any $A \in S[\theta/2, \Theta]$, we have the representation

$$A = \sum_{i=1}^{N} \widetilde{\beta}_i \widetilde{\gamma}_i \otimes \widetilde{\gamma}_i \qquad (\widetilde{\beta}_i \geqslant 0).$$

For $A \in S[\theta, \Theta]$, it is clear that

$$A - \sum_{i=1}^{N} \frac{\theta}{2N} \widetilde{\gamma}_i \otimes \widetilde{\gamma}_i \in S[\theta/2, \Theta].$$

It follows that

$$A - \sum_{i=1}^{N} \frac{\theta}{2N} \widetilde{\gamma}_i \otimes \widetilde{\gamma}_i = \sum_{i=1}^{N} \widetilde{\beta}_i \widetilde{\gamma}_i \otimes \widetilde{\gamma}_i.$$

By choosing $\beta_i(A) = \widetilde{\beta}_i + \dfrac{\theta}{2N}$, we establish (6.4). □

Remark. Clearly, in Lemma 6.1, we can require $\{\gamma_1, \gamma_2, \cdots, \gamma_N\}$ to include the vectors

$$e_i, \quad (e_i + e_j)/\sqrt{2} \qquad (i, j = 1, 2, \cdots, n, \quad i \neq j).$$

Then it is easy to verify that there exists C (depending only on n, θ, Θ) such that

$$\frac{1}{C}|D^2 u|^2 \leqslant \sum_{i=1}^{N} |D_{\gamma_k \gamma_k} u|^2 \leqslant C|D^2 u|^2,$$

where $D_{\gamma_k \gamma_k} u$ is the second derivative in the γ_k direction.

Proof of Theorem 6.1. For the set $S[1, \mu_1]$, we apply Lemma 6.2, where $\mu_1 = \mu_1(|u|_0)$ is a constant determined in the structure condition (F1). Let $\gamma_1, \gamma_2, \cdots, \gamma_N$ be the unit vectors determined in Lemma 6.2, which include e_i and $(e_i + e_j)/\sqrt{2}$ $(i, j = 1, 2, \cdots, n, \ i \neq j)$. For simplicity, we drop the subscript in γ_k if there is no confusion and denote it by γ.

Differentiating (6.1) twice with respect to γ, we obtain

$$\frac{\partial F}{\partial r_{ij}} D_{ij}(D_{\gamma\gamma} u) + \frac{\partial^2 F}{\partial r_{ij} \partial r_{kl}} D_{ij\gamma} u D_{kl\gamma} u = 0.$$

Using the concavity condition (F4), we get

$$-\frac{\partial F}{\partial r_{ij}} D_{ij}(D_{\gamma\gamma} u) \leqslant 0. \tag{6.6}$$

For $B_{2R} \subset \Omega$, we set

$$\begin{gathered} M_k(R) = \sup_{B_R} D_{\gamma_k \gamma_k} u, \qquad m_k(R) = \inf_{B_R} D_{\gamma_k \gamma_k} u, \\ \omega_k(R) = M_k(R) - m_k(R). \end{gathered} \tag{6.7}$$

Since we have (6.6), we can apply the weak Harnack inequality (Theorem 2.3 in Chapter 6) to the function $M_k(2R) - D_{\gamma_k \gamma_k} u$ to derive

$$\begin{aligned} \Phi_k(R) &\triangleq \left[⨍_{B_R} [M_k(2R) - D_{\gamma_k \gamma_k} u]^\kappa dx \right]^{1/\kappa} \\ &\leqslant C\Big(M_k(2R) - M_k(R)\Big) \qquad (k = 1, 2, \cdots, N). \end{aligned} \tag{6.8}$$

By Lemma 6.2 and the equation (6.1), we can write

$$\begin{aligned} 0 &= F(D^2u(x)) - F(D^2u(y)) \\ &= \frac{\partial F(r^*)}{\partial r_{ij}}(D_{ij}u(x) - D_{ij}u(y)) \\ &= \lambda \sum_{k=1}^{N} \beta_k (D_{\gamma_k\gamma_k}u(x) - D_{\gamma_k\gamma_k}u(y)). \end{aligned}$$

If we fix $1 \leqslant l \leqslant N$, then for $x, y \in B_{2R}$,

$$\begin{aligned} \beta_l(D_{\gamma_l\gamma_l}u(x) - D_{\gamma_l\gamma_l}u(y)) &= -\sum_{k\neq l} \beta_k(D_{\gamma_k\gamma_k}u(x) - D_{\gamma_k\gamma_k}u(y)) \\ &\geqslant -\Theta' \sum_{k\neq l}(M_k(2R) - D_{\gamma_k\gamma_k}u(y)), \end{aligned}$$

which implies that

$$\theta'(D_{\gamma_l\gamma_l}u(y) - m_l(2R)) \leqslant \Theta' \sum_{k\neq l}(M_k(2R) - D_{\gamma_k\gamma_k}u(y)) \quad \text{for } y \in B_{2R}.$$

Using (6.8), we now obtain

$$\begin{aligned} \left[⨍_{B_R} [D_{\gamma_l\gamma_l}u(y) - m_l(2R)]^\kappa dy \right]^{1/\kappa} &\leqslant C \sum_k \Phi_k(R) \\ &\leqslant C \sum_{k=1}^{N} [M_k(2R) - M_k(R)]. \end{aligned}$$

Combining this estimate with (6.8), we get

$$\begin{aligned} M_l(2R) - m_l(2R) &\leqslant C \left\{ \left[⨍_{B_R} [D_{\gamma_l\gamma_l}u(y) - m_l(2R)]^\kappa dy \right]^{1/\kappa} \right. \\ &\quad \left. + \left[⨍_{B_R} [M_l(2R) - D_{\gamma_l\gamma_l}u(y)]^\kappa dy \right]^{1/\kappa} \right\} \\ &\leqslant C \sum_{k=1}^{N} [M_k(2R) - M_k(R)]. \end{aligned}$$

Summing over $l = 1, 2, \cdots, N$, we get,

$$\sum_{l=1}^{N} \omega_l(2R) \leqslant C \Big[\sum_{k=1}^{N} \omega_k(2R) - \sum_{k=1}^{N} \omega_k(R) \Big],$$

i.e.,

$$\sum_{k=1}^{N} \omega_k(R) \leqslant \frac{C-1}{C} \sum_{k=1}^{N} \omega_k(2R).$$

By Lemma 2.1 in Chapter 4,

$$\sum_{k=1}^{N} \omega_k(\sigma R) \leqslant C\sigma^\alpha \sum_{k=1}^{N} \omega_k(R), \qquad \forall 0 < \sigma \leqslant 1, \tag{6.9}$$

where $C \geqslant 1$ and $0 < \alpha < 1$ depend only on n and μ_1. The above inequality implies the interior Hölder estimate; i.e., for any $x, y \in \Omega$, we have

$$|D^2u(x) - D^2u(y)| \leqslant C\Big[\frac{|x-y|}{\min\{d_x, d_y\}}\Big]^\alpha |D^2u|_{0;\Omega}. \tag{6.10}$$

Next, we study the Hölder norm estimate for second order derivatives near the boundary. Let $v = u - \varphi$. Then $D_k v$ satisfies

$$\begin{aligned} -\frac{\partial F}{\partial r_{ij}} D_{ij}(D_k v) &= f_k(x) \qquad (k = 1, 2, \cdots, n), \\ \sup \Big|\frac{f_k(x)}{\lambda}\Big| &\leqslant C|\varphi|_{3;\Omega}. \end{aligned} \tag{6.11}$$

By assumption, $\partial\Omega \in C^3$. For each $x_0 \in \partial\Omega$, there exists a C^3 diffeomorphism which maps Ω in a neighborhood of x_0 to B_1^+ and maps $\partial\Omega$ in a neighborhood of x_0 to a subset of $\{x_n = 0\}$. Therefore we can assume without loss of generality that $D_k v$ satisfies an equation in the form of (6.11) on B_1^+. For $k = 1, 2, \cdots, n-1$, $D_k v\Big|_{x_n=0} = 0$. By Krylov's Lemma 3.4,

$$[D_n D_k v]_{\alpha;\partial B^+_{1/2}\cap\{x_n=0\}} \leqslant C[|u|_2 + |\varphi|_3] \qquad (k = 1, 2 \cdots, n-1). \tag{6.12}$$

Utilizing the equation (6.1), we conclude that

$$[D^2 v]_{\alpha;\partial B^+_{1/2}\cap\{x_n=0\}} \leqslant C[|u|_2 + |\varphi|_3].$$

Changing back to the original domain and using a finite covering technique, we get

$$[D^2 u]_{\alpha;\partial\Omega} \leqslant C[|u|_2 + |\varphi|_3]. \tag{6.13}$$

The boundary estimate (6.13) will be used for the estimate near the boundary, which will then be combined with the interior estimate (6.9) for the global estimate.

Applying Lemma 3.1 in Chapter 6 to the differential inequality (6.6), we obtain, for $x_0 \in \partial\Omega$, $x \in \Omega$,

$$D_{\gamma\gamma}u(x) - D_{\gamma\gamma}u(x_0) \leqslant C|x - x_0|^{\alpha/(1+\alpha)}\Big\{|u|_2 + [D^2u]_{\alpha;\partial\Omega}\Big\}. \tag{6.14}$$

On the other hand, by (6.1) and Lemma 6.2,

$$\begin{aligned} 0 &= F(D^2u(x)) - F(D^2u(x_0) \\ &= \frac{\partial F(r^*)}{\partial r_{ij}}(D_{ij}u(x) - D_{ij}u(x_0)) \\ &= \lambda \sum_{k=1}^N \beta_k (D_{\gamma_k\gamma_k}u(x) - D_{\gamma_k\gamma_k}u(x_0)). \end{aligned}$$

Applying the estimate (6.14), we find that, for any fixed $1 \leqslant l \leqslant n$,

$$\begin{aligned} \beta_l[D_{\gamma_l\gamma_l}u(x_0) - D_{\gamma_l\gamma_l}u(x)] &= \sum_{k\neq l} \beta_k (D_{\gamma_k\gamma_k}u(x) - D_{\gamma_k\gamma_k}u(x_0)) \\ &\leqslant C|x - x_0|^{\alpha/(1+\alpha)}\Big\{|u|_2 + [D^2u]_{\alpha;\partial\Omega}\Big\}. \end{aligned} \tag{6.15}$$

Combining (6.14) and (6.15), we find that, for $x_0 \in \partial\Omega$, $x \in \Omega$,

$$|D_{\gamma\gamma}u(x) - D_{\gamma\gamma}u(x_0)| \leqslant C|x - x_0|^{\alpha/(1+\alpha)}\Big\{|u|_2 + [D^2u]_{\alpha;\partial\Omega}\Big\}.$$

It follows that, for any $x_0 \in \partial\Omega$ and $R > 0$,

$$\operatorname*{osc}_{B_R(x_0)\cap\Omega} D^2u \leqslant CR^{\alpha/(1+\alpha)}\Big\{|u|_2 + |\varphi|_3\Big\}. \tag{6.16}$$

Combining the interior estimate (6.9) and the estimate (6.16) near the boundary, we obtain the global estimate

$$[D^2u]_{\beta;\Omega} \leqslant C\Big\{|u|_2 + |\varphi|_3\Big\},$$

where $0 < \beta < 1$. We then conclude the proof by employing the interpolation theorem. □

We now have the following existence theorem:

Theorem 6.3. *Let the structure conditions* (F1), (F4) *be in force. Suppose that $\partial\Omega \in C^3$ and $\varphi \in C^3(\overline{\Omega})$. Then there exists $\beta = \beta(n, \mu_1) \in (0,1)$ such that the Dirichlet problem* (6.1), (1.2) *admits a solution $u \in C^{2,\beta}(\overline{\Omega})$.*

If $F(r)$ is twice continuously differentiable, then this theorem is a direct corollary of Theorems 5.1 and 6.1. If $F(r)$ is not twice differentiable, then it can be approximated by twice continuously differentiable concave functions. The details of the proof are left to the reader.

7. General fully nonlinear equations

First we consider the equation of the following form:

$$F(x, D^2u) = 0. \tag{7.1}$$

We assume that $F(x,r)$ is Hölder continuous in x, i.e.,

$$|F(x,r) - F(y,r)| \leqslant \mu_0\widetilde{\lambda}\,(|r| + \widetilde{\mu})|x-y|^\beta \quad \text{for } (x,r),(y,r) \in \Omega\times\mathbb{S}^n, \tag{7.2}$$

where μ_0, $\widetilde{\mu}$, β are positive constants, $\widetilde{\lambda} = \min\{\lambda(x), \lambda(y)\}$.

Theorem 7.1. *Let the structure conditions* (F1), (F4) *be in force. Suppose that $\varphi \in C^3(\overline{\Omega})$ and that $u \in C^3(\overline{\Omega})$ is a solution of the Dirichlet problem* (7.1), (1.2), *with $|u|_{0;\Omega} \leqslant M_0$. Under these assumptions, there exists $\alpha = \alpha(n, \mu_1)$, $0 < \alpha < 1$, such that for any $\beta \in (0, \alpha)$, if $\partial\Omega \in C^{2,\beta}$ and* (7.2) *is valid, then*

$$[D^2u]_{\beta;\Omega} \leqslant C\Big[|u|_{0;\Omega} + |\varphi|_{2,\beta;\Omega} + \widetilde{\mu}\Big], \tag{7.3}$$

where C depends only on $n, \mu_1, \mu_0, (\alpha-\beta)^{-1}$ and the $C^{2,\beta}$ norm of $\partial\Omega$.

Proof. For any ball $B_R(x_0) \subset \Omega$, let v be the solution of the following Dirichlet problem:

$$\begin{cases} F(x_0, D^2v) = 0 & \text{in } B_R(x_0), \\ v = u & \text{on } \partial B_R(x_0). \end{cases} \tag{7.4}$$

By Theorem 6.3, such a solution exists, and $v \in C^{2,\alpha}(\overline{B}_R(x_0))$ for some $\alpha = \alpha(n, \mu_1)$.

Denote the quadratic polynomial by

$$q(x) = u(x_0) + D_iu(x_0)(x_i - x_{0i}) + \frac{1}{2}D_{ij}u(x_0)(x_i - x_{0i})(x_j - x_{0j}), \tag{7.5}$$

and set

$$\widehat{v}(x) = v(x) - q(x), \qquad \widehat{u}(x) = u(x) - q(x). \tag{7.6}$$

By the interior estimate (6.9),

$$[D^2 v]_{\alpha;B_\rho} \leqslant \frac{C}{(r-\rho)^\alpha} \operatorname*{osc}_{B_r} D^2 v \leqslant \frac{2C}{(r-\rho)^\alpha}[D^2 \widehat{v}]_{0;B_r}, \qquad \forall 0 < \rho < r \leqslant R.$$

Using the interpolation inequality, we get

$$[D^2 v]_{\alpha;B_\rho} \leqslant \frac{C}{(r-\rho)^\alpha}\left[\varepsilon^\alpha [D^2 \widehat{v}]_{\alpha;B_r} + \frac{C}{\varepsilon^2}|\widehat{v}|_{0;B_r}\right].$$

If we take ε such that $\varepsilon^\alpha C/(r-\rho)^\alpha = 1/2$, then Lemma 4.1 in Chapter 2 implies that

$$[D^2 v]_{\alpha;B_\rho} \leqslant \frac{C}{(R-\rho)^{2+\alpha}}|\widehat{v}|_{0;B_R}, \quad \forall 0 < \rho \leqslant R.$$

It follows from (2.7) in Lemma 2.2 in Chapter 2 that

$$\tau^{1-\alpha}|D^3 \widetilde{v}(x_0,\tau)| \leqslant C[D^2 v]_{\alpha;B_{R/2}} \leqslant \frac{C}{R^{2+\alpha}}|\widehat{v}|_{0;B_R}, \tag{7.7}$$

where $D^3 = DD_x^2$ and $\sim$ denotes the mollified function. The function $\widehat{v}$ satisfies the boundary value problem

$$\begin{cases} F(x_0, D^2\widehat{v} + D^2 u(x_0)) = 0 & \text{in } B_R, \\ \widehat{v} = \widehat{u} & \text{on } \partial B_R. \end{cases}$$

Since $F(x_0, D^2 u(x_0)) = 0$, we can rewrite the above equation and apply the maximum principle to deduce that $|\widehat{v}|_{0;B_R} \leqslant |\widehat{u}|_{0;B_R}$. Therefore, by (7.7), for $\beta < \alpha$,

$$\tau^{1-\alpha}|D^3 \widetilde{v}(x_0,\tau)| \leqslant \frac{C}{R^{2+\alpha}}|\widehat{u}|_{0;B_R} \leqslant \frac{C}{R^{\alpha-\beta}} H^\beta_{x_0}[D^2 u; B_R(x_0)], \tag{7.8}$$

where the (pointwise) Hölder norm notation is defined in (1.1) in Chapter 2.

Since

$$F(x_0, D^2 u) - F(x_0, D^2 v) = F(x_0, D^2 u) - F(x, D^2 u) \triangleq f(x),$$

the function $w = u - v$ satisfies the equation

$$a^{ij} D_{ij} w = f(x),$$

where

$$a^{ij} = \int_0^1 \frac{\partial F}{\partial r_{ij}}(x_0, \tau D^2 u + (1-\tau) D^2 v) dx.$$

By the condition (7.2),

$$|f(x)| \leqslant \widetilde{\lambda}\mu_0 \left(|D^2 u| + \widetilde{\mu}\right)|x - x_0|^\beta.$$

The Alexandroff-Bakelman-Pucci maximum principle implies that

$$\sup_{B_R(x_0)} w \leqslant CR\|f/\lambda\|_{L^n(B_R(x_0))} \leqslant C(|u|_2 + \widetilde{\mu})R^{2+\beta}. \tag{7.9}$$

(7.8), (7.9), and (2.7) in Chapter 2 imply

$$\begin{aligned} \tau^{1-\beta}|D^3 \widetilde{u}(x_0,\tau)| &\leqslant \tau^{1-\beta}|D^3 \widetilde{v}(x_0,\tau)| + \tau^{1-\beta}|D^3 \widetilde{w}(x_0,\tau)| \\ &\leqslant C\left[\left(\frac{\tau}{R}\right)^{\alpha-\beta} H^\beta_{x_0}[D^2 u; B_R(x_0)] + (|u|_2 + \widetilde{\mu})\frac{R^{2+\beta}}{\tau^{2+\beta}}\right]. \end{aligned}$$

Take $R = N\tau$, where N is to be determined. Then

$$\tau^{1-\beta}|D^3\widetilde{u}(x_0,\tau)| \leqslant C\Big\{N^{\beta-\alpha}H^{\beta}_{x_0}[D^2u; B_R(x_0)] + N^{2+\beta}(|u|_2+\widetilde{\mu})\Big\}. \tag{7.10}$$

Since Ω is a bounded domain, which introduces difficulties in estimating Hölder norms using the method of mollifiers, we should proceed very cautiously.

For any $0 < s < t < d_{x_0} = \operatorname{dist}\{x_0, \partial\Omega\}$, we take $R \leqslant t - s$. Thus for $\tau = \dfrac{1}{N}R < \dfrac{1}{N}(t-s)$, (7.10) implies that

$$\tau^{1-\beta}|D^3\widetilde{u}(x_0,\tau)| \leqslant C\Big\{N^{\beta-\alpha}H^{\beta}_{x_0}[D^2u; B_{t-s}(x_0)] + N^{2+\beta}(|u|_2+\widetilde{\mu})\Big\}.$$

In the case $\tau \geqslant \dfrac{1}{N}(t-s)$ (u is extended to 0 outside Ω),

$$\tau^{1-\beta}|D^3\widetilde{u}(x_0,\tau)| \leqslant \tau^{1-\beta}\cdot C\tau^{-3}|u|_0 \leqslant \frac{CN^{2+\beta}}{(t-s)^{2+\beta}}|u|_0.$$

It follows that

$$\begin{aligned}
&\sup_{\tau>0}\tau^{1-\beta}|D^3\widetilde{u}(x_0,\tau)| \\
&\leqslant C\Big\{N^{\beta-\alpha}H^{\beta}_{x_0}[D^2u; B_{t-s}(x_0)] + N^{2+\beta}\Big(|u|_2+\widetilde{\mu}+\frac{CN^{2+\beta}|u|_0}{(t-s)^{2+\beta}}\Big)\Big\} \\
&\leqslant C\Big\{N^{\beta-\alpha}\sup_{\tau>0,\ y\in B_{t-s}(x_0)}\Big\{\tau^{1-\beta}|D^3\widetilde{u}(y,\tau)|\Big\} \\
&\qquad + N^{2+\beta}\Big(|u|_2+\widetilde{\mu}+\frac{CN^{2+\beta}|u|_0}{(t-s)^{2+\beta}}\Big)\Big\}.
\end{aligned}$$

In deriving the last inequality, we also made use of Lemma 2.3 in Chapter 2. Thus,

$$\begin{aligned}
&\sup_{\tau>0,\ y\in B_s(x_0)}\Big\{\tau^{1-\beta}|D^3\widetilde{u}(x_0,\tau)|\Big\} \\
&\leqslant C\Big\{N^{\beta-\alpha}\sup_{\tau>0,\ y\in B_t(x_0)}\tau^{1-\beta}|D^3\widetilde{u}(y,\tau)| \\
&\qquad + N^{2+\beta}\Big(|u|_2+\widetilde{\mu}+\frac{CN^{2+\beta}|u|_0}{(t-s)^{2+\beta}}\Big)\Big\}.
\end{aligned}$$

We take N such that $CN^{\beta-\alpha} = 1/2$. By Lemma 4.1 in Chapter 2,

$$\sup_{\tau>0,\ y\in B_\rho(x_0)}\tau^{1-\beta}|D^3\widetilde{u}(y,\tau)| \leqslant C\Big(|u|_2+\widetilde{\mu}+\frac{|u|_0}{(R-\rho)^{2+\beta}}\Big),$$
$$\forall 0<\rho<R\leqslant d_{x_0}.$$

Applying Lemma 2.3 again, we obtain

$$H^{\beta}_{x_0}[D^2u; B_\rho(x_0)] \leqslant C\Big(|u|_2+\widetilde{\mu}+\frac{|u|_0}{(R-\rho)^{2+\beta}}\Big), \quad \forall 0<\rho<R\leqslant d_{x_0}, \tag{7.11}$$

which is exactly the interior Hölder estimate for second derivatives.

The method for the estimate near the boundary is similar. Assume without loss of generality that $\varphi \equiv 0$; of course we shall also replace $\widetilde{\mu}$ by $\widetilde{\mu} + |\varphi|_{2,\beta}$ in the structure condition (7.2). Let $y_0 \in \partial\Omega$. Using the $C^{2,\beta}$ diffeomorphism in a neighborhood of y_0 if necessary, we may assume without loss of generality that y_0

is the origin and that this neighborhood is $B_1^+ = B_1 \cap \{x_n > 0\}$. On $x_n = 0$, $u = 0$. For any $x_0 \in \overline{B_1^+} \setminus \partial B_1$, $B_R(x_0) \subset B_1$, we solve the boundary value problem

$$\begin{cases} F(x_0, D^2 v) = 0 & \text{in } B_R^+(x_0), \\ v = u & \text{on } \partial B_R^+(x_0), \end{cases} \tag{7.12}$$

where $B_R^+(x_0) = B_R(x_0) \cap B_1^+$. Mollifying the singularities of ∂B_R^+ if necessary, we may assume that ∂B_R^+ is sufficiently smooth. For simplicity, we still denote it by $B_R^+(x_0)$. Then, (7.12) admits a solution $v \in C^{2,\beta}(B_R^+(x_0))$. We can derive, analogously to (7.8),

$$[D^2 v]_{\alpha; B_{R/2}^+(x_0)} \leqslant \frac{C}{R^{\alpha-\beta}} H_{x_0}^{\beta}[D^2 u; B_R^+(x_0)],$$

where we can still recover $B_R^+(x_0)$ as $B_R(x_0) \cap B_1^+$ (i.e., before the mollification). We extend u and v to $x_n < 0$ by odd extension, still denoted by u and v. Then

$$[DD_{x'}v]_{\alpha; B_{R/2}(x_0)} \leqslant 2[DD_{x'}v]_{\alpha; B_{R/2}^+(x_0)} \leqslant \frac{C}{R^{\alpha-\beta}} H_{x_0}^{\beta}[D^2 u; B_R^+(x_0)], \tag{7.13}$$

where $x' = (x_1, \cdots, x_{n-1})$. Using the equation (7.1), the second derivative $D_{nn}u$ can be represented by second derivatives $DD_{x'}u$. Therefore (7.13) can be written as

$$[DD_{x'}v]_{\alpha; B_{R/2}(x_0)} \leqslant \frac{C}{R^{\alpha-\beta}} H_{x_0}^{\beta}[DD_{x'}u; B_R^+(x_0)].$$

Using a method similar to that for the interior estimates, we obtain the following estimate corresponding to (7.11):

$$H_{x_0}^{\beta}[DD_{x'}u; B_\rho(x_0)] \leqslant C\Big(|u|_2 + \widetilde{\mu} + |\varphi|_{2,\beta} + \frac{|u|_0}{(R-\rho)^{2+\beta}}\Big), \quad \forall 0 < \rho < R \leqslant \operatorname{dist}\{x_0, \partial\, B_1\}. \tag{7.14}$$

Again, by representing $D_{nn}u$ in terms of $DD_{x'}u$, we obtain

$$H_{x_0}^{\beta}[D^2 u; B_\rho(x_0)] \leqslant C\Big(|u|_2 + \widetilde{\mu} + |\varphi|_{2,\beta} + \frac{|u|_0}{(R-\rho)^{2+\beta}}\Big), \quad \forall 0 < \rho < R \leqslant \operatorname{dist}\{x_0, \partial\, B_1\}. \tag{7.15}$$

(7.11) and (7.15) give us the global estimate (7.3). □

Theorem 7.2. *Let the structure conditions* (F1) *and* (F4) *be in force. Set*

$$M_0 = |\varphi|_{0;\Omega} + \frac{d}{n\sqrt[n]{\omega_n}} \|F(\cdot, 0)/\lambda\|_{L^n(\Omega)}, \qquad \mu_1 = \mu_1(M_0).$$

Under these assumptions, there exists $\alpha = \alpha(n, \mu_1)$, $0 < \alpha < 1$ such that for any $0 < \beta < \alpha$, if $\partial\Omega \in C^{2,\beta}$, $\varphi \in C^{2,\beta}(\overline{\Omega})$ and (7.2) is valid, then the Dirichlet problem (7.1), (1.2) admits a unique solution $u \in C^{2,\beta}(\overline{\Omega})$ satisfying the estimate (7.3).

Proof. First we assume that $\partial\Omega \in C^{3,\delta}$, $\varphi \in C^{3,\delta}(\overline{\Omega})$, where $0 < \delta < 1$. Suppose that $u \in C^{2,\beta}(\overline{\Omega})$ is a solution of (7.1) and (1.2). Then by using the technique of locally flattening the boundary and finite differences, we can show that $u \in C^{3,\delta}(\overline{\Omega})$. Thus the a priori estimate (7.3) and Theorem 5.1 imply that the Dirichlet problem (7.1), (1.2) admits a solution $u \in C^{2,\beta}(\overline{\Omega})$ satisfying the estimate (7.3). In the general case $\partial\Omega \in C^{2,\beta}$, $\varphi \in C^{2,\beta}(\overline{\Omega})$, we use approximation. □

Now we study the general fully nonlinear equation (1.1).

Theorem 7.3. *Let the structure conditions* (F1), (F3) *and* (F4) *be in force. Suppose that* u *is a solution of the Dirichlet problem* (7.1), (1.2), *satisfying* $|u|_{0;\Omega} \leqslant M_0$, $[u]_{1;\Omega} \leqslant M_1$, $[u]_{1,\overline{\alpha};\Omega} \leqslant M_2$ $(0 < \overline{\alpha} \leqslant 1)$. *Under these assumptions, there exists* $\alpha = \alpha(n, \mu_1)$, $0 < \alpha \leqslant \overline{\alpha}$, *such that for any* $0 < \beta < \alpha$, *if* $\partial\Omega \in C^{2,\beta}$, $\varphi \in C^{2,\beta}(\overline{\Omega})$, *and* $u \in C^{2,\beta}(\overline{\Omega})$, *then*

$$[D^2u]_{\beta;\Omega} \leqslant C[1 + |\varphi|_{2,\beta;\Omega}], \tag{7.16}$$

where C *depends only on the quantities* $n, \mu_1, \mu_3, M_0, M_1, M_2$, $(\alpha - \beta)^{-1}$, Ω *and* $\sup\{\Lambda(x, z, p) \mid x \in \Omega, |z| \leqslant M_0, |p| \leqslant M_1\}$.

Proof. Set

$$\widetilde{F}(x, r) = F(x, u(x), Du(x), r). \tag{7.17}$$

Using the structure condition (F3) and the assumption $[u]_{1;\Omega} \leqslant M_1$, we get

$$|\widetilde{F}(x, r) - \widetilde{F}(y, r)| \leqslant C[1 + |r|]\,[1 + [u]_{1,\beta}]\,|x - y|^{\beta}.$$

Obviously, $u \in C^{2,\beta}(\overline{\Omega})$ and

$$\widetilde{F}(x, D^2u(x)) = 0. \tag{7.18}$$

Theorem 7.2 now implies the estimate (7.16). □

Theorem 7.4. *Let the structure conditions* (F1)–(F6) *be in force. Suppose that* Λ *is bounded on* $\Omega \times \mathbb{R} \times \mathbb{R}^n$. *Under these assumptions, there exists* $0 < \alpha < 1$, *such that for any* $0 < \beta < \alpha$, *if* $\partial\Omega \in C^{2,\beta}$, $\varphi \in C^{2,\beta}(\overline{\Omega})$, *then the Dirichlet problem* (1.1), (1.2) *admits a unique solution* $u \in C^{2,\beta}(\overline{\Omega})$, *where* α *depends only on* n, μ_1, μ_2, μ_3, $|\varphi|_2$ *and* Ω.

Proof. This is a direct corollary of Theorems 5.1 and 7.3. □

Finally, we study the Bellman equation arising from control theory. Set

$$L_k u = a_k^{ij} D_{ij} u + b_k^i u + c_k u \qquad (k = 1, 2, \cdots, N).$$

Consider the Dirichlet problem

$$\inf_{1 \leqslant k \leqslant N} (L_k u - f_k) = 0 \quad \text{in } \Omega, \tag{7.19}$$

$$u = \varphi \quad \text{on } \partial\Omega. \tag{7.20}$$

We assume that

(B1) there exist $0 < \lambda \leqslant \Lambda$ such that on Ω,

$$\lambda I \leqslant (a_k^{ij}) \leqslant \Lambda I, \quad \Lambda/\lambda \leqslant \mu_1 \quad (k = 1, 2, \cdots, N),$$

where μ_1 is a constant;

(B2) there exists a constant Λ_1 such that on Ω,

$$\sum_{i,j,k} |a_k^{ij}|_\beta + \sum_{i,k} |b_k^i|_\beta + \sum_k (|c_k|_\beta + |f_k|_\beta) \leqslant \Lambda_1;$$

(B3) $c_k \leqslant 0$ on Ω, $(k = 1, 2, \cdots, N)$.

We let

$$h(y_1, \cdots, y_N) = \inf_{1 \leqslant k \leqslant N} y_k.$$

The function h is Lipschitz continuous with respect to the variables $y_1, \cdots, y_N$, and

$$D_i h \geqslant 0, \quad \sum_{i=1}^{n} D_i h = 1 \quad \text{a.e. in } \mathbb{R}^n. \tag{7.21}$$

Furthermore, h is concave with respect to the variables $y_1, \cdots, y_N$. In fact, in $\mathbb{R}^{N+1}$, the lower space of the graph $\eta = y_k$ is convex. It follows that the lower space of the graph $\eta = \inf_k \{y_k\}$, as an intersection of the lower spaces of $\eta \neq y_k$ ($k = 1, 2, \cdots, N$), is convex. This means that h is concave with respect to the variable y. The Bellman equation can be written as the following fully nonlinear equation:

$$F(x, u, Du, D^2u) = h(L_1 u - f_1, \cdots . L_N u - f_N) = 0.$$

Since

$$\frac{\partial F}{\partial r_{ij}} = a_k^{ij} D_k h,$$

(7.21) and the structure condition (B1) imply

$$\lambda I \leqslant \left(\frac{\partial F}{\partial r_{ij}}\right) \leqslant \Lambda I.$$

It is easy to verify that F also satisfies the concavity condition.

Theorem 7.5. *Under assumptions* (B1) *and* (B3), *there exists* $\alpha = \alpha(n, \mu_1)$, *such that for any* $0 < \beta < \alpha$, *if* (B2) *is valid,* $\partial\Omega \in C^{2,\beta}$, $\varphi \in C^{2,\beta}(\overline{\Omega})$, *then the Dirichlet problem* (7.19), (7.20) *admits a solution* $u \in C^{2,\beta}(\overline{\Omega})$.

Proof. Let $h_\tau(y)$ be the mollified function of $h(y)$. Consider the fully nonlinear equation

$$F_\tau(x, u, Du, D^2u) = h_\tau(L_1 u - f_1, \cdots . L_N u - f_N) = 0. \tag{7.22}$$

It is easy to verify that

$$\lambda I \leqslant \left(\frac{\partial F_\tau}{\partial r_{ij}}\right) \leqslant \Lambda I$$

and that F_τ is concave in r. Set

$$\widetilde{F}_\tau(x, r) = F_\tau(x, u(x), Du(x), r).$$

Then

$$|\widetilde{F}_\tau(x, r) - \widetilde{F}_\tau(y, r)| \leqslant C\Lambda(1 + |r| + |u|_{1,\beta})|x - y|^\beta,$$

where C depends only on n. By Theorem 7.2, there exists $\alpha = \alpha(n, \Lambda/\lambda)$ such that for $0 < \beta < \alpha$,

$$[D^2u]_{\beta;\Omega} \leqslant C[1 + |u|_{1,\beta} + |\varphi|_{2,\beta}].$$

Using the interpolation inequality, we obtain

$$[D^2u]_{\beta;\Omega} \leqslant C[1 + |u|_0 + |\varphi|_{2,\beta}].$$

The existence result now follows from Theorem 5.1 and the maximum principle. We remark here that an a priori $C^{1,\beta}$ estimate is not necessary for this equation. Finally, letting $\tau \to 0+$, we obtain the existence of a solution for (7.19), (7.20). □

Part II

Second Order Elliptic Systems

CHAPTER 8

L^2 Theory for Linear Elliptic Systems of Divergence Form

Starting with this chapter, we introduce the theory for elliptic systems. Linear systems are in Chapters 8 to 10, and nonlinear systems are in Chapters 11 and 12.

There are many differences between a second order elliptic equation and an elliptic system. In general, there is no maximum principle for an elliptic system, and the De Giorgi-Nash-Moser estimate is no longer valid. Therefore, we shall introduce mainly the L^2, Schauder and L^p theories. This chapter deals with the L^2 theory, which includes the existence of a weak solution and its H^2 regularity.

From now on, we shall always denote by Ω an open domain in $\mathbb{R}^n$, $n \geqslant 2$; by $u(x)$ a vector valued function: $u(x) = (u^1(x), \cdots, n^N(x))$, $N \geqslant 1$; and by $W^{m,p}(\Omega, \mathbb{R}^N)$ the space $W^{m,p}(\Omega) \times \cdots \times W^{m,p}(\Omega) = (W^{m,p}(\Omega))^N$, where $W^{m,p}(\Omega)$ is the standard Sobolev space (cf. Appendix 1). If $p = 2$, we shall simply denote $W^{m,p}(\Omega, \mathbb{R}^N)$ by $H^m(\Omega, \mathbb{R}^N)$.

1. Existence of weak solutions

We consider the following system of divergence form:

$$-D_\alpha[A_{ij}^{\alpha\beta}(x)D_\beta u^j] + D_\alpha f_i^\alpha = 0 \qquad (i = 1, 2, \cdots N), \tag{1.1}$$

where, as usual, repeated indices mean summation; α, β are summed from 1 to n and j from 1 to N; $D_\alpha = \partial/\partial x_\alpha$ $(\alpha = 1, 2, \cdots, n)$.

We assume throughout this chapter that $A_{ij}^{\alpha\beta}(x)$ satisfy the following conditions:

$$A_{ij}^{\alpha\beta}\xi_\alpha^i\xi_\beta^j \geqslant \lambda|\xi|^2, \quad \lambda > 0, \quad \forall \xi \in \mathbb{R}^{nN}, \tag{1.2}$$

$$A_{ij}^{\alpha\beta} \in L^\infty(\Omega), \quad |A_{ij}^{\alpha\beta}| \leqslant \Lambda \quad (\alpha, \beta = 1, 2, \cdots, n; \quad i, j = 1, 2, \cdots, N), \tag{1.3}$$

where λ, Λ are constants. (1.2) is said to be an *ellipticity condition*; that is why we call (1.1) an elliptic system.

Now we give the definition of a weak solution.

Definition 1.1. If $u \in H_{loc}^1(\Omega, \mathbb{R}^N)$ satisfies

$$\int_\Omega A_{ij}^{\alpha\beta}(x)D_\beta u^j D_\alpha \varphi^i dx = \int_\Omega f_i^\alpha D_\alpha \varphi^i dx, \quad \forall \varphi \in H_0^1(\Omega, \mathbb{R}^N), \tag{1.4}$$

then u is said to be a *weak solution* of the elliptic system (1.1).

If we impose the Dirichlet boundary condition

$$u = 0 \quad \text{on } \partial\Omega, \tag{1.5}$$

then we define a weak solution as follows:

Definition 1.2. If $u \in H_0^1(\Omega, \mathbb{R}^N)$ satisfies (1.4), then u is said to be a *weak solution* of the Dirichlet problem (1.1), (1.5).

By applying the Lax-Milgram theorem (cf. Theorem 1.1 in Chapter 1), it is not difficult to establish the existence and uniqueness of a weak solution of (1.1), (1.5).

Theorem 1.1. *Let the assumptions* (1.2), (1.3) *be in force. Suppose that* $f_i^\alpha \in L^2(\Omega)$ $(\alpha = 1, \cdots, n;\ i = 1, \cdots, N)$. *Then the problem* (1.1), (1.5) *admits a unique weak solution.*

Proof. Set

$$a(u,v) = \int_\Omega A_{ij}^{\alpha\beta}(x) D_\beta u^j D_\alpha v^i dx,$$

$$\langle T, v\rangle = \int_\Omega f_i^\alpha D_\alpha v^i dx.$$

It is easy to verify that $a(u,v)$ is a bounded bilinear form on $H_0^1(\Omega, \mathbb{R}^N) \times H_0^1(\Omega, \mathbb{R}^N)$. From (1.2), we deduce that

$$a(u,u) \geqslant \lambda \int_\Omega |Du|^2 dx, \quad \forall u \in H_0^1(\Omega, \mathbb{R}^N), \tag{1.6}$$

where $|Du|^2 = \sum_{i=1}^N \sum_{\alpha=1}^n |D_\alpha u^i|^2$. Thus $a(u,v)$ is coercive. It is clear that $\langle T, \cdot\rangle$ is a bounded linear functional on $H_0^1(\Omega, \mathbb{R}^N)$. Therefore, by the Lax-Milgram theorem, there exists a unique solution $u \in H_0^1(\Omega, \mathbb{R}^N)$, such that $a(u,v) = \langle T, v\rangle$, $\forall v \in H_0^1(\Omega, \mathbb{R}^N)$. The proof is complete. □

The condition (1.2) is also called the *strong Legendre condition.*

In the proof of Theorem 1.1, we find that the strong Legendre condition (1.2) implies the coercivity condition (1.6); the difference between a system and a single equation is that the reverse in not true: i.e., (1.6) does not imply (1.2) in general. We can only show the following theorem.

Theorem 1.2. *Suppose that* $A_{hk}^{\alpha\beta}$ *satisfy* (1.3) *and* (1.6), $\alpha, \beta = 1, \cdots, n;$ $h, k = 1, \cdots, N$. *Then for a.e.* $x \in \Omega$, *we have*

$$A_{hk}^{\alpha\beta}(x)\xi_\alpha \xi_\beta \eta^h \eta^k \geqslant \lambda |\xi|^2 |\eta|^2, \quad \forall \xi \in \mathbb{R}^n, \eta \in \mathbb{R}^N. \tag{1.7}$$

Proof. (1.6) implies that, for any complex valued function $u = w + iv$ (in the proof of this theorem $i = \sqrt{-1}$),

$$\text{Re } a(u, \overline{u}) \geqslant \lambda \int_\Omega |Du|^2 dx, \tag{1.8}$$

where $w, v \in H_0^1(\Omega, \mathbb{R}^N)$.

In (1.8), we choose $u^k(x) = \eta^k \varphi(x) e^{\tau i \xi \cdot x}$ $(k = 1, \cdots, N)$, where $\tau \in \mathbb{R}$, $\xi \in \mathbb{R}^n$, $\eta \in \mathbb{R}^N$, $\varphi \in C_0^\infty(\Omega)$. Then the left-hand side of (1.8) is evaluated as

$$\begin{aligned} \text{Re } a(u, \overline{u}) &= \text{Re} \int_\Omega A_{hk}^{\alpha\beta}(x) D_\beta u^k \overline{D_\alpha u^h} dx \\ &= \int_\Omega A_{hk}^{\alpha\beta}(x)(D_\alpha \varphi D_\beta \varphi + \tau^2 \xi_\alpha \xi_\beta \varphi^2)\eta^h \eta^k dx; \end{aligned}$$

the right-hand side of (1.8) is evaluated as

$$\begin{aligned}\lambda \int_\Omega |Du|^2 dx &= \lambda \int_\Omega \sum_{k=1}^N \sum_{\alpha=1}^n \left(|\eta^k D_\alpha \varphi|^2 + |\tau \xi_\alpha \eta^k \varphi|^2\right) dx \\ &= \lambda |\eta|^2 \int_\Omega |D\varphi|^2 dx + \lambda \tau^2 |\xi|^2 |\eta|^2 \int_\Omega |\varphi|^2 dx.\end{aligned}$$

It follows that

$$\begin{aligned}&\eta^h \eta^k \int_\Omega A_{hk}^{\alpha\beta}(x) D_\alpha \varphi D_\beta \varphi dx + \tau^2 \xi_\alpha \xi_\beta \eta^h \eta^k \int_\Omega A_{hk}^{\alpha\beta}(x) |\varphi|^2 dx \\ \geqslant\ &\lambda |\eta|^2 \int_\Omega |D\varphi|^2 dx + \lambda \tau^2 |\xi|^2 |\eta|^2 \int_\Omega |\varphi|^2 dx.\end{aligned}$$

Letting $\tau \to \infty$ after dividing by τ^2, we obtain

$$\xi_\alpha \xi_\beta \eta^h \eta^k \int_\Omega A_{hk}^{\alpha\beta}(x) \varphi^2 dx \geqslant \lambda |\xi|^2 |\eta|^2 \int_\Omega |\varphi|^2 dx, \quad \forall \varphi \in C_0^\infty(\Omega). \tag{1.9}$$

Since $C_0^\infty(\Omega)$ is dense in $L^2(\Omega)$, the above inequality is also valid for all $\varphi \in L^2(\Omega)$.

For any $x^0 \in \Omega$, we choose in (1.9) $\varphi = \varphi_\varepsilon = \chi(B(x^0, \varepsilon))/|B(x^0, \varepsilon)|^{1/2}$, where $B(x^0, \varepsilon)$ is a ball centered at x^0, with radius ε, $|B(x^0, \varepsilon)|$ is its Lebesgue measure, and $\chi(B(x^0, \varepsilon))$ is the characteristic function of $B(x^0, \varepsilon)$, i.e.,

$$\chi(B(x^0, \varepsilon)) = \begin{cases} 1, & x \in B(x^0, \varepsilon), \\ 0, & x \notin B(x^0, \varepsilon). \end{cases}$$

Obviously, $\varphi_\varepsilon \in L^2(\Omega)$, and $\displaystyle\int_\Omega |\varphi_\varepsilon|^2 dx = 1$. It follows that

$$\xi_\alpha \xi_\beta \eta^h \eta^k \int_{B(x^0,\varepsilon)} A_{hk}^{\alpha\beta}(x) \frac{1}{|B(x^0, \varepsilon)|} dx \geqslant \lambda |\xi|^2 |\eta|^2.$$

By the Lebesgue differentiation theorem (cf. [RD]), we can let $\varepsilon \to 0$ to conclude, for a.e. $x^0 \in \Omega$, that

$$A_{hk}^{\alpha\beta}(x^0) \xi_\alpha \xi_\beta \eta^h \eta^k \geqslant \lambda |\xi|^2 |\eta|^2, \quad \forall \xi \in \mathbb{R}^n, \eta \in \mathbb{R}^N.$$

This is exactly what we want to prove. □

(1.7) is called the *strong Legendre-Hadamard condition.* We have established the following relationships:

$A_{hk}^{\alpha\beta}$ satisfy the strong Legendre condition	$\Longrightarrow$	$a(u,v)$ is coercive	$\Longrightarrow$	$A_{hk}^{\alpha\beta}$ satisfy the strong Legendre-Hadamard condition

In the special case $N = 1$ (single equation), the strong Legendre-Hadamard condition implies the strong Legendre condition. Thus all three conditions are equivalent in this case. In the general case $N > 1$ (system), the strong Legendre-Hadamard condition is weaker than the strong Legendre condition. Here is an example.

Example 1.1. Let $N = n = 2$, $A_{ij}^{\alpha\beta} = \varepsilon \delta_{\alpha\beta} \delta_{ij} + a_{ij}^{\alpha\beta}$ $(\alpha, \beta = 1, 2;\ i, j = 1, 2)$, where $0 < \varepsilon < 1/2$,

$$\delta_{\alpha\beta} = \begin{cases} 1, & \alpha = \beta, \\ 0, & \alpha \neq \beta, \end{cases}$$

$a_{21}^{21} = 1$, $a_{21}^{12} = -1$, and all remaining $a_{ij}^{\alpha\beta} = 0$.

Obviously,

$$\begin{aligned} A_{ij}^{\alpha\beta}\xi_\alpha\xi_\beta\eta^i\eta^j &= \varepsilon|\xi|^2|\eta|^2 + (\xi_2\xi_1\eta^2\eta^1 - \xi_2\xi_1\eta^1\eta^2) \\ &= \varepsilon|\xi|^2|\eta|^2, \qquad \forall\xi\in\mathbb{R}^2,\ \eta\in\mathbb{R}^2; \end{aligned}$$

i.e., $A_{ij}^{\alpha\beta}$ satisfy the strong Legendre-Hadamard condition. However, for any $\varepsilon \in (0,1/2)$, there exists $\xi = (\xi_1^1, \xi_2^1, \xi_1^2, \xi_2^2) = (0,1,2\varepsilon,0)$ such that

$$A_{ij}^{\alpha\beta}\xi_\alpha^i\xi_\beta^j = \varepsilon|\xi|^2 - (\xi_2^2\xi_1^1 - \xi_2^1\xi_1^2) = \varepsilon(1+4\varepsilon^2) - 2\varepsilon = \varepsilon(4\varepsilon^2-1) < 0.$$

The strong Legendre condition is not satisfied.

The strong Legendre-Hadamard condition is an ellipticity condition. The above example shows that it is weaker than the strong Legendre condition in the case $N > 1$.

2. Energy estimates and H^2 regularity

Under a higher regularity assumption on the coefficients and the right-hand side of (1.1), we shall establish in this section the H^2 and H^k ($k \geqslant 2$) regularity for a weak solution.

We first prove a local *energy estimate* — Caccioppoli's inequality.

Theorem 2.1 (Caccioppoli's inequality). *Under the assumptions of Theorem 1.1, if $u \in H^1_{loc}(\Omega, \mathbb{R}^N)$ is a weak solution of* (1.1), *then for all $x^0 \in \Omega$ and all ρ, R with $0 < \rho < R < \operatorname{dist}(x^0, \partial\Omega)$, we have*

$$\int_{B_\rho(x^0)} |Du|^2 dx \leqslant C\left[\frac{1}{(R-\rho)^2}\int_{B_R(x^0)} |u-\nu|^2 dx + \int_{B_R(x^0)} |f|^2 dx\right], \tag{2.1}$$

where $\nu \in \mathbb{R}^N$ is an arbitrary constant vector, $|f|^2 = \sum_{i=1}^N\sum_{\alpha=1}^n |f_i^\alpha|^2$, $B_\rho(x^0)$ is a ball centered at x^0 and of radius ρ, and C is a positive constant depending only on n, N, λ, Λ.

Proof. In the integral identity (1.4) we choose $\varphi = \eta^2(u-\nu)$, where $\eta \in C_0^\infty(B_R(x^0))$ is a cutoff function:

$$0 \leqslant \eta \leqslant 1 \quad \text{on } B_R(x^0), \qquad \eta \equiv 1 \quad \text{on } B_\rho(x^0), \qquad |D\eta| \leqslant \frac{C}{R-\rho}. \tag{2.2}$$

Then

$$\begin{aligned} \int_\Omega \eta^2 A_{ij}^{\alpha\beta}(x) D_\beta u^j D_\alpha u^i dx &= -\int_\Omega 2\eta(u^i-\nu^i)A_{ij}^{\alpha\beta}D_\beta u^j D_\alpha \eta dx \\ &\quad + \int_\Omega \eta^2 f_i^\alpha D_\alpha u^i dx \\ &\quad + \int_\Omega 2\eta f_i^\alpha (u^i - \nu^i) D_\alpha \eta dx. \end{aligned}$$

Therefore

$$\begin{aligned} \lambda\int_{B_R(x^0)} |\eta Du|^2 dx &\leqslant \varepsilon\int_{B_R(x^0)} |\eta Du|^2 dx + \frac{C}{(R-\rho)^2}\int_{B_R(x^0)} |u-\nu|^2 dx \\ &\quad + C\int_{B_R(x^0)} |f|^2 dx, \end{aligned}$$

where $C = C(\varepsilon, n, N, \Lambda)$. We take $\varepsilon = \lambda/2$. Then

$$\int_{B_R(x^0)} |\eta Du|^2 dx \leqslant C\left[\frac{1}{(R-\rho)^2}\int_{B_R(x^0)} |u-\nu|^2 dx + \int_{B_R(x^0)} |f|^2 dx\right].$$

Restricting the left-hand side of the above inequality to $B_\rho(x^0)$, we immediately obtain (2.1). □

Next, we study the interior H^2 regularity of weak solutions.

Theorem 2.2. *Suppose that $A_{ij}^{\alpha\beta}(x)$ satisfy (1.2), $A_{ij}^{\alpha\beta} \in W^{1,\infty}(\Omega)$ $(\alpha, \beta = 1, \cdots, n;\ i, j = 1, \cdots, N)$, $f_i^\alpha \in H^1(\Omega)$ $(\alpha, \beta = 1, \cdots, n;\ i, j = 1, \cdots, N)$. If $u \in H^1_{loc}(\Omega, \mathbb{R}^N)$ is a weak solution of (1.1), then $u \in H^2_{loc}(\Omega, \mathbb{R}^N)$.*

Proof. Denote by e_s the unit vector in the x_s direction $(s = 1, \cdots, n)$. Let $\Delta_{h,s}u(x) = \frac{1}{h}[u(x + he_s) - u(x)]$ be the difference quotient in the x_s direction. If $\varphi \in H_0^1(\Omega, \mathbb{R}^N)$ with its support $\operatorname{spt}\varphi \subset \Omega$, then for

$$|h| < \frac{1}{2}\operatorname{dist}(\operatorname{spt}\varphi, \partial\Omega),$$

we have

$$\int_\Omega A_{ij}^{\alpha\beta}(x + he_s) D_\beta u^j(x + he_s) D_\alpha \varphi^i(x) dx = \int_\Omega f_i^\alpha(x + he_s) D_\alpha \varphi^i(x) dx.$$

Subtracting (1.4) from this equality and then dividing by h, we obtain

$$\begin{aligned}&\int_\Omega A_{ij}^{\alpha\beta}(x + he_s)[D_\beta \Delta_{h,s} u^j] D_\alpha \varphi^i dx \\ &\qquad + \int_\Omega [\Delta_{h,s} A_{ij}^{\alpha\beta}] D_\beta u^j D_\alpha \varphi^i dx = \int_\Omega [\Delta_{h,s} f_i^\alpha] D_\alpha \varphi^i dx. \end{aligned} \tag{2.3}$$

In Caccioppoli's inequality (Theorem 2.1), we take $A_{ij}^{\alpha\beta}$ to be $A_{ij}^{\alpha\beta}(x + he_s)$, u to be $\Delta_{h,s}u$, and f_i^α to be $-[\Delta_{h,s}A_{ij}^{\alpha\beta}]D_\beta u^j + \Delta_{h,s} f_i^\alpha$; we also choose $\nu = 0$, $\rho = R/2$. Then

$$\begin{aligned}\int_{B_{R/2}(x^0)} |D\Delta_{h,s}u|^2 dx \leqslant\ & \frac{C}{R^2}\int_{B_R(x^0)} |\Delta_{h,s}u|^2 dx \\ & + C\int_{B_R(x^0)} \sum_{i,j=1}^N \sum_{\alpha,\beta=1}^n |\Delta_{h,s} A_{ij}^{\alpha\beta}|^2 |Du|^2 dx \\ & + C\int_{B_R(x^0)} \sum_{i=1}^N \sum_{\alpha=1}^n |\Delta_{h,s} f_i^\alpha|^2 dx.\end{aligned}$$

The assumptions of the theorem imply that the right-hand side is uniformly bounded (with the bounds independent of h). Therefore $D^2 u \in L^2_{loc}(\Omega, \mathbb{R}^{n^2 N})$. □

On higher regularities of a weak solution, we have the following results:

Theorem 2.3. *Suppose that $A_{ij}^{\alpha\beta}(x)$ satisfy (1.2), $A_{ij}^{\alpha\beta} \in W^{k+1,\infty}(\Omega)$ $(\alpha, \beta = 1, \cdots, n;\ i, j = 1, \cdots, N)$, $f_i^\alpha \in H^{k+1}(\Omega)$ $(\alpha = 1, \cdots, n;\ i = 1, \cdots, N)$. If $u \in H^1_{loc}(\Omega, \mathbb{R}^N)$ is a weak solution of (1.1), then $u \in H^{k+2}_{loc}(\Omega, \mathbb{R}^N)$.*

Corollary 2.4. *Suppose that $A_{ij}^{\alpha\beta}(x)$ satisfy (1.2), $A_{ij}^{\alpha\beta} \in C^\infty(\Omega)$ $(\alpha, \beta = 1, \cdots, n;\ i, j = 1, \cdots, N)$, $f_i^\alpha \in C^\infty(\Omega)$ $(\alpha = 1, \cdots, n;\ i = 1, \cdots, N)$. If $u \in H^1_{loc}(\Omega, \mathbb{R}^N)$ is a weak solution of (1.1), then $u \in C^\infty(\Omega, \mathbb{R}^N)$.*

Corollary 2.5. *Suppose that $A_{ij}^{\alpha\beta}$ are constants satisfying* (1.2), (1.3) $(\alpha, \beta = 1, \cdots, n;\ i, j = 1, \cdots, N)$, $f_i^\alpha \equiv 0$ $(\alpha = 1, \cdots, n;\ i = 1, \cdots, N)$. *If* $u \in H_{loc}^1(\Omega, \mathbb{R}^N)$ *is a weak solution of* (1.1), *then for any* $B_R(x^0) \subset\subset \Omega$ *and any positive integer* k,

$$\|u\|_{H^k(B_{R/2}(x^0), \mathbb{R}^N)} \leqslant C\|u\|_{L^2(B_R(x^0), \mathbb{R}^N)},$$

where C *depends only on* $n, N, \lambda, \Lambda, k, R$.

Proof. Under our assumptions, Corollary 2.4 implies that $u \in C^\infty(\Omega, \mathbb{R}^N)$. It satisfies

$$\int_\Omega A_{ij}^{\alpha\beta} D_\beta u^j D_\alpha \varphi^i dx = 0, \qquad \forall \varphi \in H_0^1(\Omega, \mathbb{R}^N).$$

We choose a sequence of R_k as follows: $R_0 = R$, $R_k = \frac{1}{2}\left(\frac{R}{2} + R_{k-1}\right)$ $(k = 1, 2, \cdots)$. Applying Caccioppoli's inequality (Theorem 2.1) to u, we obtain

$$\int_{B_{R_1}(x^0)} |Du|^2 dx \leqslant \frac{C}{(R - R_1)^2} \int_{B_R(x^0)} |u|^2 dx,$$

where C depends only on n, N, λ, Λ.

It is easy to verify that $D_s u^j$ $(s = 1, \cdots, n)$ satisfies

$$\int_\Omega A_{ij}^{\alpha\beta} D_\beta (D_s u^j) D_\alpha \varphi^i dx = 0, \qquad \forall \varphi \in H_0^1(\Omega, \mathbb{R}^N), \ \operatorname{spt} \varphi \subset \Omega.$$

Applying Caccioppoli's inequality to Du, we obtain

$$\begin{aligned} \int_{B_{R_2}(x^0)} |D^2 u|^2 dx &\leqslant \frac{C}{(R_1 - R_2)^2} \int_{B_{R_1}(x^0)} |u|^2 dx \\ &\leqslant \frac{C}{(R_1 - R_2)^2 (R - R_1)^2} \int_{B_R(x^0)} |u|^2 dx. \end{aligned}$$

Applying Caccioppoli's inequality k times, we get the corollary. □

For a weak solution of the Dirichlet problem (1.1), (1.5), its interior H^2 regularity is given by Theorem 2.2; its global H^2 regularity is as follows:

Theorem 2.6. *In addition to the assumptions of Theorem 2.2, assume that* $\partial\Omega \in C^2$. *If* $u \in H_0^1(\Omega, \mathbb{R}^N)$ *is a weak solution of the Dirichlet problem* (1.1), (1.5), *then* $u \in H^2(\Omega, \mathbb{R}^N)$, *and*

$$\|u\|_{H^2(\Omega, \mathbb{R}^N)} \leqslant C\Big\{\|u\|_{L^2(\Omega, \mathbb{R}^N)} + \|f\|_{H^1(\Omega, \mathbb{R}^{nN})}\Big\}, \tag{2.4}$$

where $f = (f_i^\alpha)$, *and* C *depends only on* n, N, λ, $\|A_{ij}^{\alpha\beta}\|_{W^{1,\infty}}$ *and* $\partial\Omega$.

The proof of this theorem is similar to that of Theorem 5.2 in Chapter 1 and is omitted here. On higher regularities of a weak solution of the Dirichlet problem, we have the following result:

Theorem 2.7. *In addition to the assumptions of Theorem 2.3, assume that* $\partial\Omega \in C^{k+2}$. *If* $u \in H_0^1(\Omega, \mathbb{R}^N)$ *is a weak solution of the Dirichlet problem* (1.1), (1.5), *then* $u \in H^{k+2}(\Omega, \mathbb{R}^N)$.

CHAPTER 9

Schauder Theory for Linear Elliptic Systems of Divergence Form

In this chapter, we study the Schauder theory for linear elliptic systems of divergence form. Here, we introduce another method, which appears in [MR3] and [CP], for establishing the Schauder estimates without tedious computations of potentials; see also [GQ1] and [GQ2]. This method can also be used to study the regularity of weak solutions of nonlinear elliptic systems (cf. §5 of Chapter 12).

As a preliminary, we first introduce the Morrey and Campanato spaces.

1. Morrey and Campanato spaces

As before, $B(x^0, R)$ or $B_R(x^0)$ denotes the n dimensional ball with radius R, centered at x^0.

By the Sobolev embedding theorem (see Appendix 1), if $u \in W^{1,p}(B_R(x^0))$, where $p > n$, then $u \in C^{0,\delta}(\overline{B_R(x^0)})$, where

$$\delta = 1 - \frac{n}{p}.$$

On the other hand, if $p \leqslant n$, then u may not even be continuous. Nonetheless, the Morrey theorem on Hölder continuity tells us:

Theorem 1.1 (Morrey). *Let $u \in W^{1,p}(B_R(x^0))$, $p > 1$. If for all $x \in B_R(x^0)$ and all $\rho : 0 < \rho < d(x) = R - |x - x^0|$,*

$$\int_{B_\rho(x^0)} |Du|^p dx \leqslant C\Big(\frac{\rho}{d(x)}\Big)^{n-p+p\delta}, \qquad 0 < \delta < 1, \tag{1.1}$$

then for all r, $0 < r < R$, we have $u \in C^{0,\delta}(\overline{B_r(x^0)})$.

This theorem is a localized version of Theorem 4.2 in Chapter 5; its proof will be postponed until the end of this section. This theorem tells us that it may not be necessary to show that $u \in W^{1,p}(B_R(x^0))$ $(p > n)$ in order to establish that $u \in C^{0,\delta}(\overline{B_r(x^0)})$; all we need is $u \in W^{1,p}(B_R(x^0))$ for some $p > 1$, together with the estimate

$$\rho^{-\mu} \int_{B_\rho(x)} |Du|^p dx \leqslant C_1, \qquad \forall x \in B_R(x^0),\ \forall 0 < \rho < d(x),$$

where μ is sufficiently large $(\mu > n - p)$ and C_1 is independent of x, ρ. In this case, we can still prove that $u \in C^{0,\delta}(\overline{B_r(x^0)})$. It is therefore beneficial to introduce spaces involving two parameters p and μ.

Let Ω be a bounded domain in $\mathbb{R}^n$. We introduce the following notation:

$$\Omega(x^0, R) = \Omega \cap B(x^0, R),$$

$$u_{x^0,R} = \fint_{\Omega(x^0,R)} u(x)dx = \frac{1}{|\Omega(x^0, R)|} \int_{\Omega(x^0,R)} u(x)dx;$$

we denote by $\operatorname{diam}\Omega$ the diameter of Ω and by ω_n the Lebesgue measure of the n dimensional unit ball.

Definition 1.1. If for a bounded domain Ω there exists a positive constant A such that for all $x \in \Omega$ and all ρ with $0 < \rho < \operatorname{diam}\Omega$ the estimate

$$|\Omega(x, \rho)| \geqslant A\rho^n$$

is valid, then we say Ω is a domain of type (A).

Definition 1.2 (Morrey spaces). Let $p \geqslant 1$, $\mu \geqslant 0$. The collection of functions $u \in L^p(\Omega)$ satisfying

$$\sup_{x\in\Omega,\ 0<\rho<\operatorname{diam}\Omega} \rho^{-\mu} \int_{\Omega(x,\rho)} |u(z)|^p dz < +\infty$$

with the norm

$$\|u\|_{L^{p,\mu}(\Omega)} = \left\{ \sup_{x\in\Omega,\ 0<\rho<\operatorname{diam}\Omega} \rho^{-\mu} \int_{\Omega(x,\rho)} |u(z)|^p dz \right\}^{1/p}$$

constitutes a normed linear space and is called the *Morrey space* $L^{p,\mu}(\Omega)$.

It can be shown that $L^{p,\mu}(\Omega)$ is a Banach space.

Lemma 1.2. *$L^{p,\mu}(\Omega)$ has the following properties:*

(i) $L^{p,0}(\Omega) \simeq L^p(\Omega)$.

(ii) $L^{p,n}(\Omega) \simeq L^\infty(\Omega)$.

(iii) *If $\mu > n$, then $L^{p,\mu}(\Omega) = \{0\}$.*

(iv) *If $p \leqslant q$ and $(n - \mu)/p \geqslant (n - \nu)/q$, then $L^{q,\nu}(\Omega) \subset L^{p,\mu}(\Omega)$.*

Proof. (i) is obvious.

(ii) If $u \in L^\infty(\Omega)$, then

$$\begin{aligned} \|u\|_{L^{p,n}(\Omega)} &\leqslant \left\{ \sup_{x\in\Omega,\ 0<\rho<\operatorname{diam}\Omega} \rho^{-n} \|u\|^p_{L^\infty(\Omega)} |\Omega(x, \rho)| \right\}^{1/p} \\ &\leqslant \left\{ \sup_{x\in\Omega,\ 0<\rho<\operatorname{diam}\Omega} \rho^{-n} \|u\|^p_{L^\infty(\Omega)} \omega_n \rho^n \right\}^{1/p} \\ &\leqslant \omega_n^{1/p} \|u\|_{L^\infty(\Omega)}. \end{aligned}$$

It follows that $u \in L^{p,n}(\Omega)$, and $\|u\|_{L^{p,n}(\Omega)} \leqslant \omega_n^{1/p} \|u\|_{L^\infty(\Omega)}$.

Conversely, if $u \in L^{p,n}(\Omega)$, then by the Lebesgue differentiation theorem, for a.e. $x \in \Omega$, we have

$$\begin{aligned} u(x) &= \lim_{\rho\to 0} \frac{1}{|\Omega(x, \rho)|} \int_{\Omega(x,\rho)} u(z)dz \\ &= \lim_{\rho\to 0} \frac{1}{|B(x, \rho)|} \int_{\Omega(x,\rho)} u(z)dz. \end{aligned}$$

Therefore

$$\begin{aligned}
|u(x)| &\leqslant \sup_\rho \left\{ \frac{1}{\omega_n \rho^n} \int_{\Omega(x,\rho)} |u(z)| dz \right\} \\
&\leqslant \sup_\rho \left\{ \frac{1}{\omega_n \rho^n} |\Omega(x,\rho)|^{1-1/p} \left(\int_{\Omega(x,\rho)} |u|^p dz \right)^{1/p} \right\} \\
&\leqslant \omega_n^{-1/p} \|u\|_{L^{p,n}(\Omega)}.
\end{aligned}$$

It follows that $u \in L^\infty(\Omega)$, and $\|u\|_{L^\infty(\Omega)} \leqslant \omega_n^{-1/p} \|u\|_{L^{p,n}(\Omega)}$.

(iii) If $u \in L^{p,\mu}(\Omega)$ for some $\mu > n$, then

$$\begin{aligned}
&\left| \frac{1}{|B(x,\rho)|} \int_{\Omega(x,\rho)} u(z) dz \right| \\
&\leqslant \frac{1}{\omega_n \rho^n} (\omega_n \rho^n)^{1-1/p} \rho^{\mu/p} \left\{ \rho^{-\mu} \int_{\Omega(x,\rho)} |u|^p dz \right\}^{1/p} \\
&\leqslant C \rho^{(\mu-n)/p} \|u\|_{L^{p,\mu}(\Omega)} \to 0 \quad \text{as } \rho \to 0.
\end{aligned}$$

It follows that $u(x) = 0$, for a.e. $x \in \Omega$.

(iv) If $u \in L^{q,\nu}(\Omega)$, then

$$\begin{aligned}
\rho^{-\mu} \int_{\Omega(x,\rho)} |u|^p dz &\leqslant \rho^{-\mu} |\Omega(x,\rho)|^{1-p/q} \left\{ \int_{\Omega(x,\rho)} |u|^q dz \right\}^{p/q} \\
&\leqslant C \rho^{p[(n-\mu)/p-(n-\nu)/q]} \left\{ \rho^{-\nu} \int_{\Omega(x,\rho)} |u|^q dz \right\}^{p/q}.
\end{aligned}$$

The bound on the right-hand side of the above inequality is independent of x and ρ, by our assumptions. Hence $u \in L^{p,\mu}(\Omega)$. □

Now we introduce the Campanato spaces.

Definition 1.3. Let $p \geqslant 1$, $\mu \geqslant 0$. The collection of functions $u \in L^p(\Omega)$ satisfying

$$[u]_{p,\mu;\Omega} \triangleq \left\{ \sup_{x \in \Omega,\ 0<\rho<\operatorname{diam}\Omega} \rho^{-\mu} \int_{\Omega(x,\rho)} |u(z) - u_{x,\rho}|^p dz \right\}^{1/p} < +\infty$$

with the norm

$$\|u\|_{\mathscr{L}^{p,\mu}(\Omega)} = \|u\|_{L^p(\Omega)} + [u]_{p,\mu;\Omega}$$

constitutes a normed linear space and is called the *Campanato space* $\mathscr{L}^{p,\mu}(\Omega)$.

It can be shown that $\mathscr{L}^{p,\mu}(\Omega)$ is a Banach space.

Just like the space $L^{p,\mu}(\Omega)$, if $p \leqslant q$ and $(n-\mu)/p \geqslant (n-\nu)/q$, then $\mathscr{L}^{q,\nu}(\Omega) \subset \mathscr{L}^{p,\mu}(\Omega)$. The proof is left as an exercise for the reader.

Next, we shall prove: if Ω is a domain of type (A), then

$$\begin{aligned}
&\mathscr{L}^{p,\mu}(\Omega) \simeq L^{p,\mu}(\Omega) \quad \text{for } 0 \leqslant \mu < n; \\
&\mathscr{L}^{p,\mu}(\Omega) \simeq C^{0,\delta}(\overline{\Omega}),\ \delta = \frac{\mu-n}{p}, \quad \text{for } n < \mu \leqslant n+p; \\
&\mathscr{L}^{p,\mu}(\Omega) = \{\text{constant}\} \quad \text{for } \mu > n+p.
\end{aligned}$$

In order to prove these results, we need the following lemma.

Lemma 1.3. *Let Ω be a domain of type* (A). *If $u \in \mathscr{L}^{p,\mu}(\Omega)$, where $p \geqslant 1$, $\mu \geqslant 0$, then for all $x \in \overline{\Omega}$ and all $\widetilde{r}, \widetilde{R}$ with $0 < \widetilde{r} < \widetilde{R}$, we have*

$$|u_{x,\widetilde{R}} - u_{x,\widetilde{r}}| \leqslant C(p, A)[u]_{p,\mu}\widetilde{r}^{-n/p}\widetilde{R}^{\mu/p}, \tag{1.2}$$

where $[u]_{p,\mu} = [u]_{p,\mu;\Omega}$.

Proof. For simplicity, we use R and r in place of $\widetilde{R}$ and $\widetilde{r}$ in our proof. Clearly,

$$|u_{x,R} - u_{x,r}|^p \leqslant 2^{p-1}(|u(z) - u_{x,R}|^p + |u(z) - u_{x,r}|^p).$$

Integrating this inequality over $\Omega(x, r)$, we obtain

$$\begin{aligned} &|u_{x,R} - u_{x,r}|^p|\Omega(x,r)| \\ &\leqslant 2^{p-1}\left[\int_{\Omega(x,R)} |u(z) - u_{x,R}|^p dz + \int_{\Omega(x,r)} |u(z) - u_{x,r}|^p dz\right]. \end{aligned}$$

Since Ω is a domain of type (A), $|\Omega(x,r)| \geqslant Ar^n$; it follows that

$$\begin{aligned} |u_{x,R} - u_{x,r}|^p &\leqslant 2^{p-1}A^{-1}r^{-n}(R^\mu[u]^p_{p,\mu} + r^\mu[u]^p_{p,\mu}) \\ &\leqslant C(p, A)[u]^p_{p,\mu}r^{-n}R^\mu. \end{aligned}$$

Taking the pth root on both sides, we immediately get (1.2). □

Now we give the relationship between $\mathscr{L}^{p,\mu}(\Omega)$ and $L^{p,\mu}(\Omega)$ for $\mu < n$. We always use $[u]_{p,\mu}$ to denote $[u]_{p,\mu;\Omega}$, if there is no confusion.

Theorem 1.4. *Let Ω be a domain of type (A). Then*

$$\mathscr{L}^{p,\mu}(\Omega) \simeq L^{p,\mu}(\Omega) \qquad \text{for } 0 \leqslant \mu < n.$$

Proof. First we show that for $\mu \geqslant 0$, if $u \in L^{p,\mu}(\Omega)$, then $u \in \mathscr{L}^{p,\mu}(\Omega)$, and

$$\|u\|_{\mathscr{L}^{p,\mu}(\Omega)} \leqslant C\|u\|_{L^{p,\mu}(\Omega)}.$$

Suppose that $u \in L^{p,\mu}(\Omega)$. For all $x \in \Omega$ and all $\rho > 0$, we have

$$\begin{aligned} \int_{\Omega(x,\rho)} |u(z) - u_{x,\rho}|^p dz &\leqslant 2^{p-1}\int_{\Omega(x,\rho)} (|u|^p + |u_{x,\rho}|^p)dz \\ &= 2^{p-1}\left[\int_{\Omega(x,\rho)} |u|^p dz + |\Omega(x,\rho)| \cdot |u_{x,\rho}|^p\right]. \end{aligned} \tag{1.3}$$

Clearly

$$\begin{aligned} |u_{x,\rho}|^p &\leqslant \frac{1}{|\Omega(x,\rho)|^p}\left[\int_{\Omega(x,\rho)} |u|^p dz\right]^p \\ &\leqslant \frac{1}{|\Omega(x,\rho)|^p}\left[|\Omega(x,\rho)|^{1-1/p}\left(\int_{\Omega(x,\rho)} |u|^p dz\right)^{1/p}\right]^p \\ &= |\Omega(x,\rho)|^{-1}\int_{\Omega(x,\rho)} |u|^p dz. \end{aligned}$$

Substituting this inequality into (1.3), we get

$$\int_{\Omega(x,\rho)} |u(z) - u_{x,\rho}|^p dz \leqslant 2^p \int_{\Omega(x,\rho)} |u|^p dz.$$

It follows that $[u]_{p,\mu} \leqslant 2\|u\|_{L^{p,\mu}(\Omega)}$, and $u \in \mathscr{L}^{p,\mu}(\Omega)$; furthermore,

$$\|u\|_{\mathscr{L}^{p,\mu}(\Omega)} \leqslant C\|u\|_{L^{p,\mu}(\Omega)},$$

where the constant C depends on $\operatorname{diam}\Omega$.

Next we show that for $0 \leqslant \mu < n$, if $u \in \mathscr{L}^{p,\mu}(\Omega)$, then $u \in L^{p,\mu}(\Omega)$, and

$$\|u\|_{L^{p,\mu}(\Omega)} \leqslant C\|u\|_{\mathscr{L}^{p,\mu}(\Omega)}.$$

Suppose that $u \in \mathscr{L}^{p,\mu}(\Omega)$. For all $x \in \Omega$ and all $\rho > 0$, we have

$$\begin{aligned}\rho^{-\mu}&\int_{\Omega(x,\rho)}|u|^p dz \\ &\leqslant 2^{p-1}\left\{\rho^{-\mu}\int_{\Omega(x,\rho)}|u(z)-u_{x,\rho}|^p dz + \omega_n\rho^{n-\mu}|u_{x,\rho}|^p\right\}.\end{aligned} \tag{1.4}$$

Clearly, for $R > \rho > 0$, we have

$$|u_{x,\rho}|^p \leqslant 2^{p-1}[|u_{x,R}|^p + |u_{x,R}-u_{x,\rho}|^p]. \tag{1.5}$$

In order to estimate $|u_{x,R}-u_{x,\rho}|$, we choose in Lemma 1.3

$$\widetilde{R} = R_i = \frac{R}{2^i}, \quad \widetilde{r} = R_{i+1} = \frac{R}{2^{i+1}} \qquad (i = 0, 1, 2, \cdots).$$

Then

$$|u_{x,R_i} - u_{x,R_{i+1}}| \leqslant C(p,A)[u]_{p,\mu}\left(\frac{R}{2^i}\right)^{\mu/p}\left(\frac{R}{2^{i+1}}\right)^{-n/p}.$$

It follows that for any integer $h > 0$,

$$\begin{aligned}|u_{x,R} - u_{x,R_{h+1}}| &\leqslant C(p,A)[u]_{p,\mu}R^{(\mu-n)/p}2^{n/p}\sum_{i=0}^{h}2^{i(n-\mu)/p} \\ &\leqslant C(p,A,n)[u]_{p,\mu}R^{(\mu-n)/p}[2^{(n-\mu)(h+1)/p}-1]/[2^{(n-\mu)/p}-1] \\ &\leqslant C(p,A,n,\mu)[u]_{p,\mu}R^{(\mu-n)/p}2^{(n-\mu)(h+1)/p} \\ &= C(p,A,n,\mu)[u]_{p,\mu}R_{h+1}^{(\mu-n)/p}.\end{aligned}$$

For any fixed ρ $(0 < \rho < \operatorname{diam}\Omega)$, we choose h and R such that

$$\operatorname{diam}\Omega \leqslant 2^{h+1}\rho < 2\operatorname{diam}\Omega, \qquad R = 2^{h+1}\rho.$$

Then

$$|u_{x,R} - u_{x,\rho}| \leqslant C(p,A,n,\mu)[u]_{p,\mu}\rho^{(\mu-n)/p}, \tag{1.6}$$

and

$$|u_{x,R}| = \frac{1}{|\Omega|}\left|\int_\Omega u(z)dz\right| \leqslant |\Omega|^{-1/p}\|u\|_{L^p(\Omega)}. \tag{1.7}$$

Substituting (1.6) and (1.7) into (1.5), we get

$$|u_{x,\rho}|^p \leqslant 2^{p-1}\Big[|\Omega|^{-1}\|u\|^p_{L^p(\Omega)} + C^p[u]^p_{p,\mu}\rho^{\mu-n}\Big].$$

Substituting this inequality into (1.4), we derive

$$\begin{aligned}\rho^{-\mu}&\int_{\Omega(x,\rho)}|u|^p dx \\ &\leqslant 2^{p-1}[u]^p_{p,\mu} + \omega_n\cdot 2^{2p-2}\Big[|\Omega|^{-1}(\operatorname{diam}\Omega)^{n-\mu}\|u\|^p_{L^p(\Omega)} + C^p[u]^p_{p,\mu}\Big] \\ &\leqslant C\Big[\|u\|^p_{L^p(\Omega)} + [u]^p_{p,\mu}\Big],\end{aligned}$$

where C depends on p, A, n, μ, diam Ω. Therefore, $u \in L^{p,\mu}(\Omega)$ and

$$\|u\|_{L^{p,\mu}(\Omega)} \leqslant C\|u\|_{\mathscr{L}^{p,\mu}(\Omega)}. \qquad \square$$

Now we give the relationship between $\mathscr{L}^{p,\mu}(\Omega)$ and $C^{0,\delta}(\overline{\Omega})$ for $\mu > n$.

Theorem 1.5 (Integral characterization of Hölder continuity). *Let Ω be a domain of type* (A). *If $n < \mu \leqslant n+p$, then $\mathscr{L}^{p,\mu}(\Omega) \simeq C^{0,\delta}(\overline{\Omega})$, $\delta = (\mu - n)/p$; if $\mu > n+p$, then $\mathscr{L}^{p,\mu}(\Omega) = \{\text{constant}\}$.*

Proof. We first show that: if $u \in C^{0,\delta}(\overline{\Omega})$, $0 < \delta \leqslant 1$, then $u \in \mathscr{L}^{p,\mu}(\Omega)$, $\mu = n + p\delta$, and $\|u\|_{\mathscr{L}^{p,\mu}(\Omega)} \leqslant C\|u\|_{C^{0,\delta}(\overline{\Omega})}$. Here $\|u\|_{C^{0,\delta}(\overline{\Omega})} = |u|_{0;\Omega} + [u]_{0,\delta;\Omega}$ (see Definition 1.1 in Chapter 2). For simplicity, we often drop the subscript Ω and use $|u|_0$ and $[u]_{0,\delta}$ instead of $|u|_{0;\Omega}$ and $[u]_{0,\delta;\Omega}$.

Suppose that $u \in C^{0,\delta}(\overline{\Omega})$, $0 < \delta \leqslant 1$. Then $\forall x \in \Omega$, $\forall z \in \Omega(x,\rho)$, we have

$$\begin{aligned} |u(z) - u_{x,\rho}| &\leqslant \frac{1}{|\Omega(x,\rho)|}\int_{\Omega(x,\rho)} |u(z) - u(t)|dt \\ &\leqslant \frac{1}{A\rho^n}\int_{\Omega(x,\rho)} [u]_{0,\delta}|z-t|^\delta dt \\ &\leqslant \frac{C[u]_{0,\delta}}{A\rho^n}\int_0^{2\rho} r^\delta r^{n-1}dr \\ &\leqslant C(n,A,\delta)[u]_{0,\delta}\,\rho^\delta. \end{aligned}$$

It follows that

$$\begin{aligned} \rho^{-\mu}\int_{\Omega(x,\rho)} |u(z) - u_{x,\rho}|^p dz &\leqslant C(n,A,\delta,p)[u]_{0,\delta}^p\,\rho^{n+p\delta-\mu} \\ &= C(n,A,\delta,p)[u]_{0,\delta}^p \qquad (\mu = n + p\delta). \end{aligned}$$

Therefore

$$[u]_{p,\mu} \leqslant C[u]_{0,\delta}\,,$$

where $C = C(n,A,\delta,p)$. Since $\|u\|_{L^p(\Omega)} \leqslant |\Omega|^{1/p}|u|_0$, we conclude that $u \in \mathscr{L}^{p,\mu}(\Omega)$, and

$$\|u\|_{\mathscr{L}^{p,\mu}(\Omega)} \leqslant C\|u\|_{C^{0,\delta}(\overline{\Omega})},$$

where $C = C(n,A,\delta,p,|\Omega|)$.

Next we show that: if $u \in \mathscr{L}^{p,\mu}(\Omega)$, $n < \mu \leqslant n+p$, then $u \in C^{0,\delta}(\overline{\Omega})$, $\delta = (\mu - n)/p$, and $\|u\|_{C^{0,\delta}(\overline{\Omega})} \leqslant C\|u\|_{\mathscr{L}^{p,\mu}(\Omega)}$.

The conclusion should be understood as saying that if $u \in \mathscr{L}^{p,\mu}(\Omega)$, then there exists $\widetilde{u} \in C^{0,\delta}(\overline{\Omega})$ such that $\widetilde{u}(x) = u(x)$ for a.e. $x \in \Omega$, and

$$\|\widetilde{u}\|_{C^{0,\delta}(\overline{\Omega})} \leqslant C\|u\|_{\mathscr{L}^{p,\mu}(\Omega)}.$$

The proof is divided into four steps.

Step 1. For any fixed $x \in \overline{\Omega}$ and any $R > 0$, we set $R_h = R/2^h$ $(h = 0,1,2,\cdots)$. We claim that the sequence $\{u_{x,R_h}\}$ converges as $h \to \infty$. It suffices to show that $\{u_{x,R_h}\}$ is a Cauchy sequence.

Choose in Lemma 1.3 $\widetilde{R} = R_i$, $\widetilde{r} = R_{i+1}$ $(i = 0,1,2,\cdots)$. Then

$$|u_{x,R_i} - u_{x,R_{i+1}}| \leqslant C[u]_{p,\mu}R^{(\mu-n)/p}2^{i(n-\mu)/p+n/p}.$$

Therefore, for $0 \leqslant k < h$, we have

$$\begin{aligned} |u_{x,R_k} - u_{x,R_h}| &\leqslant \sum_{i=k}^{h-1} |u_{x,R_i} - u_{x,R_{i+1}}| \\ &\leqslant \sum_{i=k}^{h-1} C[u]_{p,\mu} R^{(\mu-n)/p} 2^{i(n-\mu)/p+n/p} \\ &= C \cdot 2^{n/p} [u]_{p,\mu} R^{(\mu-n)/p} 2^{k(n-\mu)/p} \frac{1 - 2^{(h-k)(n-\mu)/p}}{1 - 2^{(n-\mu)/p}}, \end{aligned}$$

which can be rewritten as

$$|u_{x,R_k} - u_{x,R_h}| \leqslant C[u]_{p,\mu} R_k^{(\mu-n)/p}. \tag{1.8}$$

This implies that for fixed $x \in \overline{\Omega}$, $\{u_{x,R_h}\}$ is a Cauchy sequence. We set

$$\widetilde{u}_R(x) = \lim_{h\to\infty} u_{x,R_h}.$$

Step 2. We claim that $\widetilde{u}_R(x)$ is independent of the choice of R, i.e., $\widetilde{u}_r(x) = \widetilde{u}_R(x)$ for $R > r > 0$. In fact, by Lemma 1.3,

$$\begin{aligned} |u_{x,R_h} - u_{x,r_h}| &\leqslant C(p,A)[u]_{p,\mu} r_h^{-n/p} R_h^{\mu/p} \\ &= C(p,A)[u]_{p,\mu} \left(\frac{R_h}{r_h}\right)^{n/p} R_h^{(\mu-n)/p} \\ &= C(p,A)[u]_{p,\mu} \left(\frac{R}{r}\right)^{n/p} R_h^{(\mu-n)/p}, \end{aligned}$$

which implies that $\lim_{h\to\infty} |u_{x,R_h} - u_{x,r_h}| = 0$, i.e., $\widetilde{u}_r(x) = \widetilde{u}_R(x)$. From now on, we shall simply denote $\widetilde{u}(x) = \widetilde{u}_R(x)$.

Step 3. We claim that $\widetilde{u}(x) = u(x)$ for a.e. $x \in \Omega$. Choosing $k = 0$ in (1.8) and then letting $h \to \infty$, we obtain

$$|u_{x,R} - \widetilde{u}(x)| \leqslant C[u]_{p,\mu} R^{(\mu-n)/p}, \tag{1.9}$$

where $C = C(p, A, n, \mu)$. This shows that $u_{x,R}$ converges uniformly to $\widetilde{u}(x)$ on $\overline{\Omega}$ as $R \to 0$. On the other hand, the Lebesgue differentiation theorem implies that $\lim_{R\to 0} u_{x,R} = u(x)$ for a.e. $x \in \Omega$. Hence $\widetilde{u}(x) = u(x)$ for a.e. $x \in \Omega$.

Step 4. We show that $\|\widetilde{u}\|_{C^{0,\delta}(\overline{\Omega})} \leqslant C\|u\|_{\mathscr{L}^{p,\mu}(\Omega)}$.

For any $x, y \in \overline{\Omega}$, $x \neq y$, we set $R = |x - y|$. Then

$$\begin{aligned} &|\widetilde{u}(x) - \widetilde{u}(y)| \\ &\leqslant |u_{x,2R} - \widetilde{u}(x)| + |u_{x,2R} - u_{y,2R}| + |u_{y,2R} - \widetilde{u}(y)| \\ &\triangleq I + II + III. \end{aligned} \tag{1.10}$$

From (1.9) we easily deduce that

$$I + III \leqslant C[u]_{p,\mu} R^{(\mu-n)/p}. \tag{1.11}$$

Now we estimate II. Set $G = \Omega(x, 2R) \cap \Omega(y, 2R)$. Then

$$\begin{aligned} &\int_G |u_{x,2R} - u_{y,2R}| dz \\ &\leqslant \int_{\Omega(x,2R)} |u_{x,2R} - u(z)| dz + \int_{\Omega(y,2R)} |u(z) - u_{y,2R}| dz. \end{aligned}$$

Clearly,

$$\begin{aligned}
&\int_{\Omega(x,\rho)} |u(z) - u_{x,\rho}|dz \\
&\leqslant \quad |\Omega(x,\rho)|^{1-1/p}\left\{\int_{\Omega(x,\rho)} |u(z) - u_{x,\rho}|^p dz\right\}^{1/p} \\
&= \quad |\Omega(x,\rho)|^{1-1/p}\rho^{\mu/p}\left\{\rho^{-\mu}\int_{\Omega(x,\rho)} |u(z) - u_{x,\rho}|^p dz\right\}^{1/p} \\
&\leqslant \quad C[u]_{p,\mu}\, \rho^{(\mu-n)/p+n},
\end{aligned}$$

where $C = C(p, A, n, \mu)$. It follows that

$$|u_{x,2R} - u_{y,2R}| \cdot |G| \leqslant 2C[u]_{p,\mu}\, R^{(\mu-n)/p+n}.$$

The definition of G implies that $G \supset \Omega(x, R)$ and therefore $|G| \geqslant |\Omega(x, R)| \geqslant AR^n$. Thus

$$|u_{x,2R} - u_{y,2R}| \leqslant 2A^{-1}C[u]_{p,\mu}\, R^{(\mu-n)/p}. \tag{1.12}$$

From (1.10)–(1.12), we deduce that

$$|\widetilde{u}(x) - \widetilde{u}(y)| \leqslant C[u]_{p,\mu}\, R^{(\mu-n)/p} = C[u]_{p,\mu}\, |x-y|^\delta, \tag{1.13}$$

where $\delta = (\mu - n)/p > 0$, $C = C(p, A, n, \mu)$. This implies that

$$[\widetilde{u}]_{0,\delta} \leqslant C[u]_{p,\mu}\, .$$

Finally, we show that

$$\sup_{x\in\overline{\Omega}} |\widetilde{u}(x)| \leqslant C\|u\|_{\mathscr{L}^{p,\mu}(\Omega)}.$$

$\widetilde{u}(x)$ is continuous on $\overline{\Omega}$, by (1.13). Therefore there exists $y \in \overline{\Omega}$ such that

$$\widetilde{u}(y) = \widetilde{u}_\Omega \triangleq \frac{1}{|\Omega|}\int_\Omega \widetilde{u}(z)dz.$$

It follows that, for all $x \in \overline{\Omega}$, we have

$$\begin{aligned}
|\widetilde{u}(x)| &\leqslant |\widetilde{u}_\Omega| + |\widetilde{u}(x) - \widetilde{u}(y)| \\
&\leqslant |\Omega|^{-1/p}\|u\|_{L^p(\Omega)} + C(\operatorname{diam}\Omega)^{(\mu-n)/p}[u]_{p,\mu}\, ,
\end{aligned}$$

which implies that

$$\sup_{x\in\overline{\Omega}} |\widetilde{u}(x)| \leqslant C\|u\|_{\mathscr{L}^{p,\mu}(\Omega)},$$

where $C = C(p, A, n, \mu, \operatorname{diam}\Omega)$.

It is obvious from the above proof that $\mathscr{L}^{p,\mu}(\Omega) = \{\text{constant}\}$ when $\mu > n + p$. □

With these results and Poincaré's inequality (cf. Appendix 1), we can easily prove Theorem 1.1.

Proof of Theorem 1.1. By assumption, $u \in W^{1,p}(B_R(x^0))$, $p \geqslant 1$; furthermore, for all $x \in B_R(x^0)$ and all ρ with $0 < \rho < d(x) = R - |x - x^0|$, we have

$$\int_{B_\rho(x)} |Du|^p dz \leqslant C\Big(\frac{\rho}{d(x)}\Big)^{n-p+p\delta}, \quad 0 < \delta < 1.$$

For each fixed r $(0 < r < R)$, for all $x \in B_r(x^0)$ and all ρ with $0 < \rho < R - r$, we have

$$\int_{B_\rho(x)} |Du|^p dz \leqslant C\Big(\frac{\rho}{R-r}\Big)^{n-p+p\delta}.$$

By Poincaré's inequality,

$$\begin{aligned}\int_{B_\rho(x)} |u(z) - u_{x,\rho}|^p dz &\leqslant C\rho^p \int_{B_\rho(x)} |Du|^p dz \\ &\leqslant C\rho^{n+p\delta},\end{aligned}$$

where C depends only on n, p, δ, $R-r$ and is independent of x, ρ. By Theorem 1.5, $u \in C^{0,\delta}(\overline{B_r(x^0)})$. □

By following the above proof, the reader may also give a proof of Theorem 4.2 in Chapter 5.

2. Schauder theory

Consider the elliptic system of divergence form

$$-D_\alpha[A_{ij}^{\alpha\beta}(x)D_\beta u^j] + D_\alpha f_i^\alpha = 0 \qquad (i = 1, \cdots, N), \tag{2.1}$$

where the coefficients $A_{ij}^{\alpha\beta}$ satisfy the strong Legendre condition

$$A_{ij}^{\alpha\beta}\xi_\alpha^i\xi_\beta^j \geqslant \lambda|\xi|^2, \quad \forall\xi \in \mathbb{R}^{nN}, \quad \lambda > 0. \tag{2.2}$$

We shall prove that if $A_{ij}^{\alpha\beta}$, $f_i^\alpha \in C^{0,\delta}$, then a weak solution of (2.1) belongs to $C^{1,\delta}$.

2.1. Two lemmas.

We shall first establish two very useful lemmas.

Lemma 2.1 (Iteration lemma). *Let $\Phi(\rho)$ be a nonnegative and nondecreasing function. Suppose that*

$$\Phi(\rho) \leqslant A\Big[\Big(\frac{\rho}{R}\Big)^\alpha + \varepsilon\Big]\Phi(R) + BR^\beta, \quad \forall 0 < \rho \leqslant R \leqslant R_0,$$

where A, α, β, R_0 are nonnegative constants, $\beta < \alpha$. Then there exist positive constants $\varepsilon_0 = \varepsilon_0(A, \alpha, \beta)$ and $C = C(A, \alpha, \beta)$ such that if $\varepsilon < \varepsilon_0$, then

$$\Phi(\rho) \leqslant C\Big[\Big(\frac{\rho}{R}\Big)^\beta + B\rho^\beta\Big], \quad \forall 0 < \rho \leqslant R \leqslant R_0.$$

Proof. Under our assumption, we have, for all $\tau \in (0, 1)$,

$$\Phi(\tau R) \leqslant A\tau^\alpha[1 + \varepsilon\tau^{-\alpha}]\Phi(R) + BR^\beta, \qquad R \leqslant R_0,$$

where we assume without loss of generality that $A \geqslant 1$.

First, we take γ such that $\beta < \gamma < \alpha$. Then we choose τ such that $2A\tau^\alpha = \tau^\gamma$, i.e., $\tau = \exp\Big(-\log(2A)/(\alpha - \gamma)\Big)$ (it is clear that $0 < \tau < 1$). Finally, we take ε_0 such that $\varepsilon_0\tau^{-\alpha} < 1$, i.e., $\varepsilon_0 < \exp\Big(-\alpha\log(2A)/(\alpha - \gamma)\Big)$. With these choices of γ, τ and ε_0, we have, for $\varepsilon < \varepsilon_0$,

$$\begin{aligned}\Phi(\tau R) &\leqslant 2A\tau^\alpha\Phi(R) + BR^\beta \\ &\leqslant \tau^\gamma\Phi(R) + BR^\beta, \quad R \leqslant R_0.\end{aligned}$$

Now we iterate. For any positive integer k, we have

$$\begin{aligned}
\Phi(\tau^{k+1}R) &\leqslant \tau^{\gamma}\Phi(\tau^k R) + B\tau^{k\beta}R^{\beta} \\
&\leqslant \cdots \leqslant \\
&\leqslant \tau^{(k+1)\gamma}\Phi(R) + B\tau^{k\beta}R^{\beta}\sum_{j=0}^{k}\tau^{j(\gamma-\beta)} \\
&\leqslant \tau^{(k+1)\beta}\Phi(R) + B\tau^{k\beta}R^{\beta}\frac{1-\tau^{(\gamma-\beta)(k+1)}}{1-\tau^{\gamma-\beta}} \\
&\leqslant \tau^{(k+1)\beta}\Big[\Phi(R) + BR^{\beta}\frac{1}{\tau^{\beta}-\tau^{\gamma}}\Big],
\end{aligned}$$

i.e.,

$$\Phi(\tau^{k+1}R) \leqslant C\tau^{(k+1)\beta}[\Phi(R) + BR^{\beta}], \tag{2.3}$$

where $C = C(A, \alpha, \beta)$.

For any $\rho \leqslant R$, we choose k such that $\tau^{k+1}R < \rho \leqslant \tau^k R$. The inequality (2.3), together with the nondecreasing property of Φ and our choice of k, implies that

$$\begin{aligned}
\Phi(\rho) &\leqslant \Phi(\tau^k R) \\
&\leqslant C\tau^{k\beta}[\Phi(R) + BR^{\beta}] \\
&\leqslant C\tau^{-\beta}\Big(\frac{\rho}{R}\Big)^{\beta}[\Phi(R) + BR^{\beta}] \\
&\leqslant C\tau^{-\beta}\Big[\Big(\frac{\rho}{R}\Big)^{\beta}\Phi(R) + B\rho^{\beta}\Big] \\
&\leqslant C_1\Big[\Big(\frac{\rho}{R}\Big)^{\beta}\Phi(R) + B\rho^{\beta}\Big],
\end{aligned}$$

where $C_1 = C_1(A, \alpha, \beta)$. □

Next, we establish two important estimates for weak solutions of an elliptic system with constant coefficients.

Lemma 2.2. *Let u be a weak solution of (2.1), where we assume that $A_{ij}^{\alpha\beta}$ are constants satisfying (2.2) and $|A_{ij}^{\alpha\beta}| \leqslant \Lambda$ $(i, j = 1, \cdots, N;\ \alpha, \beta = 1, \cdots, n)$, $f_i^{\alpha} \equiv 0$ $(i = 1, \cdots, N;\ \alpha = 1, \cdots, n)$. Then there exists a constant $C > 0$ such that for all $x^0 \in \Omega$ and all ρ, R with $0 < \rho \leqslant R < \operatorname{dist}(x^0, \partial\Omega)$, the following two estimates are valid:*

$$\int_{B_\rho(x^0)} |u|^2 dx \leqslant C\Big(\frac{\rho}{R}\Big)^{n}\int_{B_R(x^0)} |u|^2 dx, \tag{2.4}$$

$$\int_{B_\rho(x^0)} |u - u_{x^0,\rho}|^2 dx \leqslant C\Big(\frac{\rho}{R}\Big)^{n+2}\int_{B_R(x^0)} |u - u_{x^0,R}|^2 dx, \tag{2.5}$$

where C depends on n, N, λ, Λ and $\operatorname{dist}(x^0, \partial\Omega)$.

Proof. First we prove (2.4).

Corollary 2.5 in Chapter 8 implies that $u \in H^k(B_{R/2}(x^0), \mathbb{R}^N)$ and that

$$\|u\|_{H^k(B_{R/2}(x^0),\mathbb{R}^N)} \leqslant C\|u\|_{L^2(B_R(x^0),\mathbb{R}^N)} \qquad (k = 1, 2, \cdots), \tag{2.6}$$

where $C = C(n, N, \lambda, \Lambda, k, R)$.

We choose $2k > n$. By the Sobolev embedding theorem,

$$\sup_{B_{R/2}(x^0)} |u| \leqslant C\|u\|_{H^k(B_{R/2}(x^0),\mathbb{R}^N)}. \tag{2.7}$$

It follows from (2.7) that, for $0 < \rho < R/2$,

$$\int_{B_\rho(x^0)} |u|^2 dx \leqslant \omega_n \rho^n \sup_{B_{R/2}(x^0)} |u|^2 \leqslant C\rho^n \|u\|^2_{H^k(B_{R/2}(x^0),\mathbb{R}^N)}.$$

By (2.6),

$$\int_{B_\rho(x^0)} |u|^2 dx \leqslant C(R)\rho^n \int_{B_R(x^0)} |u|^2 dx, \tag{2.8}$$

where $C(R)$ depends also on R, in addition to the dependence on $n, N, \lambda, \Lambda, d$, where $d = \operatorname{dist}(x^0, \partial\Omega)$. Using a rescaling technique, we next show that $C(R) = C_1 R^{-n}$, where $C_1 = C_1(n, N, \lambda, \Lambda, d)$. In fact, we can first fix a constant $a < \operatorname{dist}(x^0, \partial\Omega)$. Then (2.8) implies that, for any $r < a/2$,

$$\int_{B_r(x^0)} |u|^2 dx \leqslant C(a) r^n \int_{B_a(x^0)} |u|^2 dx.$$

Set $y = a(x - x_0)/R$, $r = a\rho/R$. Then $|x - x_0| < \rho$ implies that $|y| < r < a/2$; therefore,

$$\begin{aligned}
\int_{B_\rho(x^0)} |u(x)|^2 dx &= \left(\frac{R}{a}\right)^n \int_{B_r(O)} \left|u\left(x_0 + \frac{R}{a}y\right)\right|^2 dy \\
&\leqslant \left(\frac{R}{a}\right)^n C(a) r^n \int_{B_a(O)} \left|u\left(x_0 + \frac{R}{a}y\right)\right|^2 dy \\
&= C_1 \left(\frac{\rho}{R}\right)^n \int_{B_R(x^0)} |u(x)|^2 dx,
\end{aligned}$$

where $C_1 = C_1(n, N, \lambda, \Lambda, d)$. This proves that (2.4) is valid for $0 < \rho < R/2$. It is obvious that (2.4) is valid for $R/2 \leqslant \rho \leqslant R$.

Now we prove the inequality (2.5).

Since u satisfies a homogeneous elliptic system with constant coefficients, Du also satisfies the same system. It follows from (2.4) that, for $\rho < R/2$,

$$\int_{B_\rho(x^0)} |Du|^2 dx \leqslant C\left(\frac{\rho}{R}\right)^n \int_{B_R(x^0)} |Du|^2 dx. \tag{2.9}$$

By Poincaré's inequality, the left-hand side of the inequality (2.9) is no less than

$$\frac{1}{C(n)\rho^2} \int_{B_\rho(x^0)} |u - u_{x^0,\rho}|^2 dx,$$

while the right-hand side of the inequality (2.9) is bounded by

$$C\left(\frac{\rho}{R}\right)^n \frac{C}{R^2} \int_{B_R(x^0)} |u - u_{x^0,R}|^2 dx,$$

by Caccioppoli's inequality. Thus (2.5) is valid for $0 < \rho < R/2$.

For $R/2 \leqslant \rho \leqslant R$, we have

$$\begin{aligned}\left(\frac{\rho}{R}\right)^{n+2} & \int_{B_R(x^0)} |u - u_{x^0,R}|^2 dx \\ \geqslant & \left(\frac{1}{2}\right)^{n+2} \int_{B_R(x^0)} |u - u_{x^0,R}|^2 dx \\ \geqslant & C(n) \int_{B_\rho(x^0)} |u - u_{x^0,\rho}|^2 dx.\end{aligned}$$

In deriving the last inequality, we made use of the fact that the function

$$\Phi(\rho) = \int_{B_\rho(x^0)} |u(x) - u_{x^0,\rho}|^2 dx$$

is a nondecreasing function in ρ. In fact, setting

$$\Psi(\nu) = \int_{B_\rho(x^0)} |u(x) - \nu|^2 dx, \quad \nu \in \mathbb{R}^N,$$

we can easily derive that $\Phi(\rho) = \min\limits_{\nu \in \mathbb{R}^N} \Psi(\nu)$. Thus for all $\rho \leqslant R$, we have

$$\begin{aligned}\int_{B_\rho(x^0)} |u(x) - u_{x^0,\rho}|^2 dx & \leqslant \int_{B_\rho(x^0)} |u(x) - u_{x^0,R}|^2 dx \\ & \leqslant \int_{B_R(x^0)} |u(x) - u_{x^0,R}|^2 dx.\end{aligned}$$ □

2.2. Constant coefficient elliptic systems.

Theorem 2.3. *Let u be a weak solution of (2.1), where we assume that $A_{ij}^{\alpha\beta}$ are constants satisfying (2.2) and $|A_{ij}^{\alpha\beta}| \leqslant \Lambda$ $(i,j = 1,\cdots,N;\ \alpha,\beta = 1,\cdots,n)$. If $f_i^\alpha \in \mathscr{L}^{2,\mu}_{loc}(\Omega)$, $0 \leqslant \mu < n+2$ $(i = 1,\cdots,N;\ \alpha = 1,\cdots,n)$, then $Du \in \mathscr{L}^{2,\mu}_{loc}(\Omega, \mathbb{R}^{nN})$. In particular, if $f_i^\alpha \in C^{0,\delta}_{loc}(\Omega)$ $(i = \cdots,N;\ \alpha = 1,\cdots,n)$, $0 < \delta < 1$, then $Du \in C^{0,\delta}_{loc}(\Omega, \mathbb{R}^{nN})$.*

Proof. For all $\widetilde{\Omega} \subset\subset \Omega$, all $x^0 \in \widetilde{\Omega}$, and all R, $0 < R < d/2$, $d = \operatorname{dist}(\widetilde{\Omega}, \partial\Omega)$, we let $v \in H^1(B_R(x^0), \mathbb{R}^N)$ be the solution of the following elliptic system:

$$\begin{cases} \displaystyle\int_{B_R(x^0)} A_{ij}^{\alpha\beta} D_\beta v^j D_\alpha \varphi^i dx = 0 \qquad \forall \varphi \in H_0^1(B_R(x^0), \mathbb{R}^N), \\ v - u \in H_0^1(B_R(x^0), \mathbb{R}^N). \end{cases} \tag{2.10}$$

The existence of such a solution is guaranteed by the Lax-Milgram theorem.

It is easy to see that Dv satisfies a homogeneous elliptic system with constant coefficients. It follows that (2.5) is valid for Dv, i.e.,

$$\int_{B_\rho(x^0)} |Dv - (Dv)_{x^0,\rho}|^2 dx \leqslant C\left(\frac{\rho}{R}\right)^{n+2} \int_{B_R(x^0)} |Dv - (Dv)_{x^0,R}|^2 dx, \tag{2.11}$$

where $0 < \rho \leqslant R$, $C = C(n, N, \lambda, \Lambda, d)$.

For simplicity, we shall denote $B_\rho(x^0)$ by B_ρ and $(Du)_{x^0,\rho}$ by $(Du)_\rho$.

Set $w = u - v$. Then

$$\begin{aligned}
&\int_{B_\rho} |Du - (Du)_\rho|^2 dx \\
&\leqslant 2\int_{B_\rho} |Dv - (Dv)_\rho|^2 dx + 2\int_{B_\rho} |Dw - (Dw)_\rho|^2 dx \\
&\leqslant C\Big(\frac{\rho}{R}\Big)^{n+2}\int_{B_R} |Dv - (Dv)_R|^2 dx \\
&\qquad + C\int_{B_\rho} |Dw|^2 dx + C\int_{B_\rho} |(Dw)_\rho|^2 dx \\
&\leqslant C\Big(\frac{\rho}{R}\Big)^{n+2}\int_{B_R} |Du - (Du)_R|^2 dx \\
&\qquad + C\Big[\int_{B_R} |Dw|^2 dx + \int_{B_R} |(Dw)_R|^2 dx \\
&\qquad\quad + \int_{B_\rho} |Dw|^2 dx + \int_{B_\rho} |(Dw)_\rho|^2 dx\Big].
\end{aligned}$$

Using Hölder's inequality, we easily obtain

$$\int_{B_\rho} |(Dw)_\rho|^2 dx \leqslant \int_{B_\rho} |Dw|^2 dx \quad \forall \rho > 0.$$

Thus

$$\begin{aligned}
&\int_{B_\rho} |Du - (Du)_\rho|^2 dx \\
&\leqslant C\Big(\frac{\rho}{R}\Big)^{n+2}\int_{B_R} |Du - (Du)_R|^2 dx + C\int_{B_R} |Dw|^2 dx.
\end{aligned} \tag{2.12}$$

Now we estimate $\int_{B_R} |Dw|^2 dx$. It is clear that $w = u - v \in H_0^1(B_R, \mathbb{R}^N)$ satisfies

$$\int_{B_R} A_{ij}^{\alpha\beta} D_\beta w^j D_\alpha \varphi^i dx = \int_{B_R} \Big[f_i^\alpha - (f_i^\alpha)_R\Big] D_\alpha \varphi^i dx, \qquad \forall \varphi \in H_0^1(B_R, \mathbb{R}^N).$$

By choosing $\varphi = w$ in the above equation and using the ellipticity condition and Hölder's inequality, we derive

$$\int_{B_R} |Dw|^2 dx \leqslant C\int_{B_R} |f - f_R|^2 dx, \tag{2.13}$$

where $f = (f_i^\alpha)$. Since $f_i^\alpha \in \mathscr{L}^{2,\mu}_{loc}(\Omega)$ $(i = 1, \cdots, N;\ \alpha = 1, \cdots, n)$, we have

$$\int_{B_R} |f - f_R|^2 dx \leqslant [f]^2_{2,\mu;\Omega'} R^\mu,$$

where $\Omega' = \{x \in \Omega |\, \mathrm{dist}(x, \partial\Omega) > \frac{1}{2}\,\mathrm{dist}(\widetilde{\Omega}, \partial\Omega)\}$. Substituting this inequality into (2.13), we get

$$\int_{B_R} |Dw|^2 dx \leqslant C[f]^2_{2,\mu;\Omega'} R^\mu. \tag{2.14}$$

Substituting (2.14) into (2.12), we conclude that

$$\int_{B_\rho} |Du-(Du)_\rho|^2 dx \leqslant C\Big(\frac{\rho}{R}\Big)^{n+2}\int_{B_R}|Du-(Du)_R|^2dx + C[f]^2_{2,\mu;\Omega'}R^\mu,$$

where $C = C(n, N, \lambda, \Lambda, d)$.

In the iteration lemma (Lemma 2.1), we choose $\Phi(\rho) = \int_{B_\rho} |Du-(Du)_\rho|^2 dx$, $A = C$, $B = C[f]^2_{2,\mu;\Omega'}$, $\alpha = n+2$, $\beta = \mu$, $\varepsilon = 0$. Then

$$\int_{B_\rho} |Du-(Du)_\rho|^2 dx \leqslant C\Big\{\frac{1}{R^\mu}\int_{B_R}|Du-(Du)_R|^2dx + [f]^2_{2,\mu;\Omega'}\Big\}\rho^\mu.$$

It follows that

$$\rho^{-\mu}\int_{B_\rho} |Du-(Du)_\rho|^2 dx \leqslant C\Big\{\|Du\|^2_{L^2(\Omega',\mathbb{R}^{nN})} + [f]^2_{2,\mu;\Omega'}\Big\},$$

where $C = C(n, N, \lambda, \Lambda, \mu, \operatorname{dist}(\widetilde{\Omega}, \partial\Omega))$. This implies that

$$Du \in \mathscr{L}^{2,\mu}_{loc}(\Omega, \mathbb{R}^{nN}).$$

The proof is complete. $\square$

We also have the following result on the global regularity.

Theorem 2.4. *Suppose that $A^{\alpha\beta}_{ij}$ are constants satisfying (2.2), $|A^{\alpha\beta}_{ij}| \leqslant \Lambda$ $(i,j = 1,\cdots,N;\ \alpha,\beta = 1,\cdots,n)$, $f^\alpha_i \in \mathscr{L}^{2,\mu}(\Omega)$, $0 \leqslant \mu < n+2$ $(i = 1,\cdots,N;\ \alpha = 1,\cdots,n)$, and $\partial\Omega$ is smooth. If $u \in H^1_0(\Omega,\mathbb{R}^N)$ is a weak solution of (2.1) with the Dirichlet boundary condition*

$$u\Big|_{\partial\Omega} = 0, \tag{2.15}$$

then $Du \in \mathscr{L}^{2,\mu}(\Omega,\mathbb{R}^{nN})$; furthermore,

$$\|Du\|_{\mathscr{L}^{2,\mu}(\Omega,\mathbb{R}^{nN})} \leqslant C\|f\|_{\mathscr{L}^{2,\mu}(\Omega,\mathbb{R}^{nN})},$$

where $C = C(n, N, \lambda, \Lambda, \mu, \operatorname{diam}\Omega)$.

Remark 2.1. Since $L^{2,\mu}(\Omega) \subset \mathscr{L}^{2,\mu}(\Omega)$ for $0 \leqslant \mu < n+2$ and $\|\cdot\|_{\mathscr{L}^{2,\mu}(\Omega)} \leqslant C\|\cdot\|_{L^{2,\mu}(\Omega)}$, the assumption on f^α_i in Theorem 2.4 can be changed to $f^\alpha_i \in L^{2,\mu}(\Omega)$, $0 \leqslant \mu < n+2$. In this case, the weak solution u of the Dirichlet problem (2.1), (2.15) satisfies $Du \in \mathscr{L}^{2,\mu}(\Omega,\mathbb{R}^{nN})$, and

$$\|Du\|_{\mathscr{L}^{2,\mu}(\Omega,\mathbb{R}^{nN})} \leqslant C\|f\|_{L^{2,\mu}(\Omega,\mathbb{R}^{nN})},$$

where $C = C(n, N, \lambda, \Lambda, \mu, \operatorname{diam}\Omega)$.

2.3. Variable coefficient elliptic systems.

Theorem 2.5. *Suppose that $A^{\alpha\beta}_{ij}(x)$ satisfy (2.2), $A^{\alpha\beta}_{ij} \in C^0(\overline{\Omega})$, $|A^{\alpha\beta}_{ij}| \leqslant \Lambda$ $(i,j = 1,\cdots,N;\ \alpha,\beta = 1,\cdots,n)$, and $f^\alpha_i \in L^{2,\mu}(\Omega)$, $0 < \mu < n$ $(i = 1,\cdots,N;\ \alpha = 1,\cdots,n)$. If $u \in H^1(\Omega,\mathbb{R}^N)$ is a weak solution of (2.1), then $Du \in L^{2,\mu}_{loc}(\Omega,\mathbb{R}^{nN})$, and for any $\widetilde{\Omega} \subset\subset \Omega$, we have the estimate*

$$\|Du\|_{L^{2,\mu}(\widetilde{\Omega},\mathbb{R}^{nN})} \leqslant C\Big[\|Du\|_{L^2(\Omega,\mathbb{R}^{nN})} + \|f\|_{L^{2,\mu}(\Omega,\mathbb{R}^{nN})}\Big],$$

where C depends only on $n, N, \lambda, \Lambda, \mu$, $\operatorname{dist}(\widetilde{\Omega}, \partial\Omega)$, and the modulus of continuity of $A_{ij}^{\alpha\beta}$.

Proof. We use the technique of freezing coefficients.

For all $\widetilde{\Omega} \subset\subset \Omega$, all $x^0 \in \widetilde{\Omega}$, and all R with $0 < R < d$, $d = \operatorname{dist}(\widetilde{\Omega}, \partial\Omega)$, we rewrite (2.1) as

$$-D_\alpha(A_{ij}^{\alpha\beta}(x^0)D_\beta u^j) + D_\alpha F_i^\alpha = 0 \qquad (i = 1, \cdots, N),$$

where $F_i^\alpha = f_i^\alpha + [A_{ij}^{\alpha\beta}(x^0) - A_{ij}^{\alpha\beta}(x)]D_\beta u^j \quad (\alpha = 1, \cdots, n;\ i = 1, \cdots, N)$.

Let v be the solution of the following elliptic system:

$$\begin{cases} \displaystyle\int_{B_R(x^0)} A_{ij}^{\alpha\beta}(x^0)D_\beta v^j D_\alpha \varphi^i dx = 0, & \forall \varphi \in H_0^1(B_R(x^0), \mathbb{R}^N), \\ v - u \in H_0^1(B_R(x^0), \mathbb{R}^N). \end{cases}$$

Obviously, Dv satisfies a homogeneous elliptic system with constant coefficients. It follows that (2.4) is valid for Dv; i.e., for all ρ, $0 < \rho \leqslant R$, we have

$$\int_{B_\rho(x^0)} |Dv|^2 dx \leqslant C\Big(\frac{\rho}{R}\Big)^n \int_{B_R(x^0)} |Dv|^2 dx,$$

where $C = C(n, N, \lambda, \Lambda, d)$.

Set $w = u - v$. Then

$$\int_{B_\rho(x^0)} |Du|^2 dx \leqslant 2\int_{B_\rho(x^0)} |Dv|^2 dx + 2\int_{B_\rho(x^0)} |Dw|^2 dx.$$

Thus for all $\rho, 0 < \rho \leqslant R$, we have

$$\text{(2.16)} \qquad \int_{B_\rho(x^0)} |Du|^2 dx \leqslant C\Big(\frac{\rho}{R}\Big)^n \int_{B_R(x^0)} |Du|^2 dx + C\int_{B_R(x^0)} |Dw|^2 dx.$$

Now we estimate $\displaystyle\int_{B_R(x^0)} |Dw|^2 dx$.

Clearly, $w \in H_0^1(B_R(x^0), \mathbb{R}^N)$ satisfies the equation

$$\begin{aligned} &\int_{B_R(x^0)} A_{ij}^{\alpha\beta}(x^0)D_\beta w^j D_\alpha \varphi^i dx \\ &\quad = \int_{B_R(x^0)} \Big\{f_i^\alpha + [A_{ij}^{\alpha\beta}(x^0) - A_{ij}^{\alpha\beta}(x)]D_\beta u^j\Big\} D_\alpha \varphi^i dx, \\ &\qquad\qquad \forall \varphi \in H_0^1(B_R(x^0), \mathbb{R}^N). \end{aligned}$$

Choosing $\varphi = w$ in the above equation, we obtain

$$\text{(2.17)} \qquad \int_{B_R(x^0)} |Dw|^2 dx \leqslant C\int_{B_R(x^0)} |f|^2 dx + C\omega^2(R)\int_{B_R(x^0)} |Du|^2 dx,$$

where

$$\omega(R) = \sup_{x \in B_R(x^0)} \Big\{\sum_{i,j=1}^N \sum_{\alpha,\beta=1}^n |A_{ij}^{\alpha\beta}(x^0) - A_{ij}^{\alpha\beta}(x)|^2\Big\}^{1/2}.$$

Substituting (2.17) into (2.16), we get

$$\int_{B_\rho(x^0)} |Du|^2 dx \leqslant C\Big[\Big(\frac{\rho}{R}\Big)^n + \omega^2(R)\Big]\int_{B_R(x^0)} |Du|^2 dx + C\|f\|_{L^{2,\mu}}^2 \cdot R^\mu.$$

Since $A_{ij}^{\alpha\beta} \in C^0(\overline{\Omega})$, there exists R_0, $0 < R_0 < d$, such that for $0 < \rho \leqslant R_0$,

$$\int_{B_\rho(x^0)} |Du|^2 dx \leqslant C\left[\left(\frac{\rho}{R}\right)^n + \varepsilon\right] \int_{B_R(x^0)} |Du|^2 dx + C\|f\|_{L^{2,\mu}}^2 \cdot R^\mu,$$

where $C = C(n, N, \lambda, \Lambda, d)$.

We choose, in the iteration lemma (Lemma 2.1), $\Phi(\rho) = \displaystyle\int_{B_\rho(x^0)} |Du|^2 dx$, $A = C$, $B = C\|f\|_{L^{2,\mu}}^2$, $\alpha = n$, $\beta = \mu$; then there exists an $\widetilde{R}_0$, $0 < \widetilde{R}_0 < R_0$, such that for $0 < \rho \leqslant R \leqslant \widetilde{R}_0$,

$$\int_{B_\rho(x^0)} |Du|^2 dx \leqslant C\left[R^{-\mu} \int_\Omega |Du|^2 dx + \|f\|_{L^{2,\mu}(\Omega, \mathbb{R}^{nN})}^2\right] \rho^\mu.$$

It follows that

$$\|Du\|_{L^{2,\mu}(\widetilde{\Omega}, \mathbb{R}^{nN})} \leqslant C\Big[\|Du\|_{L^2(\Omega, \mathbb{R}^{nN})} + \|f\|_{L^{2,\mu}(\Omega, \mathbb{R}^{nN})}\Big],$$

where C depends on $n, N, \lambda, \Lambda, \mu, \operatorname{dist}(\widetilde{\Omega}, \partial\Omega)$, and the modulus of continuity of $A_{ij}^{\alpha\beta}$. The proof is complete. $\square$

Theorem 2.6. *Suppose that $A_{ij}^{\alpha\beta}(x)$ satisfy (2.2), $A_{ij}^{\alpha\beta} \in C^{0,\delta}(\Omega)$, $|A_{ij}^{\alpha\beta}| \leqslant \Lambda$ $(i, j = 1, \cdots, N;\ \alpha, \beta = 1, \cdots, n)$, and $f_i^\alpha \in C^{0,\delta}(\Omega)$, $0 < \delta < 1$ $(i = 1, \cdots, N;\ \alpha = 1, \cdots, n)$. If $u \in H^1(\Omega, \mathbb{R}^N)$ is a weak solution of (2.1), then*

$$Du \in C_{loc}^{0,\delta}(\Omega, \mathbb{R}^{nN});$$

furthermore, for all $\widetilde{\Omega} \subset\subset \Omega$, we have the estimate

$$\|Du\|_{C^{0,\delta}(\overline{\widetilde{\Omega}}, \mathbb{R}^{nN})} \leqslant C\Big\{\|Du\|_{L^2(\Omega', \mathbb{R}^{nN})} + \|f\|_{C^{0,\delta}(\overline{\Omega'}, \mathbb{R}^{nN})}\Big\},$$

where C depends only on $n, N, \lambda, \Lambda, \delta$, $[A_{ij}^{\alpha\beta}]_{0,\delta;\Omega'}$, $\operatorname{dist}(\widetilde{\Omega}, \partial\Omega)$ and $\operatorname{diam}\Omega$. Here $\Omega' = \{x \in \Omega \mid \operatorname{dist}(x, \partial\Omega) > \frac{1}{2}\operatorname{dist}(\widetilde{\Omega}, \partial\Omega)\}$.

Proof. For all $\widetilde{\Omega} \subset\subset \Omega$, all $x^0 \in \widetilde{\Omega}$, and all R, $0 < R < \frac{1}{2}\operatorname{dist}(\widetilde{\Omega}, \partial\Omega)$, we rewrite (2.1) as

$$-D_\alpha(A_{ij}^{\alpha\beta}(x^0) D_\beta u^j) + D_\alpha F_i^\alpha = 0 \qquad (i = 1, \cdots, N),$$

where $F_i^\alpha = f_i^\alpha - (f_i^\alpha)_{x^0,R} + [A_{ij}^{\alpha\beta}(x^0) - A_{ij}^{\alpha\beta}(x)] D_\beta u^j$.

Let v be the solution of the following elliptic system:

$$\begin{cases} \displaystyle\int_{B_R(x^0)} A_{ij}^{\alpha\beta}(x^0) D_\beta v^j D_\alpha \varphi^i dx = 0, & \forall \varphi \in H_0^1(B_R(x^0), \mathbb{R}^N), \\ v - u \in H_0^1(B_R(x^0), \mathbb{R}^N). \end{cases}$$

Applying estimate (2.5) to Dv, we obtain

$$\int_{B_\rho(x^0)} |Dv - (Dv)_{x^0,\rho}|^2 dx \leqslant C\left(\frac{\rho}{R}\right)^{n+2} \int_{B_R(x^0)} |Dv - (Dv)_{x^0,R}|^2 dx,$$

where $C = C(n, N, \lambda, \Lambda, \operatorname{dist}(\widetilde{\Omega}, \partial\Omega))$.

Set $w = u - v$. Analogously to the proof of Theorem 2.3, we derive

$$\begin{aligned}
&\int_{B_\rho(x^0)} |Du - (Du)_{x^0,\rho}|^2 dx \\
&\leqslant C\Big(\frac{\rho}{R}\Big)^{n+2} \int_{B_R(x^0)} |Du - (Du)_{x^0,R}|^2 dx + C\int_{B_R(x^0)} |Dw|^2 dx \\
&\leqslant C\Big(\frac{\rho}{R}\Big)^{n+2} \int_{B_R(x^0)} |Du - (Du)_{x^0,R}|^2 dx + C\int_{B_R(x^0)} |F - F_{x^0,R}|^2 dx \\
&\leqslant C\Big(\frac{\rho}{R}\Big)^{n+2} \int_{B_R(x^0)} |Du - (Du)_{x^0,R}|^2 dx + C\int_{B_R(x^0)} |f - f_{x^0,R}|^2 dx \\
&\qquad + C \sup_{x\in B_R(x^0)} \sum_{i,j=1}^{N} \sum_{\alpha,\beta=1}^{n} |A_{ij}^{\alpha\beta}(x) - A_{ij}^{\alpha\beta}(x^0)|^2 \cdot \int_{B_R(x^0)} |Du|^2 dx.
\end{aligned}$$

Since $A_{ij}^{\alpha\beta} \in C^{0,\delta}(\Omega)$,

$$\sup_{x\in B_R(x^0)} \sum |A_{ij}^{\alpha\beta}(x) - A_{ij}^{\alpha\beta}(x^0)|^2 \leqslant \sum [A_{ij}^{\alpha\beta}]^2_{o,\delta;\Omega'} R^{2\delta},$$

where $\Omega' = \Big\{x \in \Omega \;\Big|\; \operatorname{dist}(x, \partial\Omega) > \frac{1}{2}\operatorname{dist}(\widetilde{\Omega}, \partial\Omega)\Big\}$. It follows that

$$\begin{aligned}
&\int_{B_\rho(x^0)} |Du - (Du)_{x^0,\rho}|^2 dx \\
&\leqslant C\Big(\frac{\rho}{R}\Big)^{n+2} \int_{B_R(x^0)} |Du - (Du)_{x^0,R}|^2 dx \\
&\qquad + C[f]^2_{2,n+2\delta;\Omega'} R^{n+2\delta} \\
&\qquad + C\sum [A_{ij}^{\alpha\beta}]^2_{0,\delta;\Omega'} R^{2\delta} \int_{B_R(x^0)} |Du|^2 dx.
\end{aligned} \tag{2.18}$$

For any $\varepsilon > 0$, we have $f \in C^{0,\delta} \simeq \mathscr{L}^{2,n+2\delta} \subset \mathscr{L}^{2,n-\varepsilon} \subset L^{2,n-\varepsilon}$. Therefore, $f \in L^{2,n-\varepsilon}(\Omega', \mathbb{R}^{nN})$. By Theorem 2.5,

$$Du \in L^{2,n-\varepsilon}(\widetilde{\Omega}, \mathbb{R}^{nN}),$$

and

$$\int_{B_R(x^0)} |Du|^2 dx \leqslant C\Big[\|Du\|^2_{L^2(\Omega',\mathbb{R}^{nN})} + \|f\|^2_{L^{2,n-\varepsilon}(\Omega',\mathbb{R}^{nN})}\Big] R^{n-\varepsilon},$$

where ε is a small positive constant; C depends on $n, N, \lambda, \Lambda, \delta, \varepsilon$, $\operatorname{dist}(\widetilde{\Omega}, \partial\Omega)$, and the modulus of continuity of $A_{ij}^{\alpha\beta}$. Substituting this inequality into (2.18), we get

$$\begin{aligned}
&\int_{B_\rho(x^0)} |Du - (Du)_{x^0,\rho}|^2 dx \\
&\leqslant C\Big(\frac{\rho}{R}\Big)^{n+2} \int_{B_R(x^0)} |Du - (Du)_{x^0,R}|^2 dx + C[f]^2_{2,n+2\delta;\Omega'} R^{n+2\delta} \\
&\qquad + C\sum [A_{ij}^{\alpha\beta}]^2_{0,\delta;\Omega'} \Big\{\|Du\|^2_{L^2(\Omega',\mathbb{R}^{nN})} + \|f\|^2_{L^{2,n-\varepsilon}(\Omega',\mathbb{R}^{nN})}\Big\} R^{n+2\delta-\varepsilon}.
\end{aligned}$$

Applying the iteration lemma, we now obtain

$$\int_{B_\rho(x^0)} |Du - (Du)_{x^0,\rho}|^2 dx \leqslant C\Big\{\|Du\|^2_{L^2(\Omega',\mathbb{R}^{nN})} + \|f\|^2_{L^{2,n-\varepsilon}(\Omega',\mathbb{R}^{nN})}\Big\} \rho^{n+2\delta-\varepsilon},$$

where C depends on $n, N, \lambda, \Lambda, \delta, \varepsilon$, $\|A_{ij}^{\alpha\beta}\|_{C^{0,\delta}}$, $\operatorname{dist}(\widetilde{\Omega}, \partial\Omega)$ and $\operatorname{diam}\Omega$. It follows that

$$Du \in C_{loc}^{0,\delta-\varepsilon/2}(\Omega, \mathbb{R}^{nN}), \quad \forall \varepsilon > 0,$$

and that

$$\|Du\|_{C^{0,\delta-\varepsilon/2}(\overline{\widetilde{\Omega}},\mathbb{R}^{nN})} \leqslant C\Big\{\|Du\|_{L^2(\Omega',\mathbb{R}^{nN})} + \|f\|_{\mathscr{L}^{2,n+2\delta}(\Omega',\mathbb{R}^{nN})}\Big\}.$$

In particular, Du is locally bounded and

$$\int_{B_R(x^0)} |Du|^2 dx \leqslant C\Big[\|Du\|^2_{L^2(\Omega',\mathbb{R}^{nN})} + \|f\|^2_{C^{0,\delta}(\overline{\Omega}',\mathbb{R}^{nN})}\Big] R^n,$$

where C depends on $n, N, \lambda, \Lambda, \delta$, $\|A_{ij}^{\alpha\beta}\|_{C^{0,\delta}}$, $\operatorname{dist}(\widetilde{\Omega}, \partial\Omega)$ and $\operatorname{diam}\Omega$. Substituting this inequality into (2.18), we derive

$$\begin{aligned}\int_{B_\rho(x^0)} &|Du - (Du)_{x^0,\rho}|^2 dx \\ \leqslant\ & C\Big(\frac{\rho}{R}\Big)^{n+2} \int_{B_R(x^0)} |Du - (Du)_{x^0,R}|^2 dx \\ &+C\Big[\|Du\|^2_{L^2(\Omega',\mathbb{R}^{nN})} + \|f\|^2_{C^{0,\delta}(\overline{\Omega}',\mathbb{R}^{nN})}\Big] R^{n+2\delta}.\end{aligned}$$

Applying again the iteration lemma and Theorem 1.5, we conclude that

$$Du \in C_{loc}^{0,\delta}(\Omega, \mathbb{R}^{nN}),$$

together with the desired estimate. The proof is complete. □

For the global regularity of a weak solution of the Dirichlet problem (2.1), (2.15), we have the following result:

Theorem 2.7. *Suppose that $A_{ij}^{\alpha\beta}(x)$ satisfy (2.2), $A_{ij}^{\alpha\beta} \in C^{0,\delta}(\overline{\Omega})$ $(i, j = 1, \cdots, N;\ \alpha, \beta = 1, \cdots, n)$, $f_i^\alpha \in C^{0,\delta}(\overline{\Omega})$ $(i = 1, \cdots, N;\ \alpha = 1, \cdots, n)$, and $\partial\Omega \in C^{1,\delta}$, $0 < \delta < 1$. If $u \in H_0^1(\Omega, \mathbb{R}^N)$ is a weak solution of the Dirichlet problem (2.1), (2.15), then*

$$Du \in C^{0,\delta}(\overline{\Omega}, \mathbb{R}^{nN}).$$

For higher regularities, we have

Theorem 2.8. *Suppose that $A_{ij}^{\alpha\beta}(x)$ satisfy (2.2), $A_{ij}^{\alpha\beta} \in C^{k,\delta}(\overline{\Omega})$ $(i, j = 1, \cdots, N;\ \alpha, \beta = 1, \cdots, n)$, $f_i^\alpha \in C^{k,\delta}(\overline{\Omega})$ $(i = 1, \cdots, N;\ \alpha = 1, \cdots, n)$, and $\partial\Omega \in C^{k+1,\delta}$, $k \geqslant 1$, $0 < \delta < 1$. If $u \in H_0^1(\Omega, \mathbb{R}^N)$ is a weak solution of the Dirichlet problem (2.1), (2.15), then*

$$Du \in C^{k,\delta}(\overline{\Omega}, \mathbb{R}^{nN}).$$

CHAPTER 10

L^p Theory for Linear Elliptic Systems of Divergence Form

In this chapter, we introduce the L^p theory for elliptic systems in divergence form. We shall first use the Stampacchia interpolation theorem to establish L^p estimates for weak solutions of elliptic systems with constant coefficients. Then we shall use the technique of freezing coefficients to establish L^p estimates for weak solutions of elliptic systems with continuous coefficients.

As a preliminary, we first introduce the BMO spaces and the Stampacchia interpolation theorem.

1. BMO spaces and the Stampacchia interpolation theorem

As in the previous chapters, we use $u_\Omega = \fint_\Omega u(z)dz$ to denote the average of u on Ω.

Definition 1.1. Suppose that Q_0 is a cube in $\mathbb{R}^n$. If $u \in L^1(Q_0)$ satisfies

$$|u|_{*,Q_0} \triangleq \sup_{Q\subset Q_0} \fint_Q |u - u_Q|dx < +\infty,$$

where the supremum is taken over all subcubes parallel to Q_0, then we say that $u \in \mathrm{BMO}(Q_0)$.

The norm of an element in $\mathrm{BMO}(Q_0)$ is defined to be

$$\|u\|_{\mathrm{BMO}(Q_0)} = \|u\|_{L^1(Q_0)} + |u|_{*,Q_0}.$$

It can be shown that $\mathrm{BMO}(Q_0)$ is a Banach space under this norm.

From the definition, it is easily seen that $\mathrm{BMO}(Q_0) \simeq \mathscr{L}^{1,n}(Q_0)$, where $\mathscr{L}^{1,n}(Q_0)$ is a Campanato space (see Definition 1.3 in Chapter 9).

In 1961, F. John and L. Nirenberg ([JN]) proved a very deep result:

Theorem 1.1 (John and Nirenberg). *There exist two constants C_1 and C_2, depending only on n, such that for all $Q \subset Q_0$,*

$$\mathrm{meas}\{x \in Q_0 \mid |u(x) - u_Q| > t\} \leqslant C_1|Q| \exp\Big\{ -\frac{C_2}{|u|_{*,Q_0}} t\Big\}, \quad \forall u \in \mathrm{BMO}(Q_0).$$

The proof of this theorem is given in Appendix 3 (or see [JN]). Using this theorem, we can easily prove that $\mathrm{BMO}(Q_0) \simeq \mathscr{L}^{p,n}(Q_0)$, $\forall p \geqslant 1$. In fact, if $u \in \mathscr{L}^{p,n}(Q_0)$, $p \geqslant 1$, then it is obvious that $u \in \mathrm{BMO}(Q_0)$, and $\|u\|_{\mathrm{BMO}(Q_0)} \leqslant$

$C\|u\|_{\mathscr{L}^{p,n}(Q_0)}$. On the other hand, if $u \in \mathrm{BMO}(Q_0)$, then for $p \geqslant 1$ and $Q \subset Q_0$, we have

$$\begin{aligned}\int_Q |u-u_Q|^p dx &= p\int_0^\infty t^{p-1}\operatorname{meas}\{x\in Q \mid |u(x)-u_Q|>t\}dt\\ &\leqslant pC_1\int_0^\infty t^{p-1}|Q|\exp\Big\{-\frac{C_2}{|u|_{*,Q_0}}t\Big\}dt\\ &= pC_1\Big(\frac{|u|_{*,Q_0}}{C_2}\Big)^p|Q|\int_0^\infty e^{-t}t^{p-1}dt\\ &\leqslant C(p,n)|u|^p_{*,Q_0}|Q|.\end{aligned}$$

It follows that $\displaystyle ⨍_Q |u-u_Q|^p dx \leqslant C(p,n)|u|^p_{*,Q_0}$. Therefore, $u \in \mathscr{L}^{p,n}(Q_0)$, and $[u]_{p,n;Q_0} \leqslant C(p,n)|u|_{*,Q_0}$. Clearly,

$$\begin{aligned}\|u\|^p_{L^p(Q_0)} &\leqslant C\Big[\int_{Q_0}|u-u_{Q_0}|^p dx + |u_{Q_0}|^p|Q_0|\Big]\\ &\leqslant C(p,n,|Q_0|)\|u\|^p_{\mathrm{BMO}(Q_0)}.\end{aligned}$$

Hence

$$\|u\|_{\mathscr{L}^{p,n}(Q_0)} \leqslant C\|u\|_{\mathrm{BMO}(Q_0)}.$$

Now we state an interpolation theorem between L^q and BMO spaces.

Theorem 1.2 (Stampacchia interpolation theorem). *Let $1 < q < +\infty$. Suppose that T is both a bounded linear operator: $L^q(Q_0) \to L^q(Q_0)$, and a bounded linear operator: $L^\infty(Q_0) \to \mathrm{BMO}(Q_0)$; i.e.,*

$$\begin{aligned}&\|Tu\|_{L^q(Q_0)} \leqslant C_1\|u\|_{L^q(Q_0)}, \quad \forall u \in L^q(Q_0),\\ &\|Tu\|_{\mathrm{BMO}(Q_0)} \leqslant C_2\|u\|_{L^\infty(Q_0)}, \quad \forall u \in L^\infty(Q_0).\end{aligned}$$

Then for $p \in [q,+\infty)$, T is a bounded linear operator: $L^p(Q_0) \to L^p(Q_0)$; i.e.,

$$\|Tu\|_{L^p(Q_0)} \leqslant C\|u\|_{L^p(Q_0)}, \quad \forall u \in L^p(Q_0),$$

where C depends on n, q, p, C_1, C_2.

For a proof of this theorem, see Appendix 4.

Remark 1.1. In the above definitions and theorems, the cubes can be replaced by balls.

2. L^p theory

First we prove

Theorem 2.1. *Suppose that $u \in H_0^1(B_R, \mathbb{R}^N)$ satisfies the equation*

$$\int_{B_R} A_{ij}^{\alpha\beta} D_\beta u^j D_\alpha \varphi^i dx = \int_{B_R} f_i^\alpha D_\alpha \varphi^i dx, \quad \forall \varphi \in H_0^1(B_R,\mathbb{R}^N), \tag{2.1}$$

where B_R is a ball of radius R, $A_{ij}^{\alpha\beta}$ are constants, $|A_{ij}^{\alpha\beta}| \leqslant \Lambda$ $(i, j = 1, \cdots, N;$ $\alpha, \beta = 1, \cdots, n)$, $A_{ij}^{\alpha\beta}\xi_\alpha^i\xi_\beta^j \geqslant \lambda|\xi|^2$, $\lambda > 0$, and $f_i^\alpha \in L^p(B_R)$ $(i = 1, \cdots, N;$ $\alpha = 1, \cdots, n)$, $p \geqslant 2$. Under these assumptions, $Du \in L^p(B_R, \mathbb{R}^{nN})$, and

$$\|Du\|_{L^p(B_R,\mathbb{R}^{nN})} \leqslant C\|f\|_{L^p(B_R,\mathbb{R}^{nN})},$$

where $f = (f_i^\alpha)$ and $C = C(n, N, \lambda, \Lambda, p)$ is independent of R.

Proof. Let $u \in H_0^1(B_R, \mathbb{R}^N)$ be the unique solution of (2.1) corresponding to f. We define the operator T as follows:

$$Tf \triangleq Du.$$

By choosing $\varphi = u$ in (2.1), we obtain the estimate

$$\|Du\|_{L^2(B_R,\mathbb{R}^{nN})} \leqslant C\|f\|_{L^2(B_R,\mathbb{R}^{nN})},$$

where $C = C(n, N, \lambda)$. This shows that T is a bounded linear operator from $L^2(B_R, \mathbb{R}^{nN})$ to $L^2(B_R, \mathbb{R}^{nN})$.

On the other hand, Remark 2.1 in Chapter 9 implies that if $f \in L^{2,n}(B_R, \mathbb{R}^{nN})$, then $Du \in \mathscr{L}^{2,n}(B_R, \mathbb{R}^{nN})$, and

$$\|Du\|_{\mathscr{L}^{2,n}(B_R,\mathbb{R}^{nN})} \leqslant C\|f\|_{L^{2,n}(B_R,\mathbb{R}^{nN})},$$

where $C = C(n, N, \lambda, \Lambda)$. Using Lemma 1.2 in Chapter 9 and §1 of this chapter, we conclude that T is a bounded linear operator from $L^\infty(B_R, \mathbb{R}^{nN})$ to $\mathrm{BMO}(B_R, \mathbb{R}^{nN})$.

By the Stampacchia interpolation theorem, for $2 \leqslant p < \infty$, T is a bounded linear operator from $L^p(B_R, \mathbb{R}^{nN})$ to $L^p(B_R, \mathbb{R}^{nN})$; i.e., if $f \in L^p(B_R, \mathbb{R}^{nN})$, then $Du \in L^p(B_R, \mathbb{R}^{nN})$, and

$$\|Du\|_{L^p(B_R,\mathbb{R}^{nN})} \leqslant C\|f\|_{L^p(B_R,\mathbb{R}^{nN})},$$

where $C = C(n, N, \lambda, \Lambda, p)$. The proof is complete. □

Next, we study the local L^p estimates for weak solutions of elliptic systems of divergence form.

Theorem 2.2. *Suppose that $u \in H^1(\Omega, \mathbb{R}^N)$ satisfies the equation*

$$\int_\Omega A_{ij}^{\alpha\beta}(x)D_\beta u^j D_\alpha\varphi^i dx = \int_\Omega f_i^\alpha D_\alpha\varphi^i dx, \quad \forall\varphi \in H_0^1(\Omega, \mathbb{R}^N),$$

where $A_{ij}^{\alpha\beta} \in C^0(\overline{\Omega})$, $|A_{ij}^{\alpha\beta}| \leqslant \Lambda$ $(i, j = 1, \cdots, N;$ $\alpha, \beta = 1, \cdots, n)$, $A_{ij}^{\alpha\beta}\xi_\alpha^i\xi_\beta^j \geqslant \lambda|\xi|^2$, $\lambda > 0$, and $f_i^\alpha \in L^p(\Omega)$ $(i = 1, \cdots, N;$ $\alpha = 1, \cdots, n)$, $p \geqslant 2$. Under these assumptions, $Du \in L^p_{loc}(\Omega, \mathbb{R}^{nN})$, and for any $\widetilde{\Omega} \subset\subset \Omega$,

$$\|Du\|_{L^p(\widetilde{\Omega},\mathbb{R}^{nN})} \leqslant C\Big[\|u\|_{H^1(\Omega,\mathbb{R}^N)} + \|f\|_{L^p(\Omega,\mathbb{R}^{nN})}\Big],$$

where $f = (f_i^\alpha)$ and C depends on $n, N, \lambda, \Lambda, p$, $\operatorname{dist}(\widetilde{\Omega}, \partial\Omega)$, and the modulus of continuity of $A_{ij}^{\alpha\beta}$.

Proof. It suffices to prove, $\forall \widetilde{\Omega} \subset\subset \Omega$, $\forall x^0 \in \widetilde{\Omega}$, $\forall R : 0 < R < \operatorname{dist}(\widetilde{\Omega}, \partial\Omega)$, that $Du \in L^p(B_{R/2}(x^0), \mathbb{R}^{nN})$.

By our assumption, $u \in H^1(B_R(x^0), \mathbb{R}^N)$ satisfies the equation

$$\int_{B_R(x^0)} A_{ij}^{\alpha\beta}(x) D_\beta u^j D_\alpha \varphi^i dx = \int_{B_R(x^0)} f_i^\alpha D_\alpha \varphi^i dx, \quad \forall \varphi \in H_0^1(B_R(x^0), \mathbb{R}^N). \tag{2.2}$$

Freezing the coefficients, we can rewrite (2.2) as

$$\begin{aligned} &\int_{B_R(x^0)} A_{ij}^{\alpha\beta}(x^0) D_\beta u^j D_\alpha \varphi^i dx \\ &= \int_{B_R(x^0)} \Big\{ [A_{ij}^{\alpha\beta}(x^0) - A_{ij}^{\alpha\beta}(x)] D_\beta u^j D_\alpha \varphi^i + f_i^\alpha D_\alpha \varphi^i \Big\} dx, \\ &\qquad \forall \varphi \in H_0^1(B_R(x^0), \mathbb{R}^N). \end{aligned}$$

The above system looks like an elliptic system with constant coefficients, to which Theorem 2.1 can be applied. However, here u does not vanish on ∂B_R, and therefore we use a cutoff function η to multiply u and consider ηu instead. The cutoff function $\eta \in C_0^\infty(B_R)$ satisfies

$$0 \leqslant \eta \leqslant 1 \quad \text{on } B_R, \qquad \eta \equiv 1 \quad \text{in } B_\rho, \qquad |D\eta| \leqslant \frac{C}{R-\rho},$$

where $0 < \rho < R$.

The proof is divided into four steps.

Step 1. Derive the equation for $\eta u \in H_0^1(B_R, \mathbb{R}^N)$.

For all $\varphi \in C_0^\infty(B_R, \mathbb{R}^N)$, we have

$$\begin{aligned} &\int_{B_R} A_{ij}^{\alpha\beta}(x^0) D_\beta(\eta u^j) D_\alpha \varphi^i dx \\ &= \int_{B_R} [A_{ij}^{\alpha\beta}(x^0) - A_{ij}^{\alpha\beta}(x)] D_\beta(\eta u^j) D_\alpha \varphi^i dx + \int_{B_R} A_{ij}^{\alpha\beta}(x) D_\beta(\eta u^j) D_\alpha \varphi^i dx \\ &= \int_{B_R} [A_{ij}^{\alpha\beta}(x^0) - A_{ij}^{\alpha\beta}(x)] D_\beta(\eta u^j) D_\alpha \varphi^i dx + \int_{B_R} A_{ij}^{\alpha\beta}(x) \eta\, D_\beta u^j D_\alpha \varphi^i dx \\ &\quad + \int_{B_R} A_{ij}^{\alpha\beta}(x) u^j\, D_\beta \eta D_\alpha \varphi^i dx \\ &= \int_{B_R} [A_{ij}^{\alpha\beta}(x^0) - A_{ij}^{\alpha\beta}(x)] D_\beta(\eta u^j) D_\alpha \varphi^i dx + \int_{B_R} A_{ij}^{\alpha\beta}(x) D_\beta u^j D_\alpha(\eta \varphi^i) dx \\ &\quad - \int_{B_R} \varphi^i A_{ij}^{\alpha\beta}(x) D_\beta u^j D_\alpha \eta dx + \int_{B_R} A_{ij}^{\alpha\beta}(x) u^j\, D_\beta \eta D_\alpha \varphi^i dx. \end{aligned}$$

Using (2.2) for the second term, we derive

$$\int_{B_R} A_{ij}^{\alpha\beta}(x^0)D_\beta(\eta u^j)D_\alpha\varphi^i dx$$
$$= \int_{B_R}[A_{ij}^{\alpha\beta}(x^0)-A_{ij}^{\alpha\beta}(x)]D_\beta(\eta u^j)D_\alpha\varphi^i dx + \int_{B_R} f_i^\alpha D_\alpha(\eta\varphi^i)dx$$
$$- \int_{B_R}\varphi^i A_{ij}^{\alpha\beta}(x)D_\beta u^j D_\alpha\eta dx + \int_{B_R} A_{ij}^{\alpha\beta}(x)u^j\, D_\beta\eta D_\alpha\varphi^i dx.$$

Setting

$$G_i = f_i^\alpha D_\alpha\eta - A_{ij}^{\alpha\beta}(x)D_\beta u^j D_\alpha\eta, \qquad G=(G_i),$$
$$F_i^\alpha = f_i^\alpha\eta + A_{ij}^{\alpha\beta}(x)u^j D_\beta\eta, \qquad F=(F_i^\alpha),$$

we then obtain the equation for ηu:

$$\int_{B_R} A_{ij}^{\alpha\beta}(x^0)D_\beta(\eta u^j)D_\alpha\varphi^i dx$$
$$= \int_{B_R}[A_{ij}^{\alpha\beta}(x^0)-A_{ij}^{\alpha\beta}(x)]D_\beta(\eta u^j)D_\alpha\varphi^i dx \tag{2.3}$$
$$+\int_{B_R} G_i\varphi^i dx + \int_{B_R} F_i^\alpha D_\alpha\varphi^i dx, \quad \forall\varphi\in C_0^\infty(B_R,\mathbb{R}^{nN}).$$

Finally, we rewrite (2.3) as

$$\int_{B_R} A_{ij}^{\alpha\beta}(x^0)D_\beta(\eta u^j)D_\alpha\varphi^i dx$$
$$= \int_{B_R}[A_{ij}^{\alpha\beta}(x^0)-A_{ij}^{\alpha\beta}(x)]D_\beta(\eta u^j)D_\alpha\varphi^i dx \tag{2.4}$$
$$+\int_{B_R}\widetilde{F}_i^\alpha D_\alpha\varphi^i dx, \qquad \forall\varphi\in C_0^\infty(B_R,\mathbb{R}^{nN}),$$

where $\widetilde{F}_i^\alpha = F_i^\alpha + w_\alpha^i$, and the $w_s\in H_0^1(B_R,\mathbb{R}^N)$ satisfy

$$-\int_{B_R}\delta^{\alpha\beta}\delta_{ij}D_\beta w_s^j D_\alpha\phi^i dx = \int_{B_R}\delta^{\alpha s}G_i D_\alpha\phi^i dx,$$
$$\forall\phi\in C_0^\infty(B_R,\mathbb{R}^N),$$

where $s=1,\cdots,n$.

Step 2. Analyze the relationship between the regularities of $\widetilde{F}=(\widetilde{F}_i^\alpha)$ and the regularities of the solution u.

First, it is easily seen that w_s $(s=1,\cdots,n)$ are weak solutions of a system of the same type as (2.1). By Theorem 2.1, if $G\in L^r(B_R,\mathbb{R}^N)$, $r\geqslant 2$, then $Dw_s\in L^r(B_R,\mathbb{R}^{nN})$, and

$$\|Dw_s\|_{L^r(B_R,\mathbb{R}^{nN})}\leqslant C\|G\|_{L^r(B_R,\mathbb{R}^N)} \quad (s=1,\cdots,n).$$

By the Sobolev embedding theorem,

$$\|w\|_{L^{r^*}(B_R,\mathbb{R}^{nN})}\leqslant C\|G\|_{L^r(B_R,\mathbb{R}^N)}, \tag{2.5}$$

where $C=C(n,N,\lambda,\Lambda,r)$, $w=(w_s)=(w_s^i)$, and r^* is the Sobolev conjugate of r.

Now suppose that $u \in W^{1,l}(B_R, \mathbb{R}^N)$, $l \geqslant 2$. What space does $\widetilde{F}$ belong to? For simplicity, we shall use $l \wedge p$ to denote $\min(l,p)$.

It is assumed that $f \in L^p(B_R, \mathbb{R}^{nN})$, $p \geqslant 2$. If $u \in W^{1,l}(B_R, \mathbb{R}^N)$, $l \geqslant 2$, then $G \in L^{l\wedge p}(B_R, \mathbb{R}^N)$, $F \in L^{l^*\wedge p}(B_R, \mathbb{R}^N)$. By (2.5), we have $w_s \in L^{(l\wedge p)^*}(B_R, \mathbb{R}^N)$ $(s = 1, \cdots, n)$. Therefore, $\widetilde{F} \in L^{l^*\wedge p}(B_R, \mathbb{R}^{nN})$.

Step 3. Let $q = l^* \wedge p$. We want to show that if $u \in W^{1,l}(B_R, \mathbb{R}^N)$, $l \geqslant 2$, then for sufficiently small R, we have $\eta u \in W_0^{1,q}(B_R, \mathbb{R}^N)$.

In order to show this, we define a map $\widetilde{T} : W_0^{1,q}(B_R, \mathbb{R}^N) \to W_0^{1,q}(B_R, \mathbb{R}^N)$ as follows: for any $V \in W_0^{1,q}(B_R, \mathbb{R}^N)$, we set $\widetilde{T}V = v$, where $v \in W_0^{1,q}(B_R, \mathbb{R}^N)$ is the solution of the elliptic system

$$\begin{aligned}\int_{B_R} & A_{ij}^{\alpha\beta}(x^0) D_\beta v^j D_\alpha \varphi^i dx \\ &= \int_{B_R} [A_{ij}^{\alpha\beta}(x^0) - A_{ij}^{\alpha\beta}(x)] D_\beta V^j D_\alpha \varphi^i dx \\ &\quad + \int_{B_R} \widetilde{F}_i^\alpha D_\alpha \varphi^i dx, \qquad \forall \varphi \in C_0^\infty(B_R, \mathbb{R}^{nN}).\end{aligned}$$

Theorem 2.1 implies that

$$\|Dv\|_{L^q(B_R,\mathbb{R}^{nN})} \leqslant C\|[A(x^0) - A(x)]DV\|_{L^q(B_R,\mathbb{R}^{nN})} + C\|\widetilde{F}\|_{L^q(B_R,\mathbb{R}^{nN})}, \tag{2.6}$$

where $A(x)DV$ denotes the matrix $(A_{ij}^{\alpha\beta}(x) D_\beta V^j)$ and $C = C(n, N, \lambda, \Lambda, q)$.

Next, we show that $\widetilde{T}$ is a contraction if R is small enough and therefore has a unique fixed point.

For $V_1, V_2 \in W_0^{1,q}(B_R, \mathbb{R}^N)$, we denote by v_1 and v_2 the solutions of the corresponding elliptic system. By Poincaré's inequality and (2.6),

$$\begin{aligned}\|\widetilde{T}V_1 - \widetilde{T}V_2\|_{W_0^{1,q}(B_R,\mathbb{R}^N)} &= \|v_1 - v_2\|_{W_0^{1,q}(B_R,\mathbb{R}^N)} \\ &\leqslant C\|Dv_1 - Dv_2\|_{L^q(B_R,\mathbb{R}^{nN})} \\ &\leqslant C\|[A(x^0) - A(x)]D(V_1 - V_2)\|_{L^q(B_R,\mathbb{R}^{nN})} \\ &\leqslant C \max_{x\in B_R(x^0)} |A(x^0) - A(x)| \, \|V_1 - V_2\|_{W_0^{1,q}(B_R,\mathbb{R}^N)}.\end{aligned}$$

Since $A_{ij}^{\alpha\beta} \in C^0(\overline{\Omega})$, for sufficiently small R, we have

$$\|\widetilde{T}V_1 - \widetilde{T}V_2\|_{W_0^{1,q}(B_R,\mathbb{R}^N)} \leqslant \theta\|V_1 - V_2\|_{W_0^{1,q}(B_R,\mathbb{R}^N)},$$

where $0 < \theta < 1$. Thus $\widetilde{T}$ is a contraction and has a unique fixed point V, provided R is small enough.

It is obvious that V is a solution of (2.4). By uniqueness, $V = \eta u$. From (2.6), we deduce that

$$\|D(\eta u)\|_{L^q(B_R,\mathbb{R}^{nN})} \leqslant C\|[A(x^0) - A(x)]D(\eta u)\|_{L^q(B_R,\mathbb{R}^{nN})} + C\|\widetilde{F}\|_{L^q(B_R,\mathbb{R}^{nN})}.$$

It follows that, for sufficiently small R,

$$\begin{aligned}
&\|D(\eta u)\|_{L^q(B_R,\mathbb{R}^{nN})}\\
\leqslant\ & C\|\widetilde{F}\|_{L^q(B_R,\mathbb{R}^{nN})}\\
\leqslant\ & C\Big\{\|w\|_{L^q(B_R,\mathbb{R}^{nN})}+\|F\|_{L^q(B_R,\mathbb{R}^{nN})}\Big\}\\
\leqslant\ & C\Big\{\|G\|_{L^{l\wedge p}(B_R,\mathbb{R}^{N})}+\|F\|_{L^q(B_R,\mathbb{R}^{nN})}\Big\}\\
\leqslant\ & \frac{C}{R-\rho}\Big\{\|f\|_{L^p(B_R,\mathbb{R}^{nN})}+\|Du\|_{L^l(B_R,\mathbb{R}^{nN})}+\|u\|_{L^{l^*}(B_R,\mathbb{R}^{N})}\Big\}\\
\leqslant\ & \frac{C}{R-\rho}\Big\{\|f\|_{L^p(B_R,\mathbb{R}^{nN})}+\|u\|_{W^{1,l}(B_R,\mathbb{R}^{N})}\Big\},
\end{aligned}$$

where C depends on $n, N, \lambda, \Lambda, l, p$, and the modulus of continuity of $A_{ij}^{\alpha\beta}$. We now have

$$\|Du\|_{L^q(B_\rho,\mathbb{R}^{nN})}\leqslant\frac{C}{R-\rho}\Big\{\|f\|_{L^p(B_R,\mathbb{R}^{nN})}+\|u\|_{W^{1,l}(B_R,\mathbb{R}^{N})}\Big\},\tag{2.7}$$

where $q=l^*\wedge p$ and C depends on the quantities listed above.

Step 4. We show that $Du\in L^p(B_{R/2}(x^0),\mathbb{R}^{nN})$.

Take a sequence of concentric balls whose radii are

$$\frac{R}{2}<\cdots<R_k<\cdots R_2<R_1<R.$$

Take in (2.7) $\rho=R_1$, $l=2$. Then

$$\|Du\|_{L^{2^*\wedge p}(B_{R_1},\mathbb{R}^{nN})}\leqslant\frac{C}{R-R_1}\Big\{\|f\|_{L^p(B_R,\mathbb{R}^{nN})}+\|u\|_{W^{1,2}(B_R,\mathbb{R}^{N})}\Big\}.$$

If $p\leqslant 2^*$, then $Du\in L^p(B_{R_1},\mathbb{R}^{nN})$, and therefore $Du\in L^p(B_{R/2},\mathbb{R}^{nN})$.

If $p>2^*$, then $Du\in L^{2^*}(B_{R_1},\mathbb{R}^{nN})$, and we have the estimate

$$\|Du\|_{L^{2^*}(B_{R_1},\mathbb{R}^{nN})}\leqslant\frac{C}{R-R_1}\Big\{\|f\|_{L^p(B_R,\mathbb{R}^{nN})}+\|u\|_{W^{1,2}(B_R,\mathbb{R}^{N})}\Big\}.$$

Take in (2.7) $R=R_1$, $\rho=R_2$, $l=2^*$. Then

$$\begin{aligned}
&\|Du\|_{L^{2^{**}\wedge p}(B_{R_2},\mathbb{R}^{nN})}\\
\leqslant\ & \frac{C}{R_1-R_2}\Big\{\|f\|_{L^p(B_R,\mathbb{R}^{nN})}+\|u\|_{W^{1,2^*}(B_{R_1},\mathbb{R}^{N})}\Big\}\\
\leqslant\ & \frac{C}{(R-R_1)(R_1-R_2)}\Big\{\|f\|_{L^p(B_R,\mathbb{R}^{nN})}+\|u\|_{W^{1,2}(B_R,\mathbb{R}^{N})}\Big\}.
\end{aligned}$$

Now, if $p\leqslant 2^{**}$, then $Du\in L^p(B_{R_2},\mathbb{R}^{nN})$, and therefore $Du\in L^p(B_{R/2},\mathbb{R}^{nN})$.

If $p>2^{**}$, then $Du\in L^{2^{**}}(B_{R_2},\mathbb{R}^{nN})$. We can repeat the above procedure. After finitely many iterations, we can always conclude that $Du\in L^p(B_{R/2},\mathbb{R}^{nN})$ and

$$\|Du\|_{L^p(B_{R/2},\mathbb{R}^{nN})}\leqslant C\Big\{\|f\|_{L^p(\Omega,\mathbb{R}^{nN})}+\|u\|_{H^1(\Omega,\mathbb{R}^{N})}\Big\},$$

where C depends on $n, N, \lambda, \Lambda, l, p, \operatorname{dist}(\widetilde{\Omega},\partial\Omega)$, and the modulus of continuity of $A_{ij}^{\alpha\beta}$. □

Remark 2.1. In Theorem 2.2, the assumption $A_{ij}^{\alpha\beta} \in C^0(\overline{\Omega})$ cannot be dropped. See the example in §4 of Chapter 12, given by N. G. Meyers.

CHAPTER 11

Existence of Weak Solutions of Nonlinear Elliptic Systems

1. Introduction

In this chapter, we study the existence of weak solutions of the following nonlinear elliptic systems of divergence form:

$$-D_\alpha A_i^\alpha(x,u,Du)+B_i(x,u,Du)=0 \qquad (i=1,\cdots,N), \tag{1.1}$$

where $A_i^\alpha, B_i : \Omega\times\mathbb{R}^N\times\mathbb{R}^{nN}\to\mathbb{R}$, Ω being a bounded domain in $\mathbb{R}^n$. The meaning of *ellipticity* will be explained.

The definition of a weak solution of (1.1) depends on the structure conditions on A_i^α, B_i.

Definition 1.1. If A_i^α, B_i satisfy the following structure conditions:

$$A_i^\alpha(x,u,p)p_\alpha^i \geqslant \lambda|p|^2-\Lambda|u|^r-f^2(x), \tag{1.2}$$

$$\begin{cases} |A_i^\alpha(x,u,p)|\leqslant \Lambda_1(|p|+|u|^{r/2}+f_i^\alpha(x)), \\ |B_i(x,u,p)|\leqslant \Lambda\Big(|p|^{2(1-1/r)}+|u|^{r-1}+f_i(x)\Big), \end{cases} \tag{1.3}$$

where $\lambda,\Lambda,\Lambda_1$ are positive constants, $f_i^\alpha, f_i\geqslant 0$, $f, f_i^\alpha\in L^2(\Omega)$, $f_i\in L^{r/(r-1)}(\Omega)$, and $r=2^*$ is the Sobolev conjugate of 2 (it equals $2n/(n-2)$ when $n>2$ and is an arbitrary real number in $[2,+\infty)$ when $n=2$), then we say that A_i^α, B_i satisfy the *controllable structure conditions.* Here, (1.2) is an *ellipticity condition* and (1.3) is a *growth condition*, which is referred to as the *controllable growth condition.*

We will seek a weak solution of (1.1) in the space $H^1(\Omega,\mathbb{R}^N)$ when A_i^α, B_i satisfy the controllable structure conditions.

Definition 1.2. Suppose that A_i^α, B_i satisfy (1.2), (1.3). If $u\in H^1(\Omega,\mathbb{R}^N)$ satisfies

$$\int_\Omega [A_i^\alpha(x,u,Du)D_\alpha\varphi^i+B_i(x,u,Du)\varphi^i]dx=0, \quad \forall\varphi\in H_0^1(\Omega,\mathbb{R}^N), \tag{1.4}$$

then u is said to be a *weak solution* of the elliptic system (1.1).

The controllable growth condition (1.3) guarantees that the integral identity (1.4) makes sense. However, this set of growth conditions does not seem to be very natural. When $N=1$ (single equation case), we know that the natural growth condition is

$$\begin{cases} |A^\alpha(x,u,p)|\leqslant \Lambda_1(|p|+g(x)), \\ |B(x,u,p)|\leqslant \Lambda(|p|^2+f(x)), \end{cases}$$

where $f, g \geqslant 0$ and $f, g \in L^1(\Omega)$.

Now we consider the corresponding natural growth condition for systems.

Definition 1.3. Suppose that A_i^α, B_i satisfy

$$A_i^\alpha(x,u,p)p_\alpha^i \geqslant \lambda|p|^2 - \Lambda_1 f^2(x) \quad \text{for } |u| \leqslant M, \tag{1.5}$$

$$\begin{cases} |A_i^\alpha(x,u,p)| \leqslant \Lambda_1(|p| + f_i^\alpha(x)) \\ |B_i(x,u,p)| \leqslant \Lambda(|p|^2 + f_i(x)) \end{cases} \quad \text{for } |u| \leqslant M, \tag{1.6}$$

where $\lambda, \Lambda, \Lambda_1$ are positive constants (they may depend on M), $f_i^\alpha, f_i \geqslant 0$, $f, f_i^\alpha \in L^2(\Omega)$, $f_i \in L^1(\Omega)$. Then we say that A_i^α, B_i satisfy the *natural structure conditions*; (1.6) is a *natural growth condition.*

Under the natural growth condition, the integral identity (1.4) may not make sense for functions $u \in H^1(\Omega, \mathbb{R}^N)$ and $\varphi \in H_0^1(\Omega, \mathbb{R}^N)$. Thus we will seek solutions in $H^1 \cap L^\infty(\Omega, \mathbb{R}^N)$.

Definition 1.4. Suppose that A_i^α, B_i satisfy the natural structure conditions. If $u \in H^1 \cap L^\infty(\Omega, \mathbb{R}^N)$ satisfies the integral identity (1.4) for any $\varphi \in H_0^1 \cap L^\infty(\Omega, \mathbb{R}^N)$, then u is said to be a *weak solution* of (1.1).

There are deeper reasons for such a definition of weak solutions; we refer the readers to the book by O. A. Ladyženskaja and N. N. Ural′ceva [LU, Chapter 1], and the report by S. Hildebrandt [HB, §2].

There are many methods of dealing with the existence of weak solutions for elliptic systems. Among them are the variational method, the monotone operator method, the topological degree method, and the parabolic systems approximation method, etc. Each method has its own merits and limitations. We shall introduce the variational method in this book, which has the advantage of not requiring a priori estimates. However, the elliptic system considered must be associated with a variational problem (i.e., has a *variational structure*). For the existence of weak solutions of those elliptic systems not associated with a variational problem, we refer the readers to [NC], [ZK].

2. The variational method

2.1. Existence of weak solutions for regular variational problems.

In this subsection, we first consider the existence of a solution of a regular variational problem. We will establish its relationship with the weak solution of some elliptic system in the next two subsections.

In a variational problem, we seek a minimum point of the functional

$$J[u] = \int_\Omega F(x, u, Du)dx$$

in the *admissible set* $\mathscr{K}$, where $F(x,u,p) : \Omega \times \mathbb{R}^N \times \mathbb{R}^{nN} \to \mathbb{R}$. If F is convex in p, then the variational problem is said to be a *regular variational problem*, and J is said to be a *regular integral functional.*

We take the admissible set to be a subset of the Sobolev space $H^1(\Omega, \mathbb{R}^N)$:

$$\mathscr{K} = \{u \in H^1(\Omega, \mathbb{R}^N) \mid u - g \in H_0^1(\Omega, \mathbb{R}^N)\}, \tag{2.1}$$

where $g \in H^1(\Omega, \mathbb{R}^N)$ is such that $J[g] < +\infty$.

Obviously, $\mathscr{K}$ is a closed, convex subset of $H^1(\Omega, \mathbb{R}^N)$.

A more precise statement of the variational problem is

$$\text{Find } u \in \mathscr{K} \text{ such that } J[u] \leqslant J[v], \quad \forall v \in \mathscr{K}. \tag{2.2}$$

In order to clearly present the proof, we prove an existence theorem under a stronger condition (cf. [GQ2]). We will briefly introduce a more general result at the end of this subsection.

Theorem 2.1. *Suppose that $F(x, u, p)$ and $F_p(x, u, p)$ are continuous on $\Omega \times \mathbb{R}^N \times \mathbb{R}^{nN}$, F is convex in p, and $F(x, u, p) \geqslant \lambda |p|^2$, $\lambda > 0$. Then the problem* (2.2) *admits a solution.*

Proof. The proof is divided into two steps.

Step 1. Show that the functional $J[u]$ is weakly sequential lower semicontinuous in $H^1_{loc}(\Omega, \mathbb{R}^N)$, i.e., if $u_k, u \in H^1(\Omega, \mathbb{R}^N)$ and u_k converges weakly to u in $H^1(D, \mathbb{R}^N)$ ($\forall D \subset\subset \Omega$, ∂D smooth), then

$$J[u] \leqslant \liminf_{k \to \infty} J[u_k]. \tag{2.3}$$

First, under our assumption, we have $J[u_k] \geqslant 0$; therefore, $\liminf\limits_{k \to \infty} J[u_k]$ exists. If $\liminf\limits_{k \to \infty} J[u_k] = +\infty$, there is nothing to prove. In the case $\liminf\limits_{k \to \infty} J[u_k] < +\infty$, we choose a subsequence $\{u_{k_j}\}$ such that

$$\liminf_{k \to \infty} J[u_k] = \lim_{j \to \infty} J[u_{k_j}]. \tag{2.4}$$

Since $\{u_{k_j}\}$ is weakly convergent to u in $W^{1,2}(D, \mathbb{R}^N)$, $\{u_{k_j}\}$ must be weakly bounded in $W^{1,1}(D, \mathbb{R}^N)$. By the uniform boundedness theorem in Banach spaces, $\{u_{k_j}\}$ is bounded in $W^{1,1}(D, \mathbb{R}^N)$. By the compact embedding theorem, some subsequence of $\{u_{k_j}\}$, still denoted by $\{u_{k_j}\}$, converges strongly in $L^1(D, \mathbb{R}^N)$. Therefore, a further subsequence, still denoted by $\{u_{k_j}\}$, converges almost everywhere in D:

$$u_{k_j}(x) \to u(x), \quad \text{a.e. } x \in D.$$

Using the Egorov theorem[1] and the Lusin theorem[2], we deduce that, for any $\varepsilon > 0$, there exists a compact subset $K \subset D$ such that $\operatorname{meas}(D \setminus K) < \varepsilon$, and

(i) $u_{k_j} \to u$ uniformly on K,

(ii) u and Du are continuous on K.

---

[1]**Egorov theorem.** *Let E be a Lebesgue measurable set in $\mathbb{R}^n$ with* $\operatorname{meas} E < \infty$. *Suppose that $f(x)$ and $f_k(x)$ $(k = 1, 2, \cdots)$ are measurable functions which are finite almost everywhere. If $f_k(x)$ converges to $f(x)$ almost everywhere on E, then for any $\varepsilon > 0$, there exists a measurable subset B of E with* $\operatorname{meas}(E \setminus B) < \varepsilon$, *such that $\{f_k(x)\}$ converges uniformly to $f(x)$ on B.*

[2]**Lusin theorem.** *Let E be a Lebesgue measurable set in $\mathbb{R}^n$ with* $\operatorname{meas} E < \infty$. *If $f(x)$ is a measurable function which is finite almost everywhere on E, then for any $\varepsilon > 0$, there exists a compact subset C of E with* $\operatorname{meas}(E \setminus C) < \varepsilon$, *such that $f(x)$ is continuous on C.*

Next, using the convexity of F in p, we obtain

$$\begin{aligned}&\int_K F(x,u_{k_j},Du_{k_j})dx\\ &\quad\geqslant \int_K F(x,u_{k_j},Du)dx+\int_K F_{p^i_\alpha}(x,u_{k_j},Du)(D_\alpha u^i_{k_j}-D_\alpha u^i)dx.\end{aligned}$$

We rewrite this inequality as

$$\begin{aligned}&\int_K F(x,u_{k_j},Du_{k_j})dx\\ &\quad\geqslant \int_K F(x,u_{k_j},Du)dx+\int_K F_{p^i_\alpha}(x,u,Du)(D_\alpha u^i_{k_j}-D_\alpha u^i)dx\\ &\qquad+\int_K \Big[F_{p^i_\alpha}(x,u_{k_j},Du)-F_{p^i_\alpha}(x,u,Du)\Big](D_\alpha u^i_{k_j}-D_\alpha u^i)dx.\end{aligned}\tag{2.5}$$

Since F is continuous and u_{k_j} converges to u uniformly on K, we have

$$\int_K F(x,u_{k_j},Du)dx\to\int_K F(x,u,Du)dx\quad (k_j\to\infty).$$

Since $F_p(x,u(x),Du(x))$ is bounded on K and Du_{k_j} converges weakly to Du in $L^1(D,\mathbb{R}^N)$, we have

$$\int_K F_{p^i_\alpha}(x,u,Du)(D_\alpha u^i_{k_j}-D_\alpha u^i)dx\to 0\quad (k_j\to\infty).$$

Since $F_{p^i_\alpha}(x,u_{k_j}(x),Du(x))\to F_{p^i_\alpha}(x,u,Du(x))$ uniformly on K and the L^1 norm of $D_\alpha u^i_{k_j}-D_\alpha u^i$ is bounded, we have

$$\int_K \Big[F_{p^i_\alpha}(x,u_{k_j},Du)-F_{p^i_\alpha}(x,u,Du)\Big](D_\alpha u^i_{k_j}-D_\alpha u^i)dx\to 0\quad (k_j\to\infty).$$

Now we can let $k_j\to\infty$ in (2.5), and conclude that

$$\liminf_{k_j\to\infty}\int_K F(x,u_{k_j},Du_{k_j})dx\geqslant\int_K F(x,u,Du)dx.$$

Since $F\geqslant 0$ and $K\subset D\subset\subset\Omega$, we get

$$\liminf_{k_j\to\infty}\int_\Omega F(x,u_{k_j},Du_{k_j})dx\geqslant\int_\Omega F(x,u,Du)\chi_K(x)dx,$$

where $\chi_K(x)$ is the characteristic function on K. Notice that the left-hand side of the above inequality is independent of K. Letting $\varepsilon=1/m\to 0$ $(m\to\infty)$, we can find a sequence of compact sets $K_m\subset D$ such that $\operatorname{meas}(D\setminus K_m)\to 0$ $(m\to\infty)$. Replacing K_m by $\bigcup_{i=1}^m K_i$ if necessary, we may assume without loss of generality that $\{K_m\}$ is monotonically increasing. By the monotone convergence theorem for Lebesgue integrals,

$$\liminf_{k_j\to\infty}\int_\Omega F(x,u_{k_j},Du_{k_j})dx\geqslant\int_D F(x,u,Du)dx.$$

This inequality is valid for all $D\subset\subset\Omega$. It follows that

$$\liminf_{k_j\to\infty}\int_\Omega F(x,u_{k_j},Du_{k_j})dx\geqslant\int_\Omega F(x,u,Du)dx.$$

Using (2.4), we get (2.3).

Step 2. Establish the existence of a solution for the variational problem (2.2).

Under our assumptions, $J[u] \geqslant 0$. Since $J[u]$ is bounded below in $\mathscr{K}$, there is a greatest lower bound μ. Since $J[g] < \infty$, we have $\mu < \infty$.

Let $\{u_k\}$ be a minimizing sequence, i.e., $u_k \in \mathscr{K}$, and

$$\lim_{k\to\infty} J[u_k] = \mu.$$

Using the assumption, we have, for sufficiently large k,

$$\lambda \int_\Omega |Du_k|^2 dx \leqslant J[u_k] \leqslant \mu + 1. \tag{2.6}$$

Clearly,

$$\int_\Omega |u_k|^2 dx \leqslant 2\left\{ \int_\Omega |u_k - g|^2 dx + \int_\Omega |g|^2 dx \right\}.$$

By Poincaré's inequality,

$$\int_\Omega |u_k - g|^2 dx \leqslant C \int_\Omega |D(u_k - g)|^2 dx.$$

It follows that

$$\begin{aligned} \int_\Omega |u_k|^2 dx &\leqslant C \int_\Omega |D(u_k - g)|^2 dx + C \int_\Omega |g|^2 dx \\ &\leqslant C \int_\Omega |Du_k|^2 dx + C\|g\|^2_{H^1(\Omega,\mathbb{R}^N)}. \end{aligned} \tag{2.7}$$

Combining (2.6) and (2.7), we obtain

$$\|u_k\|_{H^1(\Omega,\mathbb{R}^N)} \leqslant C,$$

where C depends on $n, \lambda, \mu, \Omega, \|g\|^2_{H^1(\Omega,\mathbb{R}^N)}$ and is independent of k. Therefore there is a subsequence $\{u_{k_j}\}$ which converges weakly to some function $u \in H^1(\Omega, \mathbb{R}^N)$. It is clear that $u \in \mathscr{K}$. Using the weakly sequential lower semicontinuity of J (Step 1), we conclude that

$$J[u] \leqslant \liminf_{k_j\to\infty} J[u_{k_j}] = \lim_{k_j\to\infty} J[u_{k_j}] = \mu;$$

therefore,

$$J[u] = \mu;$$

i.e., u is a minimum point of J in $\mathscr{K}$. □

In Theorem 2.1, the assumption on F can be relaxed. A better result in this area is

Theorem 2.2 (E. Acerbi, N Fusco). *Suppose that for any $(u,p) \in \mathbb{R}^N \times \mathbb{R}^{nN}$ the function $F(x,u,p) : \Omega \times \mathbb{R}^N \times \mathbb{R}^{nN} \to \mathbb{R}$ is measurable in x, for almost every $x \in \Omega$ it is continuous in (u,p), and it is* quasi-convex, *i.e., for a.e. $x^0 \in \Omega$, all $u_0 \in \mathbb{R}^N$ and all $p_0 \in \mathbb{R}^{nN}$, we have*

$$\int_\Omega F(x^0, u_0, p_0) dx \leqslant \int_\Omega F(x^0, u_0, p_0 + D\varphi) dx, \quad \forall \varphi \in C_0^\infty(\Omega, \mathbb{R}^N),$$

and

$$\lambda |p|^s \leqslant F(x,u,p) \leqslant \Lambda(|p|^2 + 1)^{s/2},$$

where $s \geqslant 1$, $\lambda > 0$, $\Lambda > 0$. *Under these assumptions, the functional*

$$J[u] = \int_\Omega F(x, u, Du)dx$$

admits a minimum point in the set

$$\mathscr{K} = \left\{ u \in W^{1,s}(\Omega, \mathbb{R}^N) \,\middle|\, u - g \in W_0^{1,s}(\Omega, \mathbb{R}^N) \right\},$$

where $g \in W^{1,s}(\Omega, \mathbb{R}^N)$.

We refer readers to [AF1] for a proof.

2.2. The relationship between variational problems and boundary value problems for elliptic systems.

In this subsection, we study the relationship between a regular variational problem in the classical calculus of variations and the corresponding boundary value problem for an elliptic system. We still use

$$J[u] = \int_\Omega F(x, u, Du)dx$$

for the integral functional. In the classical calculus of variations, we assume that $F \in C^1(\overline{\Omega} \times \mathbb{R}^N \times \mathbb{R}^{nN})$ and take

$$\mathscr{C} = \left\{ u \in C^1(\overline{\Omega}, \mathbb{R}^N) \,\middle|\, u = g \quad \text{on } \partial\Omega \right\} \tag{2.8}$$

as the admissible class, where $g \in C^1(\overline{\Omega})$.

2.2.1. *First necessary condition.*

We know that, if $u \in \mathscr{C}$ is such that

$$J[u] \leqslant J[v], \quad \forall v \in \mathscr{C},$$

then

$$\frac{d}{dt} J[u + t\varphi]\Big|_{t=0} = 0, \quad \forall \varphi \in C_0^1(\Omega, \mathbb{R}^N). \tag{2.9}$$

It follows that

$$\int_\Omega \Big[F_{p_\alpha^i}(x, u, Du) D_\alpha \varphi^i + F_{u^i}(x, u, Du)\varphi^i \Big] dx = 0, \quad \forall \varphi \in C_0^1(\Omega, \mathbb{R}^N). \tag{2.10}$$

(2.10) is said to be the *weak form of the Euler system* of J at the point u.

If we further assume that $F, u, \partial\Omega$ all belong to C^2, then we can integrate by parts in (2.10) to obtain

$$\int_\Omega \Big[-D_\alpha F_{p_\alpha^i}(x, u, Du) + F_{u^i}(x, u, Du) \Big] \varphi^i dx = 0, \quad \forall \varphi \in C_0^1(\Omega, \mathbb{R}^N).$$

It follows that

$$-D_\alpha F_{p_\alpha^i}(x, u, Du) + F_{u^i}(x, u, Du) = 0 \quad (i = 1, \cdots, N), \quad x \in \Omega; \tag{2.11}$$

i.e.,

$$
(2.11)' \qquad \begin{gathered}
F_{p^i_\alpha p^j_\beta}(x,u,Du)D_\alpha D_\beta u^j + F_{p^i_\alpha u^j}(x,u,Du)D_\alpha u^j \\
+F_{p^i_\alpha x_\alpha}(x,u,Du) - F_{u^i}(x,u,Du) = 0 \\
(i = 1,\cdots,N), \quad x \in \Omega.
\end{gathered}
$$

(2.11) or (2.11)′ is said to be the *Euler system* of J at the point u.

2.2.2. *Second necessary condition.*

Under the assumption that $F \in C^2$, another necessary condition is the so-called Legendre-Hadamard condition; i.e., for a.e. $x \in \Omega$,

$$
(2.12) \qquad F_{p^h_\alpha p^k_\beta}(x,u(x),Du(x))\xi_\alpha\xi_\beta\eta^h\eta^k \geqslant 0, \quad \forall \xi \in \mathbb{R}^n,\ \eta \in \mathbb{R}^N.
$$

In fact, if $u \in \mathscr{C}$ satisfies

$$
J[u] \leqslant J[v], \quad \forall v \in \mathscr{C},
$$

then in addition to (2.9), we also have

$$
\frac{d^2}{dt^2}J[u+t\varphi]\Big|_{t=0} \geqslant 0, \quad \forall \varphi \in C_0^\infty(\Omega,\mathbb{R}^N).
$$

A simple computation shows that

$$
(2.13) \qquad \begin{aligned}
\int_\Omega \Big\{ & F_{p^h_\alpha p^k_\beta}(x,u,Du)D_\alpha\varphi^h D_\beta\varphi^k + 2F_{p^h_\alpha u^k}(x,u,Du)\varphi^k D_\alpha\varphi^h \\
& +F_{u^h u^k}(x,u,Du)\varphi^k\varphi^h\Big\}dx \geqslant 0, \qquad \forall \varphi \in C_0^\infty(\Omega,\mathbb{R}^N).
\end{aligned}
$$

For a complex valued function $\varphi = \varphi_1 + i\varphi_2$, where $\varphi_l \in C_0^\infty(\Omega,\mathbb{R}^N)$, $l = 1,2$, we can apply the above inequality to φ_1 and φ_2 to obtain

$$
(2.14) \qquad \begin{aligned}
\operatorname{Re}\int_\Omega \Big\{ & F_{p^h_\alpha p^k_\beta}(x,u,Du)\overline{D_\beta\varphi^k} D_\alpha\varphi^h + 2F_{p^h_\alpha u^k}(x,u,Du)\overline{\varphi^k} D_\alpha\varphi^h \\
& +F_{u^h u^k}(x,u,Du)\overline{\varphi^k}\varphi^h\Big\}dx \geqslant 0.
\end{aligned}
$$

We choose $\varphi^h(x) = \eta^h\phi(x)e^{\tau i x\cdot\xi}$ $(h = 1,\cdots,N)$, where $\xi \in \mathbb{R}^n$, $\eta \in \mathbb{R}^N$, $\phi \in C_0^\infty(\Omega)$. After a simple calculation, we find that

$$
\begin{aligned}
\int_\Omega \Big\{ & F_{p^h_\alpha p^k_\beta}(x,u,Du)(D_\alpha\phi \cdot D_\beta\phi + \tau^2\xi_\alpha\xi_\beta\phi^2)\eta^h\eta^k \\
& +2F_{p^h_\alpha u^k}(x,u,Du)\eta^h\eta^k\phi D_\alpha\phi + F_{u^h u^k}(x,u,Du)\eta^h\eta^k\phi^2\Big\}dx \geqslant 0.
\end{aligned}
$$

Dividing by τ^2 and then letting $\tau \to \infty$, we get

$$
(2.15) \qquad \int_\Omega F_{p^h_\alpha p^k_\beta}(x,u,Du)\xi_\alpha\xi_\beta\eta^h\eta^k\phi^2(x)dx \geqslant 0, \quad \forall \phi \in C_0^\infty(\Omega).
$$

Now we choose ϕ as in the proof of Theorem 1.2 in Chapter 8, and obtain (2.12).

2.2.3. *Conclusions.*

From the above discussion, we see that if u is a minimum point of J, then u satisfies the elliptic system (2.10); here, ellipticity means that the Legendre-Hadamard condition (2.12) is satisfied. Thus the existence of a solution of the elliptic system (2.10) is reduced to the existence of a minimum point of the functional J; this is exactly the idea of the variational method.

In order to show the existence of a minimum point of J, we assumed that F was convex in p in Theorem 2.1. This assumption is relaxed in Theorem 2.2 to the assumption that F is quasi-convex. The concept of quasi-convexity (cf. Theorem 2.2) was introduced by C. B. Morrey in 1952 ([MR2]). He pointed out that, for $N \geqslant 1$, a necessary condition for the existence of a minimum point is that F is quasi-convex, not convex (for $N = 1$, quasi-convex is equivalent to convex; for $N > 1$, quasi-convex is weaker than convex in general). One can show that when $F \in C^2$, if F is quasi-convex, then the Legendre-Hadamard condition (2.12) is satisfied. Therefore, it is more natural to call any integral functional satisfying the Legendre-Hadamard condition a regular integral functional, not just those satisfying the Legendre condition: $F_{p^i_\alpha p^j_\beta}\xi^i_\alpha\xi^j_\beta \geqslant 0$ (i.e., F is convex in p). However, in the discussion of regularity of weak solutions of elliptic systems, we shall often impose the Legendre condition, or even the strong Legendre condition:

$$F_{p^i_\alpha p^j_\beta}\xi^i_\alpha\xi^j_\beta \geqslant \lambda|\xi|^2, \quad \lambda > 0,\ \forall\xi \in \mathbb{R}^{nN}.$$

The Legendre-Hadamard condition, or even the strong Legendre-Hadamard condition

$$F_{p^i_\alpha p^j_\beta}\xi_\alpha\xi_\beta\eta^i\eta^j \geqslant \lambda|\xi|^2|\eta|^2, \quad \lambda > 0,\ \forall\xi \in \mathbb{R}^n,\ \eta \in \mathbb{R}^N,$$

will not be enough for the regularity results in Chapter 12. For more discussion, we refer readers to [GS], [FU].

Of course, when $N = 1$ (single equation), the Legendre-Hadamard condition is equivalent to the Legendre condition.

2.3. Differentiability of regular integral functionals in H^1.

In subsection 2.2, we studied the existence of the first order variation of the functional

$$J[u] = \int_\Omega F(x, u, Du)dx$$

at the point $u \in C^1(\overline{\Omega}, \mathbb{R}^N)$ (from now on, it will simply be called the differentiability of J); at the same time, we derived the Euler system of J at u. However, J need not have a minimum point in $C^1(\overline{\Omega}, \mathbb{R}^N)$ in general. We have proved in subsection 2.1 that, under appropriate assumptions, J has a minimum point in a closed convex subset $\mathscr{K}$ of $H^1(\Omega, \mathbb{R}^N)$. Now we want to study the differentiability of $J[u]$ at a minimum point $u \in H^1(\Omega, \mathbb{R}^N)$ and to derive the corresponding Euler system. It should be pointed out that, in order for the functional J to be differentiable, not only is some differentiability of F necessary, but some growth condition on F is required as well.

Theorem 2.3. *Suppose that*

1° $F:\Omega\times\mathbb{R}^N\times\mathbb{R}^{nN}\to\mathbb{R}$ *is measurable in* x, *and for a.e.* $x\in\Omega$, *it is* C^1 *in* (u,p),

2° $\lambda V^2-g_1(x)\leqslant F(x,u,p)\leqslant\Lambda V^2+g_2(x)$, *where* λ,Λ *are positive constants,* $V=(1+|u|^2+|p|^2)^{1/2}$, $g_i(x)\geqslant 0$, $g_i\in L^1(\Omega)$ $(i=1,2)$, *and*

$$\left\{\begin{array}{l}|F_{p^i_\alpha}(x,u,p)|\leqslant C(|p|+|u|^{r/2}+f_i^\alpha(x)),\\ |F_{x^i}(x,u,p)|\leqslant C(|p|^{2(1-1/r)}+|u|^{r-1}+f_i(x)),\end{array}\right. \tag{2.16}$$

where $f_i^\alpha, f_i\geqslant 0$, $f_i^\alpha\in L^2(\Omega)$, $f_i\in L^{r/(r-1)}(\Omega)$ $(i=1,\cdots,N;\ \alpha=1,\cdots,n)$, *and* $r=2^*$ *is the Sobolev conjugate of* 2.

Under these assumptions, $J[u]$ *is differentiable at* $u\in H^1(\Omega,\mathbb{R}^N)$, *and its first order variation is given by*

$$\varphi\in H_0^1(\Omega,\mathbb{R}^N)\to\int_\Omega\Big\{F_{p^i_\alpha}(x,u,Du)D_\alpha\varphi^i+F_{u^i}(x,u,Du)\varphi^i\Big\}dx. \tag{2.17}$$

Proof. Let $u\in H^1(\Omega,\mathbb{R}^N)$ be a minimum point of $J[u]$ on some subset $\mathscr{K}$ of $H^1(\Omega,\mathbb{R}^N)$. Suppose also that $u+t\varphi\in\mathscr{K}$ for all $\varphi\in H_0^1(\Omega,\mathbb{R}^N)$ and all $t\in\mathbb{R}$. The function

$$j(t)\triangleq J[u+t\varphi]$$

has its difference quotient at $t=0$:

$$\begin{aligned}\frac{j(t)-j(0)}{t}&=\frac{1}{t}\int_\Omega\Big[F(x,u+t\varphi,Du+tD\varphi)-F(x,u,Du)\Big]dx\\ &=\frac{1}{t}\int_\Omega dx\int_0^1\frac{d}{d\xi}F(x,u+\xi t\varphi,Du+\xi tD\varphi)d\xi\\ &=\int_\Omega dx\int_0^1\Big[F_{p^i_\alpha}(x,u+\xi t\varphi,Du+\xi tD\varphi)D_\alpha\varphi^i\\ &\qquad+F_{u^i}(x,u+\xi t\varphi,Du+\xi tD\varphi)\varphi^i\Big]d\xi.\end{aligned}$$

The question now is whether this difference quotient has a limit as $t\to 0$. Set

$$u^t(x)=u(x)+\xi t\varphi(x).$$

Then it is clear that $u^t(x)\to u(x)$ and $Du^t(x)\to Du(x)$ (a.e. $x\in\Omega$) as $t\to 0$. Therefore, as $t\to 0$,

$$\begin{aligned}&F_{u^i}\Big(x,u^t(x),Du^t(x)\Big)\to F_{u^i}\Big(x,u(x),Du(x)\Big),\quad\text{a.e. } x\in\Omega,\\ &F_{p^i_\alpha}\Big(x,u^t(x),Du^t(x)\Big)\to F_{p^i_\alpha}\Big(x,u(x),Du(x)\Big),\quad\text{a.e. } x\in\Omega.\end{aligned}$$

Since $F_{p^i_\alpha}$ and F_{u^i} satisfy the controllable growth condition (2.16), we can apply the Lebesgue dominated convergence theorem to conclude that

$$\lim_{t\to 0}\frac{j(t)-j(0)}{t}=\int_\Omega\Big[F_{p^i_\alpha}(x,u,Du)D_\alpha\varphi^i+F_{u^i}(x,u,Du)\varphi^i\Big]dx. \tag{2.18}$$

The proof is complete. □

Under the natural growth condition, we can also study the differentiability of $J[u]$ at $u\in H^1\cap L^\infty(\Omega,\mathbb{R}^N)$. Analogously to Theorem 2.3, we have

Theorem 2.4. *Suppose that*

1° $F : \Omega \times \mathbb{R}^N \times \mathbb{R}^{nN} \to \mathbb{R}$ *is measurable in* x, *and for a.e.* $x \in \Omega$, *it is* C^1 *in* (u, p).

2° *For* $|u| \leqslant M$, *we have*

$$\lambda|p|^2 - \mu \leqslant F(x, u, p) \leqslant \Lambda(1 + |p|^2),$$

$$\left\{ \begin{array}{l} |F_{p^i_\alpha}(x, u, p)| \leqslant \Lambda(1 + |p|), \\ |F_{x^i}(x, u, p)| \leqslant \Lambda(1 + |p|^2), \end{array} \right. \tag{2.19}$$

where λ, μ, Λ *are constants;* Λ *may depend on* M.

Under these assumptions, $J[u]$ *is differentiable at* $u \in H^1 \cap L^\infty(\Omega, \mathbb{R}^N)$, *and its first order variation is given by*

$$\varphi \in H^1_0 \cap L^\infty(\Omega, \mathbb{R}^N) \to \int_\Omega \left\{ F_{p^i_\alpha}(x, u, Du) D_\alpha \varphi^i + F_{u^i}(x, u, Du)\varphi^i \right\} dx.$$

We refer readers to [MR4] for a proof.

Therefore, it is reasonable for us to define a weak solution of the Euler system of J as follows.

Definition 2.1. Under the controllable growth condition (or, more precisely, if F satisfies the assumptions of Theorem 2.3), if $u \in H^1(\Omega, \mathbb{R}^N)$ satisfies

$$\int_\Omega \left\{ F_{p^i_\alpha}(x, u, Du) D_\alpha \varphi^i + F_{u^i}(x, u, Du)\varphi^i \right\} dx = 0, \quad \forall \varphi \in H^1_0(\Omega, \mathbb{R}^N), \tag{2.20}$$

then u is said to be a *weak solution* of the Euler system of $J[u]$.

Under the natural growth condition (or, more precisely, if F satisfies the assumptions of Theorem 2.4), if $u \in H^1 \cap L^\infty(\Omega, \mathbb{R}^N)$ satisfies (2.20) for all $\varphi \in H^1_0 \cap L^\infty(\Omega, \mathbb{R}^N)$, then u is said to be a *weak solution* of the Euler system of $J[u]$.

(2.20) is a weak form of the Euler system of J at u. If J is a regular integral functional, then (2.20) is an *elliptic system* in the sense that the Legendre condition holds.

If an elliptic system is the Euler system of some regular integral functional, then Theorems 2.1 and 2.2 give the conditions under which a weak solution exists. If this elliptic system is not the Euler system of any regular integral functional, we refer readers to [NC] and [ZK] for some existence results.

CHAPTER 12

Regularity for Weak Solutions of Nonlinear Elliptic Systems

1. H^2 regularity

In this chapter, we study the regularity of weak solutions of the elliptic system of divergence form

$$-D_\alpha A_i^\alpha(x,u,Du)+B_i(x,u,Du)=0 \qquad (i=1,\cdots,N). \tag{1.1}$$

The meaning of *ellipticity* will be explained later.

Just as with a linear system, we start with H^2 regularity. An interesting phenomenon is that H^2 regularity results depend on what structure condition is imposed on A_i^α, B_i.

In order to study the regularity, we introduce the following two sets of structure conditions. They are often referred to as the controllable and natural structure conditions, respectively.

For simplicity, we set $A=(A_i^\alpha)$, $B=(B_i)$.

First, if $A,B\in C^1$ satisfy

$$A_{i\,p_\beta^j}^\alpha(x,u,p)\xi_\alpha^i\xi_\beta^j\geqslant\lambda|\xi|^2, \tag{1.2}$$

$$\begin{cases}|A|,|A_x|,|B|,|B_x|\leqslant\Lambda(1+|u|^2+|p|^2)^{1/2},\\|A_u|,|B_u|,|A_p|,|B_p|\leqslant\Lambda,\end{cases} \tag{1.3}$$

where λ,Λ are positive constants, then A,B are said to satisfy the *controllable structure conditions.* (1.3) is a *controllable growth condition.*

Next, suppose that $A,B\in C^1$ satisfy, for $|u|\leqslant M$,

$$A_{i\,p_\beta^j}^\alpha(x,u,p)\xi_\alpha^i\xi_\beta^j\geqslant\lambda|\xi|^2, \tag{1.4}$$

$$\begin{cases}|A|,|A_x|,|A_u|\leqslant\Lambda(1+|p|),\\|A_p|\leqslant\Lambda,\\|B|,|B_x|,|B_u|\leqslant\Lambda(1+|p|^2),\\|B_p|\leqslant\Lambda(1+|p|),\end{cases} \tag{1.5}$$

where λ,Λ are positive constants, which may depend on M. In this case, A,B are said to satisfy the *natural structure conditions.* (1.5) is a *natural growth condition.*

Obviously, the growth conditions (1.3) and (1.5) are stronger than (1.3) and (1.6) of Chapter 11, respectively.

Theorem 1.1. *Let $A, B \in C^1$ satisfy the controllable structure conditions (1.2), (1.3). If $u \in H^1(\Omega, \mathbb{R}^N)$ satisfies*

$$\int_\Omega \left[A_i^\alpha(x,u,Du)D_\alpha\varphi^i + B_i(x,u,Du)\varphi^i\right]dx = 0, \quad \forall\varphi \in H_0^1(\Omega,\mathbb{R}^N), \tag{1.6}$$

then $u \in H^2_{loc}(\Omega,\mathbb{R}^N)$; furthermore, the derivatives $D_s u$ $(s = 1, \cdots, n)$ satisfy

$$\begin{aligned} \int_\Omega \Big[& A^\alpha_{i\,p^j_\beta}(x,u,Du)D_\beta D_s u^j + A^\alpha_{i\,u^j}(x,u,Du)D_s u^j \\ & + A^\alpha_{i\,x_s}(x,u,Du) - \delta_{\alpha s}B_i(x,u,Du)\Big]D_\alpha\varphi^i dx = 0, \\ & \forall\varphi \in H_0^1(\Omega,\mathbb{R}^N), \ \operatorname{spt}\varphi \subset \Omega, \end{aligned} \tag{1.7}$$

where $\delta_{\alpha s} = 1$ if $\alpha = s$, $\delta_{\alpha s} = 0$ if $\alpha \neq s$.

Proof. We first consider the simple case $A_i^\alpha(x,u,p) = A_i^\alpha(p)$, $B_i(x,u,p) \equiv 0$. In this case, (1.6) reduces to

$$\int_\Omega A_i^\alpha(Du)D_\alpha\varphi^i dx = 0, \quad \forall\varphi \in H_0^1(\Omega,\mathbb{R}^N). \tag{1.6$'$}$$

Suppose that the support $\operatorname{spt}\varphi \subset \Omega$. Then for sufficiently small $|h|$,

$$\int_\Omega \left[A_i^\alpha(Du(x+he_s)) - A_i^\alpha(Du)\right]D_\alpha\varphi^i dx = 0.$$

It follows that

$$\int_\Omega A^{\alpha\beta}_{ij(h)}(x)D_\beta\Delta_{h,s}u^j D_\alpha\varphi^i dx = 0, \quad \forall\varphi \in H_0^1(\Omega,\mathbb{R}^N), \ \operatorname{spt}\varphi \subset \Omega, \tag{1.8}$$

where

$$\begin{aligned} A^{\alpha\beta}_{ij(h)}(x) &= \int_0^1 A^\alpha_{i\,p^j_\beta}(tDu(x+he_s) + (1-t)Du(x))dt, \\ \Delta_{h,s}u^j &= \frac{u^j(x+he_s) - u^j(x)}{h}. \end{aligned}$$

Since the A_i^α satisfy (1.2) and (1.3), the $A^{\alpha\beta}_{ij(h)}$ satisfy

$$\begin{cases} A^{\alpha\beta}_{ij(h)}\xi^i_\alpha\xi^j_\beta \geqslant \lambda|\xi|^2, & \lambda > 0, \\ |A^{\alpha\beta}_{ij(h)}| \leqslant \Lambda. \end{cases} \tag{1.9}$$

Now, for all $\widetilde{\Omega} \subset\subset \Omega$, all $x^0 \in \widetilde{\Omega}$, and all R, $0 < R < \frac{1}{2}\operatorname{dist}(\widetilde{\Omega}, \partial\Omega)$, we choose $\varphi = \eta^2\Delta_{h,s}u$ in (1.8), where $\eta \in C_0^\infty(B_R(x^0))$ is a cutoff function:

$$0 \leqslant \eta \leqslant 1 \quad \text{in } B_R(x^0), \quad \eta \equiv 1 \quad \text{in } B_{R/2}(x^0), \quad |D\eta| \leqslant \frac{C}{R}. \tag{1.10}$$

After a simple computation, using also (1.9), we obtain

$$\int_{B_{R/2}(x^0)} |D\Delta_{h,s}u|^2 dx \leqslant C\|u\|^2_{H^1(\Omega,\mathbb{R}^N)}, \tag{1.11}$$

where C depends on n, N, λ, Λ and $\operatorname{dist}(\widetilde{\Omega}, \partial\Omega)$. Therefore

$$u \in H^2_{loc}(\Omega,\mathbb{R}^N).$$

Letting $h \to 0$ in (1.8), we get

$$\int_\Omega A_{i\,p_\beta^j}^\alpha(Du) D_\beta(D_s u^j) D_\alpha \varphi^i dx = 0, \quad \forall \varphi \in H_0^1(\Omega, \mathbb{R}^N),\ \operatorname{spt}\varphi \subset \Omega,$$

where $s = 1, \cdots, n$. This is exactly the integral identity (1.7) in this simple case.

In the above process, if we formally use derivatives instead of difference quotient, the computation becomes much simpler. In fact, if in $(1.6)'$ we choose $\varphi = D_s\phi$ and formally integrate by parts, we immediately obtain

$$\int_\Omega A_{i\,p_\beta^j}^\alpha(Du) D_\beta(D_s u^j) D_\alpha \phi^i dx = 0.$$

Now we choose $\phi = \eta^2 D_s u$. After a simple computation, using (1.2) and (1.3), we obtain

$$\int_{B_{R/2}(x^0)} |DD_s u|^2 dx \leqslant C\|u\|^2_{H^1(\Omega, \mathbb{R}^N)},$$

where C depends on n, N, λ, Λ and $\operatorname{dist}(\widetilde{\Omega}, \partial\Omega)$. Finally, we replace $D_s u$ in the above inequality by $\Delta_{h,s} u$ and obtain (1.11) immediately. The argument is rigorous now.

For simplicity, we shall use this method to deal with the general case.

First, we formally choose $\varphi = D_s\phi$ in (1.6) and integrate by parts to obtain

$$\begin{aligned}\int_\Omega \Big\{&[A_{i\,x_s}^\alpha + A_{i\,u^j}^\alpha D_s u^j + A_{i\,p_\beta^j}^\alpha D_s D_\beta u^j] D_\alpha \phi^i \\ &+[B_{i x_s} + B_{i u^j} D_s u^j + B_{i p_\beta^j} D_s D_\beta u^j]\phi^i\Big\} dx = 0.\end{aligned}$$

Next, for all $\widetilde{\Omega} \subset\subset \Omega$, all $x^0 \in \widetilde{\Omega}$, and all R, $0 < R < \frac{1}{2}\operatorname{dist}(\widetilde{\Omega}, \partial\Omega)$, we choose $\phi = \eta^2 D_s u$, where $\eta \in C_0^\infty(B_R(x^0))$ satisfies (1.10). Then

$$\begin{aligned}&\int_\Omega \eta^2 A_{i\,p_\beta^j}^\alpha D_\beta D_s u^j D_\alpha D_s u^i dx + \int_\Omega A_{i\,p_\beta^j}^\alpha D_\beta D_s u^j \cdot 2\eta D_\alpha\eta \cdot D_s u^i dx \\ &\quad + \int_\Omega \eta^2 A_{i\,x_s}^\alpha D_\alpha D_s u^i dx + \int_\Omega A_{i\,x_s}^\alpha \cdot 2\eta D_\alpha \eta \cdot D_s u^i dx \\ &\quad + \int_\Omega \eta^2 A_{i\,u^j}^\alpha D_s u^j D_\alpha D_s u^i dx + \int_\Omega A_{i\,u^j}^\alpha \cdot 2\eta D_\alpha\eta \cdot D_s u^j D_s u^i dx \\ &\quad + \int_\Omega \eta^2 B_{i x_s} D_s u^i dx + \int_\Omega \eta^2 B_{i u^j} D_s u^j D_s u^i dx \\ &\quad + \int_\Omega \eta^2 B_{i p_\beta^j} D_\beta D_s u^j D_s u^i dx \ = \ 0.\end{aligned} \tag{1.12}$$

Through a simple calculation, using (1.2) and (1.3), we get

$$\int_{B_R(x^0)} |\eta DD_s u|^2 dx \leqslant C \int_{B_R(x^0)} (1 + |D\eta|^2)(1 + |Du|^2) dx;$$

therefore

$$\int_{B_{R/2}(x^0)} |DD_s u|^2 dx \leqslant C\Big(\|u\|^2_{H^1(\Omega,\mathbb{R}^N)} + 1\Big),$$

where C depends on n, N, λ, Λ and $\operatorname{dist}(\widetilde{\Omega}, \partial\Omega)$.

Now, we replace $D_s u$ in the above inequality by $\Delta_{h,s}u$. Then the above procedure can be made rigorous, and we derive

$$\int_{B_{R/2}(x^0)} |D\Delta_{h,s}u|^2 dx \leqslant C\Big(\|u\|^2_{H^1(\Omega,\mathbb{R}^N)} + 1\Big),$$

where C depends on the same quantities and is independent of h. Therefore $u \in H^2_{loc}(\Omega, \mathbb{R}^N)$, and the $D_s u$ satisfy (1.7). The proof is complete. □

In general, the above result is not valid under the natural structure conditions. In the latter case, we can only show that a weak solution belongs to H^2 under additional assumptions. We have

Theorem 1.2. *Let $A, B \in C^1$ satisfy the natural structure conditions* (1.4) *and* (1.5). *Suppose that $u \in H^1 \cap L^\infty(\Omega, \mathbb{R}^N)$ satisfies*

$$\int_\Omega \Big[A_i^\alpha(x,u,Du)D_\alpha\varphi^i + B_i(x,u,Du)\varphi^i\Big]dx = 0, \quad \forall \varphi \in H_0^1 \cap L^\infty(\Omega, \mathbb{R}^N), \tag{1.13}$$

and that $u \in C^0(\Omega, \mathbb{R}^N)$. Then $u \in H^2_{loc}(\Omega, \mathbb{R}^N)$; furthermore, the derivatives $D_s u$ $(s = 1, \cdots, n)$ satisfy the integral identity (1.7) *for any $\varphi \in H_0^1 \cap L^\infty(\Omega, \mathbb{R}^N)$,* spt $\varphi \subset \Omega$.

Proof. For all $\widetilde{\Omega} \subset\subset \Omega$, all $x^0 \in \widetilde{\Omega}$, and all R, $0 < R < \frac{1}{2}\operatorname{dist}(\widetilde{\Omega}, \partial\Omega)$, we formally choose, analogously to Theorem 1.1, $\varphi = D_s(\eta^2 D_s u)$ $(s = 1, \cdots, n)$, where $\eta \in C_0^\infty(B_R(x^0))$ satisfies (1.10). After computations, we still get (1.12). However, since the structure conditions are different from those in Theorem 1.1, the final estimate is

$$\begin{aligned}\int_{B_R(x^0)} &|\eta DD_s u|^2 dx \\ \leqslant\ & C\int_{B_R(x^0)} \Big[\eta^2(1+|Du|^2) + |D\eta|^2|Du|^2\Big]dx \\ &+C\int_{B_R(x^0)} \eta^2|D_s u|^2|Du|^2 dx \qquad (s = 1, \cdots, n).\end{aligned} \tag{1.14}$$

In order to deal with the last integral in the above inequality, we formally choose $\varphi = \eta^2(u - u_R)|D_s u|^2$ in (1.13), where $\eta \in C_0^\infty(B_R(x^0))$ satisfies (1.10) as before, and $u_R = \displaystyle⨍_{B_R(x^0)} u dx$ is the average of u on $B_R(x^0)$. Then

$$\begin{aligned}&\int_{B_R(x^0)} \eta^2 A_i^\alpha D_\alpha u^i |D_s u|^2 dx + \int_{B_R(x^0)} 2\eta^2 A_i^\alpha \cdot (u^i - u_R^i) D_s u D_\alpha D_s u dx \\ &+\int_{B_R(x^0)} 2\eta D_\alpha\eta \cdot A_i^\alpha \cdot (u^i - u_R^i)|D_s u|^2 dx \\ &+\int_{B_R(x^0)} \eta^2 B_i \cdot (u^i - u_R^i)|D_s u|^2 dx = 0 \quad (s = 1, \cdots, n).\end{aligned}$$

After a simple computation, using (1.4) and (1.5), we obtain

$$
\begin{aligned}
&\int_{B_R(x^0)} \eta^2 |D_s u|^2 |Du|^2 dx \\
&\leqslant C \int_{B_R(x^0)} (\eta^2 + |D\eta|^2 |u - u_R|^2) |Du|^2 dx \\
&\quad + C \operatorname*{osc}_{B_R(x^0)} u \cdot \int_{B_R(x^0)} \eta^2 \Big\{ |DD_s u|^2 + (1 + |Du|^2) |D_s u|^2 \Big\} dx \\
&\qquad (s = 1, \cdots, n).
\end{aligned} \tag{1.15}
$$

Combining (1.14) and (1.15), we get

$$
\begin{aligned}
&\int_{B_R(x^0)} |\eta DD_s u|^2 dx + \int_{B_R(x^0)} \eta^2 |D_s u|^2 |Du|^2 dx \\
&\leqslant C \int_{B_R(x^0)} \Big[\eta^2 (1 + |Du|^2) + |D\eta|^2 |Du|^2 \Big] dx \\
&\quad + C \int_{B_R(x^0)} (\eta^2 + |D\eta|^2 |u - u_R|^2) |Du|^2 dx \\
&\quad + C \operatorname*{osc}_{B_R(x^0)} u \cdot \int_{B_R(x^0)} \eta^2 \Big\{ |DD_s u|^2 + (1 + |Du|^2) |D_s u|^2 \Big\} dx \\
&\qquad (s = 1, \cdots, n),
\end{aligned} \tag{1.16}
$$

where C depends on n, N, λ, Λ. Since $u \in C^0(\Omega, \mathbb{R}^N)$, for sufficiently small $R > 0$ we have $C \operatorname*{osc}_{B_R(x^0)} u < 1/2$. Therefore,

$$
\int_{B_{R/2}(x^0)} |DD_s u|^2 dx \leqslant C \Big(\|u\|^2_{H^1(\Omega, \mathbb{R}^N)} + 1 \Big),
$$

where C depends on n, N, λ, Λ and $\operatorname{dist}(\widetilde{\Omega}, \partial\Omega)$; λ and Λ may depend on $M = \sup_\Omega |u|$.

Replacing $D_s u$ by $\Delta_{h,s} u$, we can make the above process rigorous and obtain

$$
\int_{B_{R/2}(x^0)} |D\Delta_{h,s} u|^2 dx \leqslant C \Big(\|u\|^2_{H^1(\Omega, \mathbb{R}^N)} + 1 \Big) \quad (s = 1, \cdots, n),
$$

where C is independent of h. Therefore $u \in H^2_{loc}(\Omega, \mathbb{R}^N)$, and the $D_s u$ satisfy (1.7) for any $\varphi \in H^1_0 \cap L^\infty(\Omega, \mathbb{R}^N)$.

Notice that the integral identity (1.7) makes sense. In fact, by summing over s in (1.16), we obtain

$$
\begin{aligned}
&\int_{B_R(x^0)} \eta^2 |Du|^4 dx + \int_{B_R(x^0)} \eta^2 |D^2 u|^2 dx \\
&\leqslant C \int_{B_R(x^0)} \Big[\eta^2 + (\eta^2 + |D\eta|^2) |Du|^2 \Big] dx \\
&< +\infty,
\end{aligned}
$$

provided R is taken to be small enough. The proof is complete. □

2. Further regularity and counterexamples

Now we study further regularity of weak solutions. If A_i^α, B_i are smoother, is the weak solution smoother? In the extreme case, if A_i^α, $B_i \in C^\infty$, is it true that the weak solution u is in C^∞?

This question actually reduces to studying whether the weak solution belongs to $C^{1,\delta}$ $(0 < \delta < 1)$ when A_i^α, B_i belong to C^∞. In order to illustrate this, we examine the following simple elliptic system:

$$-D_\alpha A_i^\alpha(Du) = 0 \quad (i = 1, \cdots, N), \tag{2.1}$$

where we assume that $A_i^\alpha(p) \in C^\infty(\mathbb{R}^{nN})$ and $A^\alpha_{i\,p^j_\beta}\xi^i_\alpha\xi^j_\beta \geqslant \lambda|\xi|^2$, $\lambda > 0$. For the moment, we assume that $u \in C^{1,\delta}(\Omega, \mathbb{R}^N)$ is a weak solution of (2.1) and discuss the following two cases.

(i) Under the controllable structure conditions, it is easy to prove that $u \in C^\infty(\Omega, \mathbb{R}^N)$. In fact, Theorem 1.1 implies that $u \in H^2_{loc}(\Omega, \mathbb{R}^N)$ and the $D_s u$ satisfy

$$\int_\Omega A^\alpha_{i\,p^j_\beta}(Du)D_\beta(D_s u^j)D_\alpha\varphi^i dx = 0, \quad \forall \varphi \in H_0^1(\Omega, \mathbb{R}^N),\ \operatorname{spt}\varphi \subset \Omega. \tag{2.2}$$

Since $A^\alpha_{i\,p^j_\beta} \in C^\infty$ and $u \in C^{1,\delta}$, the coefficient $A^\alpha_{i\,p^j_\beta}(Du(x)) \in C^{0,\delta}$. By the Schauder theory (cf. Chapter 9), $D_s u^j \in C^{1,\delta}$. Therefore $A^\alpha_{i\,p^j_\beta}(Du(x)) \in C^{1,\delta}$. Applying the Schauder theory again, we have $D_s u^j \in C^{2,\delta}$, $\cdots$. Continuing this process, we conclude that $u \in C^\infty(\Omega, \mathbb{R}^N)$.

(ii) Under the natural structure conditions, we can also show that $u \in C^\infty(\Omega, \mathbb{R}^N)$. In fact, since $u \in C^{1,\delta}(\Omega, \mathbb{R}^N)$, $u \in H^1 \cap C^0(\Omega, \mathbb{R}^N)$; therefore, Theorem 1.2 implies that $u \in H^2_{loc}(\Omega, \mathbb{R}^N)$ and the $D_s u$ satisfy the integral identity (1.7) for any $\varphi \in H_0^1 \cap L^\infty(\Omega, \mathbb{R}^N)$ with $\operatorname{spt}\varphi \subset\subset \Omega$. As in (i), we can apply the Schauder theory to get higher and higher regularity for u. Thus $u \in C^\infty(\Omega, \mathbb{R}^N)$.

For a general elliptic system of divergence form (1.1), a similar result is valid. Therefore, when A_i^α, B_i are smooth, the higher regularity of weak solutions reduces to $C^{1,\delta}$ regularity problems.

In the single equation ($N = 1$) case, when A_i^α, B_i are smooth, the $C^{1,\delta}$ regularity problem for weak solutions was fully resolved by using De Giorgi's theorem in 1957 (cf. §§2–3 of Chapter 4 for his theorems). Ever since 1957, many mathematicians have tried unsuccessfully to generalize De Giorgi's results to a system ($N > 1$). Although various new proofs for the De Giorgi-Nash-Moser theorem were discovered, none of these new proofs can be generalized to the case $N > 1$. In 1968, De Giorgi himself gave a counterexample illustrating that the De Giorgi-Nash-Moser theorem is in fact not valid for $N > 1$.

Example 1 (De Giorgi [DG2], 1968). Let $\Omega = B_1(0) \subset \mathbb{R}^n$, $N = n \geqslant 3$. Consider

$$\int_{B_1(0)} A_{ij}^{\alpha\beta}(x)D_\beta u^j D_\alpha\varphi^i dx = 0, \quad \forall\varphi \in H_0^1(B_1(0), \mathbb{R}^n), \tag{2.3}$$

where

$$A_{ij}^{\alpha\beta}(x) = \delta_{\alpha\beta}\delta_{ij} + \left[(n-2)\delta_{\alpha i} + n\frac{x_i x_\alpha}{|x|^2}\right] \cdot \left[(n-2)\delta_{\beta j} + n\frac{x_j x_\beta}{|x|^2}\right]$$
$$(i,j = 1,\cdots,n;\ \alpha,\beta = 1,\cdots,n),$$

$$\delta_{\alpha\beta} = \begin{cases} 1, & \alpha = \beta, \\ 0, & \alpha \neq \beta. \end{cases}$$

Obviously, $A_{ij}^{\alpha\beta} \in L^\infty(B_1(0))$, and

$$A_{ij}^{\alpha\beta}(x)\xi_\alpha^i\xi_\beta^j = \sum_{i,\alpha=1}^n (\xi_\alpha^i)^2 + \left\{(n-2)\sum_{\alpha=1}^n \xi_\alpha^\alpha + n\sum_{i,\alpha=1}^n \frac{x_i x_\alpha \xi_\alpha^i}{|x|^2}\right\}^2 \geqslant |\xi|^2.$$

However, the vector valued function $u_\gamma(x) = \dfrac{x}{|x|^\gamma}$ is an unbounded weak solution of (2.3), where

$$\gamma = \frac{n}{2}\left\{1 - [(2n-2)^2 + 1]^{-1/2}\right\}. \tag{2.4}$$

First, $n - 2\gamma > 0$, by (2.4). Therefore

$$\int_{B_1(0)} \left[|u_\gamma|^2 + |Du_\gamma|^2\right]dx \leqslant C\int_0^1 [r^{n-2\gamma+1} + r^{n-2\gamma-1}]dr < +\infty,$$

and hence $u_\gamma \in H^1(B_1(0), \mathbb{R}^n)$.

Next, we verify the integral identity (2.3). In fact, u_γ belongs to $C^2(B_1(0)\backslash\{0\})$ and satisfies in the classical sense the equation

$$-D_\alpha(A_{ij}^{\alpha\beta}(x)D_\beta u_\gamma^j) = 0. \tag{2.3$'$}$$

Multiplying (2.3)$'$ by $\varphi \in C_0^\infty(B_1(0), \mathbb{R}^n)$ and integrating by parts over $B_1(0) \setminus B_\varepsilon(0)$, we obtain

$$\int_{B_1(0)\backslash B_\varepsilon(0)} A_{ij}^{\alpha\beta}(x)D_\beta u_\gamma^j D_\alpha\varphi^i dx - \int_{\Gamma_\varepsilon} \varphi^i A_{ij}^{\alpha\beta}(x)D_\beta u_\gamma^j \cdot \cos(\mathbf{n}, x_\alpha)dS = 0,$$

where $\Gamma_\varepsilon = \{x \in \mathbb{R}^n \mid |x| = \varepsilon\}$. Letting $\varepsilon \to 0$ and noticing that $n - \gamma - 1 > 0$ by (2.4), we get

$$\left|\int_{\Gamma_\varepsilon} \varphi^i A_{ij}^{\alpha\beta}(x)D_\beta u_\gamma^j \cdot \cos(\mathbf{n}, x_\alpha)dS\right| \leqslant C\varepsilon^{n-\gamma-1} \to 0.$$

It follows that

$$\int_{B_1(0)} A_{ij}^{\alpha\beta}(x)D_\beta u_\gamma^j D_\alpha\varphi^i dx = 0, \quad \forall\varphi \in C_0^\infty(B_1(0), \mathbb{R}^n).$$

Finally, $n \geqslant 3$ implies $\gamma > 1$, and thus u is unbounded.

This example shows that the De Giorgi-Nash-Moser theorem on Hölder continuity of weak solutions of second order divergence elliptic equations is not valid for elliptic systems ($N > 1$). Therefore the method in Chapter 5 cannot be used to study $C^{1,\delta}$ regularity of weak solutions of nonlinear divergence elliptic systems.

If, however, the coefficients are sufficiently smooth, does a weak solution of the nonlinear divergence elliptic system have $C^{1,\delta}$ regularity? Example 1 does not answer this question. In 1968, E. Giusti and M. Miranda modified De Giorgi's counterexample and answered this question negatively.

Example 2. (Giusti and Miranda [GD1], 1968). Let $\Omega = B_1(0) \subset \mathbb{R}^n$, $N = n \geqslant 3$. Consider

$$\int_{B_1(0)} A_{ij}^{\alpha\beta}(u) D_\beta u^j D_\alpha \varphi^i dx = 0, \quad \forall \varphi \in H_0^1(B_1(0), \mathbb{R}^n), \tag{2.5}$$

where

$$A_{ij}^{\alpha\beta}(u) = \delta_{\alpha\beta}\delta_{ij} + \left[\delta_{\alpha i} + \frac{4}{n-2}\frac{u^i u^\alpha}{1+|u|^2}\right] \cdot \left[\delta_{\beta j} + \frac{4}{n-2}\frac{u^j u^\beta}{1+|u|^2}\right]$$
$$(i, j = 1, \cdots, n;\ \alpha, \beta = 1, \cdots, n),$$

$$\delta_{\alpha\beta} = \begin{cases} 1, & \alpha = \beta, \\ 0, & \alpha \neq \beta. \end{cases}$$

It is not difficult to verify that $A_{ij}^{\alpha\beta}(u)\xi_\alpha^i \xi_\beta^j \geqslant |\xi|^2$ and $A_{ij}^{\alpha\beta}(u)$ is real analytic in u. However, the vector valued function $u_1(x) = \dfrac{x}{|x|} \in H^1(B_1(0), \mathbb{R}^n)$ is a weak solution of (2.5) which is not continuous.

There is also an example of a quasilinear elliptic system where the coefficients are sufficiently smooth, but the weak solution does not belong to $H^2(\Omega, \mathbb{R}^N)$. It is given by J. Nečas and J. Stará [NS].

Example 3. (Nečas and Stará, 1972). Let $\Omega = B_1(0) \subset \mathbb{R}^n$ and $N = n \geqslant 5$. Consider

$$\int_{B_1(0)} A_{ij}^{\alpha\beta}(x, u) D_\beta u^j D_\alpha \varphi^i dx = 0, \quad \forall \varphi \in H_0^1(B_1(0), \mathbb{R}^n), \tag{2.6}$$

where

$$A_{ij}^{\alpha\beta}(x, u) = \delta_{\alpha\beta}\delta_{ij} + c^2\left[\delta_{\alpha i} + b\frac{u^i u^\alpha |x|^{2\gamma-2}}{1+|u|^2|x|^{2\gamma-2}}\right] \cdot \left[\delta_{\beta j} + b\frac{u^j u^\beta |x|^{2\gamma-2}}{1+|u|^2|x|^{2\gamma-2}}\right],$$

where $\gamma \in [2, n/2)$, $b = 2n/(n-2)$, $c^2 = \gamma(n-\gamma)(n-2)^2/[(n-2\gamma)^2(n-1)^2]$.

Obviously, $A_{ij}^{\alpha\beta}(x, u)\xi_\alpha^i \xi_\beta^j \geqslant |\xi|^2$, the $A_{ij}^{\alpha\beta}(x, u)$ are sufficiently smooth, and $u_\gamma = \dfrac{x}{|x|^\gamma} \in H^1(B_1(0), \mathbb{R}^n)$ is a weak solution of (2.6). However, u_γ does not belong to $H^2(B_1(0), \mathbb{R}^n)$ when $\gamma \geqslant (n-2)/2$.

There are many other examples illustrating that the regularity problem for weak solutions of elliptic systems is much more complicated than that for second order elliptic equations; for more references, see [GQ1].

After 1968, on one hand, starting from C. B. Morrey, efforts were devoted to studying *partial regularity* (i.e., show that the weak solution has $C^{1,\delta}$ regularity on Ω_0, where Ω_0 is an open subset of Ω such that $\text{meas}(\Omega \setminus \Omega_0) = 0$). On the other hand, for many elliptic systems of special form, weak solutions have *everywhere regularity* (see [HB], [HW1], [HW2], [WI1], [WI2] and [GH]).

In §3 and §§4–5, we shall introduce the indirect and direct methods for partial regularity for weak solutions of some divergence elliptic systems, respectively.

3. Indirect method for studying regularity

In this section, we introduce the indirect method for partial regularity of a weak solution of the following divergence quasilinear elliptic system: [1]

$$-D_\alpha(A_{ij}^{\alpha\beta}(x,u)D_\beta u^j) = 0. \tag{3.1}$$

We assume throughout this section that, for $(x,u) \in \Omega \times \mathbb{R}^N$,

$$A_{ij}^{\alpha\beta}(x,u)\xi_\alpha^i\xi_\beta^j \geqslant \lambda|\xi|^2, \quad \lambda > 0, \tag{3.2}$$

$$|A_{ij}^{\alpha\beta}(x,u)| \leqslant \Lambda, \tag{3.3}$$

where λ, Λ are constants.

The main result is

Theorem 3.1. *Suppose that the $A_{ij}^{\alpha\beta}(x,u)$ satisfy (3.2) and (3.3) and are continuous uniformly on $\overline{\Omega} \times \mathbb{R}^N$. If $u \in H_{loc}^1(\Omega, \mathbb{R}^N)$ is a weak solution of (3.1), then there exists an open set $\Omega_0 \subset \Omega$ such that $u \in C^{0,\delta}(\Omega_0, \mathbb{R}^N)$, $\forall 0 < \delta < 1$, and* $\text{meas}(\Omega \setminus \Omega_0) = 0$.

In order to prove this theorem, we need several lemmas.

Lemma 3.2 (Caccioppoli's inequality). *If $u \in H_{loc}^1(\Omega, \mathbb{R}^N)$ is a weak solution of (3.1), where the $A_{ij}^{\alpha\beta}(x,u)$ satisfy (3.2) and (3.3), then for all $x^0 \in \Omega$ and all ρ and R with $0 < \rho < R < \text{dist}(x^0, \partial\Omega)$, we have*

$$\int_{B_\rho(x^0)} |Du|^2 dx \leqslant \frac{C}{(R-\rho)^2} \int_{B_R(x^0)} |u|^2 dx, \tag{3.4}$$

where $C = C(n, N, \lambda, \Lambda)$.

This lemma can be proved by following the same procedure as in the proof in Theorem 2.1 in Chapter 8.

Lemma 3.3. *Let $b_{ij}^{\alpha\beta}$ be constants satisfying $b_{ij}^{\alpha\beta}\xi_\alpha^i\xi_\beta^j \geqslant \lambda|\xi|^2$, $\lambda > 0$, $|b_{ij}^{\alpha\beta}| \leqslant \Lambda$ $(i,j = 1, \cdots, N;\ \alpha, \beta = 1, \cdots, n)$. If $u \in H_{loc}^1 \cap L^2(B_1(0), \mathbb{R}^N)$ satisfies*

$$\int_{B_1(0)} b_{ij}^{\alpha\beta} D_\beta u^j D_\alpha \varphi^i dx = 0, \quad \forall \varphi \in C_0^\infty(B_1(0), \mathbb{R}^N),$$

then there exists $c_0 = c_0(n, N, \lambda, \Lambda) > 1$ such that for all ρ, $0 < \rho < 1$,

$$\Phi(0,\rho) \leqslant c_0\rho^2\Phi(0,1), \tag{3.5}$$

where

$$\Phi(x^0,\rho) = \rho^{-n}\int_{B_\rho(x^0)} |u(x) - u_{x^0,\rho}|^2 dx, \qquad u_{x^0,\rho} = \fint_{B_\rho(x^0)} u(x)dx.$$

The proof is similar to that of (2.5) in Lemma 2.2 in Chapter 9.

[1] For this system, the higher regularity of weak solutions reduces to $C^{0,\delta}$ regularity problems.

Lemma 3.4. *Assume that*

1° $a_{ij(k)}^{\alpha\beta}(x)$ $(i,j=1,\cdots,N;\ \alpha,\beta=1,\cdots,n;\ k=1,2,\cdots)$ *satisfy*

$$a_{ij(k)}^{\alpha\beta}(x)\xi_\alpha^i\xi_\beta^j \geqslant \lambda|\xi|^2, \quad \lambda>0, \tag{3.6}$$

$$a_{ij(k)}^{\alpha\beta} \in L^\infty(B_1(0)),\ |a_{ij(k)}^{\alpha\beta}| \leqslant \Lambda, \tag{3.7}$$

$$\begin{cases} \text{as } k\to\infty,\ a_{ij(k)}^{\alpha\beta}(x)\to a_{ij}^{\alpha\beta}(x),\ \text{a.e. } x\in B_1(0), \\ \text{and the } a_{ij}^{\alpha\beta}(x) \text{ also satisfy (3.6), (3.7);} \end{cases} \tag{3.8}$$

2° $u_k \in H_{loc}^1 \cap L^2(B_1(0),\mathbb{R}^N)$ *satisfy*

$$\begin{aligned} &A_k(u_k,\varphi) \triangleq \int_{B_1(0)} a_{ij(k)}^{\alpha\beta}(x)D_\beta u_k^j D_\alpha\varphi^i dx = 0 \\ &(k=1,2,\cdots), \quad \forall\varphi\in C_0^\infty(B_1(0),\mathbb{R}^N); \end{aligned} \tag{3.9}$$

furthermore, as $k\to\infty$,

$$u_k \rightharpoonup u \quad \text{weakly in } L^2(B_1(0),\mathbb{R}^N). \tag{3.10}$$

Under these assumptions, we have $u\in H_{loc}^1(B_1(0),\mathbb{R}^N)$, *and there exists a subsequence* $\{u_{k_j}\}$ *such that, for all* ρ, $0<\rho<1$, *as* $k_j\to\infty$,

$$\begin{cases} u_{k_j}\to u \quad \text{strongly in } L^2(B_\rho(0),\mathbb{R}^N), \\ Du_{k_j}\rightharpoonup Du \quad \text{weakly in } L^2(B_\rho(0),\mathbb{R}^{nN}); \end{cases} \tag{3.11}$$

furthermore, u *satisfies*

$$A(u,\varphi)\triangleq\int_{B_1(0)} a_{ij}^{\alpha\beta}(x)D_\beta u^j D_\alpha\varphi^i dx = 0, \quad \forall\varphi\in C_0^\infty(B_1(0),\mathbb{R}^N). \tag{3.12}$$

Proof. From Lemma 3.2 and (3.10), we deduce that, for all ρ, $0<\rho<1$,

$$\int_{B_\rho(0)}|Du_k|^2dx \leqslant \frac{C}{(1-\rho)^2}\int_{B_1(0)}|u_k|^2dx\leqslant C_1, \tag{3.13}$$

where C_1 is independent of k. It follows that $\{u_k\}$ is uniformly bounded in $H^1(B_\rho(0),\mathbb{R}^N)$. Therefore there is a subsequence $\{u_{k_j}\}$ satisfying (3.11), by the compact embedding theorem.

Now we establish (3.12).

For all $\varphi\in C_0^\infty(B_1(0),\mathbb{R}^N)$, we have

$$\begin{aligned} A(u,\varphi) &= A(u-u_{k_j},\varphi)+A(u_{k_j},\varphi) \\ &= A(u-u_{k_j},\varphi)+[A(u_{k_j},\varphi)-A_{k_j}(u_{k_j},\varphi)]. \end{aligned} \tag{3.14}$$

Now we choose ρ to be sufficiently close to 1 so that $\operatorname{spt}\varphi\subset B_\rho(0)$. Then (3.11) implies that $A(u-u_{k_j},\varphi)\to 0$ as $k_j\to\infty$. Furthermore,

$$\begin{aligned} &|A(u_{k_j},\varphi)-A_{k_j}(u_{k_j},\varphi)| \\ &\leqslant C\Big(\int_{B_\rho(0)}\sum|a_{ij(k_j)}^{\alpha\beta}-a_{ij}^{\alpha\beta}|^2dx\Big)^{1/2}\Big(\int_{B_\rho(0)}|Du_{k_j}|^2dx\Big)^{1/2}. \end{aligned}$$

By (3.7), (3.8) and (3.13),

$$A(u_{k_j},\varphi)-A_{k_j}(u_{k_j},\varphi)\to 0 \quad \text{as } k_j\to\infty.$$

Letting $k_j \to \infty$ in (3.14), we now obtain (3.12). □

Proof of Theorem 3.1. The proof is divided into two steps.

Step 1. For all τ, $0<\tau<1$, we show that there exist $R_0 = R_0(\tau, n, N, \lambda, \Lambda) > 0$ and $\varepsilon_0 = \varepsilon_0(\tau, n, N, \lambda, \Lambda) > 0$ such that if for some $x^0 \in \Omega$ and some $0 < R < R_0 \wedge \operatorname{dist}(x^0, \partial\Omega)$, a weak solution $u \in H^1(\Omega, \mathbb{R}^N)$ of (3.1) satisfies

$$\Phi(x^0, R) < \varepsilon_0^2,$$

then we must have

$$\Phi(x^0, \tau R) \leqslant 2c_0\tau^2\Phi(x^0, R),$$

where $\Phi(x^0,R) = R^{-n}\displaystyle\int_{B_R(x^0)} |u(x) - u_{x^0,R}|^2dx$ is the *mean square oscillation* of u on $B_R(x^0)$, c_0 is the constant in (3.5), and $a \wedge b$ means $\min(a,b)$.

We use the indirect method to prove this conclusion. If the conclusion is not true, then there exist τ, $0<\tau<1$, $\{x_k\} \subset \Omega$, $R_k \to 0$, $\varepsilon_k \to 0$, and weak solutions u_k of (3.1), such that

$$\Phi^{(k)}(x_k, R_k) = \varepsilon_k^2,$$
$$\Phi^{(k)}(x_k, \tau R_k) > 2c_0\tau^2\varepsilon_k^2,$$

where

$$\Phi^{(k)}(x_k, R_k) = R_k^{-n}\int_{B_{R_k}(x_k)} |u_k(x) - (u_k)_{x_k,R_k}|^2dx.$$

Now we translate x_k to 0 and normalize using a rescaling (such a technique is often called the *blowup technique*):

$$x = x_k + R_k y,$$
$$v_k(y) = \frac{u_k(x_k + R_k y) - (u_k)_{x_k,R_k}}{\varepsilon_k}.$$

It is easy to verify that v_k has the following properties:

(i) $(v_k)_{0,\tau} = \dfrac{(u_k)_{x_k,\tau R_k} - (u_k)_{x_k,R_k}}{\varepsilon_k}$;

(ii) $(v_k)_{0,1} = 0$;

(iii) $\Psi^{(k)}(0,1) \triangleq \displaystyle\int_{B_1(0)} |v_k(y) - (v_k)_{0,1}|^2dy = 1$;

(iv) $\Psi^{(k)}(0,\tau) > 2c_0\tau^2$;

(v) $v_k \in H^1(B_1(0), \mathbb{R}^N)$ satisfies

$$\int_{B_1(0)} A_{ij}^{\alpha\beta}\Big(x_k + R_k y, \varepsilon_k v_k(y) + (u_k)_{x_k,R_k}\Big) D_\beta v_k^j D_\alpha \varphi^i dy = 0,$$
$$\forall \varphi \in C_0^\infty(B_1(0), \mathbb{R}^N).$$

On one hand, (ii) and (iii) imply that $\|v_k\|^2_{L^2(B_1(0),\mathbb{R}^N)} = 1$. It follows that there exists a subsequence, still denoted by $\{v_k\}$, such that as $k \to \infty$,

$$v_k \rightharpoonup v \quad \text{weakly in } L^2(B_1(0), \mathbb{R}^N). \tag{3.15}$$

Since

$$\int_{B_1(0)} |\varepsilon_k v_k(y)|^2 dy = \varepsilon_k^2 \to 0 \quad \text{as } k \to \infty,$$

there exists a subsequence, still denoted by $\{\varepsilon_k v_k\}$, which converges to 0 almost everywhere in $B_1(0)$. Set

$$b_{ij(k)}^{\alpha\beta} = A_{ij}^{\alpha\beta}(x_k, (u_k)_{x_k,R_k}).$$

By (3.2), (3.3),

$$b_{ij(k)}^{\alpha\beta}\xi_\alpha^i\xi_\beta^j \geqslant \lambda|\xi|^2, \quad \lambda > 0, \tag{3.16}$$

$$|b_{ij(k)}^{\alpha\beta}| \leqslant \Lambda. \tag{3.17}$$

Thus there exists a subsequence (still denoted by $\{b_{ij(k)}^{\alpha\beta}\}$) such that

$$b_{ij(k)}^{\alpha\beta} \to b_{ij}^{\alpha\beta} \quad \text{as } k \to \infty.$$

It is clear that the $b_{ij}^{\alpha\beta}$ satisfy (3.16), (3.17). Since the $A_{ij}^{\alpha\beta}(x,u)$ are continuous uniformly on $\overline{\Omega} \times \mathbb{R}^N$, we have, as $k \to \infty$,

$$A_{ij}^{\alpha\beta}(x_k + R_k y, \varepsilon_k v_k(y) + (u_k)_{x_k,R_k}) \to b_{ij}^{\alpha\beta} \quad \text{a.e. } y \in B_1(0).$$

Using (3.15), property (v) and Lemma 3.4, we derive $v \in H_{loc}^1(B_1(0), \mathbb{R}^N)$, and for all ρ, $0 < \rho < 1$, $\{v_k\}$ has a subsequence (still denoted by $\{v_k\}$) with the properties

$$v_k \to v \quad \text{strongly in } L^2(B_\rho(0), \mathbb{R}^N),$$
$$Dv_k \rightharpoonup Dv \quad \text{weakly in } L^2(B_\rho(0), \mathbb{R}^{nN});$$

furthermore

$$\int_{B_1(0)} b_{ij}^{\alpha\beta} D_\beta v^j D_\alpha \varphi^i dy = 0, \quad \forall \varphi \in C_0^\infty(B_1(0), \mathbb{R}^N).$$

Therefore Lemma 3.3 implies that, for all τ, $0 < \tau < 1$,

$$\Psi(0,\tau) \leqslant c_0\tau^2\Psi(0,1). \tag{3.18}$$

On the other hand, property (iv) implies that

$$\tau^{-n}\int_{B_\tau(0)} |v_k(y) - (v_k)_{0,\tau}|^2 dy > 2c_0\tau^2. \tag{3.19}$$

By Lemma 3.4 and (3.15), we have

$$v_k \to v \quad \text{strongly in } L^2(B_\tau(0), \mathbb{R}^N),$$
$$v_k \rightharpoonup v \quad \text{weakly in } L^2(B_1(0), \mathbb{R}^N).$$

Letting $k \to \infty$ in (3.19), we obtain

$$\Psi(0,\tau) = \tau^{-n}\int_{B_\tau(0)} |v(y) - v_{0,\tau}|^2 dy \geqslant 2c_0\tau^2. \tag{3.20}$$

Using the lower semicontinuity of the L^2 norm with respect to weak convergence, we get

$$\|v\|_{L^2(B_1(0),\mathbb{R}^N)}^2 \leqslant \liminf_{k\to\infty} \|v_k\|_{L^2(B_1(0),\mathbb{R}^N)}^2 = 1,$$

which implies that

$$\Psi(0,1) \leqslant 1. \tag{3.21}$$

Combining (3.20) and (3.21), we obtain

$$\Psi(0,\tau) \geqslant 2c_0\tau^2\Psi(0,1). \tag{3.22}$$

By (3.20), we have $\Psi(0,\tau) \neq 0$. Thus (3.22) contradicts (3.18).

Step 2. Show that there exists an open set $\Omega_0 \subset \Omega$ such that $u \in C^{0,\delta}_{loc}(\Omega_0, \mathbb{R}^N)$, $\forall \delta$, $0 < \delta < 1$, and $\text{meas}(\Omega \setminus \Omega_0) = 0$.

For any $\delta \in (0,1)$, we choose $\tau \in (0,1)$ such that

$$2c_0\tau^2 = \tau^{2\delta},$$

where c_0 is still the constant in (3.5). Using the result from Step 1, for all $\delta \in (0,1)$, there exist $R_0 = R_0(n, N, \lambda, \Lambda, c_0, \delta) > 0$ and $\varepsilon_0 = \varepsilon_0(n, N, \lambda, \Lambda, c_0, \delta) > 0$, such that if a weak solution of (3.1) satisfies, for some $x^0 \in \Omega$ and some $R < R_0 \wedge \text{dist}(x^0, \partial\Omega)$,

$$\Phi(x^0, R) < \varepsilon_0^2,$$

then for this x^0 and this R, we have

$$\Phi(x^0, \tau R) \leqslant 2c_0\tau^2\Phi(x^0, R) = \tau^{2\delta}\Phi(x^0, R).$$

Using iteration, we obtain

$$\Phi(x^0, \tau^k R) \leqslant \tau^{2\delta k}\Phi(x^0, R).$$

For any ρ, $0 < \rho < R$, we choose the nonnegative integer k such that

$$\tau^{k+1}R < \rho \leqslant \tau^k R.$$

Then

$$\begin{aligned}
\Phi(x^0, \rho) &= \rho^{-n}\int_{B(x^0,\rho)} |u(x) - u_{x^0,\rho}|^2 dx \\
&\leqslant (\tau^{k+1}R)^{-n}\int_{B(x^0,\tau^k R)} |u(x) - u_{x^0,\tau^k R}|^2 dx \\
&\leqslant \tau^{-n}\Phi(x^0, \tau^k R) \\
&\leqslant \tau^{-n}\tau^{2\delta k}\Phi(x^0, R) \\
&\leqslant \tau^{-n-2\delta}\Big(\frac{\rho}{R}\Big)^{2\delta}\Phi(x^0, R).
\end{aligned}$$

This shows that, if for some $x^0 \in \Omega$ and some $R < R_0 \wedge \text{dist}(x^0, \partial\Omega)$ we have $\Phi(x^0, R) < \varepsilon_0^2$, then for all ρ, $0 < \rho < R$, we have

$$\Phi(x^0, \rho) \leqslant C\Big(\frac{\rho}{R}\Big)^{2\delta}\Phi(x^0, R),$$

where $C = C(n, N, \lambda, \Lambda, \delta)$.

$\Phi(x, R)$ is continuous in x. Therefore, if $\Phi(x^0, R) < \varepsilon_0^2$ for some $x^0 \in \Omega$, then there exists a neighborhood $B_r(x^0)$ such that

$$\Phi(x, R) < \varepsilon_0^2 \quad \text{for } x \in B_r(x^0).$$

Hence for all $x \in B_r(x^0)$ and any ρ, $0 < \rho < R$, we have

$$\Phi(x, \rho) \leqslant C\Big(\frac{\rho}{R}\Big)^{2\delta}\Phi(x, R),$$

i.e.,

$$\int_{B_\rho(x)} |u(z) - u_{x,\rho}|^2 dz \leqslant C\Big(\frac{\rho}{R}\Big)^{n+2\delta}\int_{B_R(x)} |u(z) - u_{x,R}|^2 dz,$$

where $C = C(n, N, \lambda, \Lambda, \delta)$. By Theorem 1.5 in Chapter 9,

$$u \in C^{0,\delta}(B_r(x^0), \mathbb{R}^N).$$

Now, for the above δ and the corresponding $R_0 = R_0(n, N, \lambda, \Lambda, \delta)$ and $\varepsilon_0 = \varepsilon_0(n, N, \lambda, \Lambda, \delta)$, we set

$$\Omega_0 = \{x^0 \in \Omega \mid \Phi(x^0, R) < \varepsilon_0^2 \text{ for some } R < R_0\}.$$

The set Ω_0 has the following properties:

(i) $\Omega_0 \subset \Omega$, and Ω_0 is an open set.

(ii) $u \in C^{0,\delta}_{loc}(\Omega_0, \mathbb{R}^N)$.

(iii) $\operatorname{meas}(\Omega \setminus \Omega_0) = 0$. In fact, Poincaré's inequality implies that

$$R^{-n} \int_{B_R(x^0)} |u(x) - u_{x^0,R}|^2 dx \leqslant C R^{2-n} \int_{B_R(x^0)} |Du|^2 dx.$$

From the Lebesgue differentiation theorem, we deduce that, for a.e. $x^0 \in \Omega$,

$$\lim_{R \to 0} \omega_n^{-1} R^{-n} \int_{B_R(x^0)} |Du|^2 dx = |Du(x^0)|^2 < +\infty.$$

It follows that

$$\lim_{R \to 0} R^{-n} \int_{B_R(x^0)} |u(x) - u_{x^0,R}|^2 dx = 0.$$

This shows that for a.e. $x^0 \in \Omega$ and any $\delta \in (0,1)$ and the corresponding $R_0(\delta)$ and $\varepsilon_0(\delta)$, there exists $R < R_0$ such that

$$\Phi(x^0, R) = R^{-n} \int_{B_R(x^0)} |u(x) - u_{x^0,R}|^2 dx < \varepsilon_0^2,$$

which implies that $x^0 \in \Omega_0$. Therefore $\operatorname{meas}(\Omega \setminus \Omega_0) = 0$.

(iv) Ω_0 is independent of δ. In fact, we can show that $x^0 \in \Omega_0$ if and only if

$$\liminf_{\rho \to 0} \rho^{-n} \int_{B_\rho(x^0)} |u(x) - u_{x^0,\rho}|^2 dx = 0.$$

If the above equality is valid, then there exists $R < R_0$ such that

$$R^{-n} \int_{B_R(x^0)} |u(x) - u_{x^0,R}|^2 dx < \varepsilon_0^2,$$

and therefore $x^0 \in \Omega_0$. Conversely, if $x^0 \in \Omega_0$, then there exists $R < R_0$ such that

$$R^{-n} \int_{B_R(x^0)} |u(x) - u_{x^0,R}|^2 dx < \varepsilon_0^2.$$

It follows that for all $\rho < R$, we have

$$\rho^{-n} \int_{B_\rho(x^0)} |u(x) - u_{x^0,\rho}|^2 dx \leqslant C\Big(\frac{\rho}{R}\Big)^{2\delta} R^{-n} \int_{B_R(x^0)} |u(x) - u_{x^0,R}|^2 dx.$$

Fixing R and letting $\rho \to 0$, we conclude that

$$\liminf_{\rho \to 0} \rho^{-n} \int_{B_\rho(x^0)} |u(x) - u_{x^0,\rho}|^2 dx = 0.$$

The proof is complete. □

Remark 3.1. Using Poincaré's inequality and Caccioppoli's inequality, we can also prove that

$$\Omega_0 = \left\{ x \in \Omega \;\middle|\; \liminf_{\rho \to 0} \rho^{2-n} \int_{B_\rho(x^0)} |Du(z)|^2 dz = 0 \right\}.$$

4. The reverse Hölder inequality and L^p estimates for Du

We know that if $u \in L^p(\Omega)$ and $p > q \geqslant 1$, then Hölder's inequality implies that

$$\|u\|_{L^q(\Omega)} \leqslant C\|u\|_{L^p(\Omega)},$$

where $C = C(p, q, |\Omega|)$. But if $q > p$, then this inequality is not valid in general. However, we do have special cases where $q > p$ and this inequality is still valid; the Harnack inequality in Theorem 1.4 in Chapter 4 is just such an example. This indicates that the reverse Hölder inequality is valid in certain situations. We shall see later that it plays an important role in studying regularity of weak solutions.

Theorem 4.1 (Reverse Hölder inequality). *Let B be a ball in $\mathbb{R}^n$. Suppose that*

1° $g \geqslant 0$, $g \in L^q(B)$, $q > 1$; $f \geqslant 0$, $f \in L^r(B)$, $r > q$;

2° *for all $x^0 \in B$ and all R, $0 < R < \operatorname{dist}(x^0, \partial B) \wedge R_0$, we have*

$$⨍_{B_{R/2}(x^0)} g^q dx \leqslant b\left[⨍_{B_R(x^0)} g dx \right]^q + ⨍_{B_R(x^0)} f^q dx + \theta ⨍_{B_R(x^0)} g^q dx,$$

where R_0, b, θ are constants, $b > 1$, $R_0 > 0$, $0 \leqslant \theta < 1$.

Then there exist $\varepsilon > 0$ and $C > 0$ such that

$$g \in L^p_{loc}(B), \quad \forall p \in [q, q + \varepsilon),$$

and for all $B_R \subset B$, $R < R_0$, we have

$$\left[⨍_{B_{R/2}} g^p dx \right]^{1/p} \leqslant C\left\{ \left[⨍_{B_R} g^q dx \right]^{1/q} + \left[⨍_{B_R} f^p dx \right]^{1/p} \right\},$$

where C and ε depend on b, θ, n, q, r, while $B_{R/2}$ and B_R are concentric balls.

For a proof, see Appendix 5 or [GQ1].

Next, we use Theorem 4.1 to establish $W^{1,p}$ $(p > 2)$ regularity for a weak solution of an elliptic system under certain conditions. For clarity in presenting the ideas, we first study homogeneous linear elliptic systems.

Theorem 4.2. *Suppose that*

1° *the $A_{ij}^{\alpha\beta}(x)$ satisfy*

$$A_{ij}^{\alpha\beta}(x)\xi_\alpha^i \xi_\beta^j \geqslant \lambda|\xi|^2, \quad \lambda > 0; \quad A_{ij}^{\alpha\beta} \in L^\infty, \quad |A_{ij}^{\alpha\beta}| \leqslant \Lambda;$$

2° $u \in H^1(\Omega, \mathbb{R}^N)$ *satisfies*

$$\int_\Omega A_{ij}^{\alpha\beta}(x) D_\beta u^j D_\alpha \varphi^i dx = 0, \quad \forall \varphi \in H_0^1(\Omega, \mathbb{R}^N). \tag{4.1}$$

Then there exists $p > 2$ such that $|Du| \in L^p_{loc}(\Omega)$, and for $B_R \subset \Omega$, we have

$$\left[\fint_{B_{R/2}} |Du|^p dx \right]^{1/p} \leqslant C \left[\fint_{B_R} |Du|^2 dx \right]^{1/2}, \tag{4.2}$$

where C and p depend on n, N, λ, Λ.

Proof. For all $B \subset \Omega$, all $x^0 \in B$, and all R, $0 < R < \text{dist}(x^0, \partial B)$, we apply Caccioppoli's inequality (Theorem 2.1 in Chapter 8) to obtain

$$\int_{B_{R/2}(x^0)} |Du|^2 dx \leqslant \frac{C}{R^2} \int_{B_R(x^0)} |u - u_R|^2 dx, \tag{4.3}$$

where $C = C(n, N, \lambda, \Lambda)$, $u_R = \fint_{B_R(x^0)} u dx$.

Choosing $q = 2n/(n+2)$ and $v = u$ in the Sobolev-Poincaré inequality[2], we obtain

$$\left[\int_{B_R(x^0)} |u - u_R|^2 dx \right]^{1/2} \leqslant C(n) \left[\int_{B_R(x^0)} |Du|^{2n/(n+2)} dx \right]^{(n+2)/(2n)}. \tag{4.4}$$

Combining (4.3) and (4.4), we get

$$\int_{B_{R/2}(x^0)} |Du|^2 dx \leqslant \frac{C}{R^2} \left[\int_{B_R(x^0)} |Du|^{2n/(n+2)} dx \right]^{(n+2)/n}.$$

Dividing by R^n on both sides, we obtain

$$\fint_{B_{R/2}(x^0)} |Du|^2 dx \leqslant C \left[\fint_{B_R(x^0)} |Du|^{2n/(n+2)} dx \right]^{(n+2)/n},$$

where $C = C(n, N, \lambda, \Lambda)$.

Now in Theorem 4.1 we choose $g = |Du|^{2n/(n+2)}$, $q = (n+2)/n$, $f \equiv 0$, $\theta = 0$. Then

$$|Du|^{2n/(n+2)} \in L^r_{loc}(B), \quad \forall r \in \left[\frac{n+2}{n}, \frac{n+2}{n} + \varepsilon \right],$$

and for any $B_R \subset B$, we have the estimate

$$\left[\fint_{B_{R/2}} |Du|^{2nr/(n+2)} dx \right]^{1/r} \leqslant C \left[\fint_{B_R} |Du|^2 dx \right]^{n/(n+2)}.$$

---

[2]**Sobolev-Poincaré inequality.** *If $v \in W^{1,q}(B_R)$, $1 \leqslant q < n$, then $v \in L^p(B_R)$, $p = nq/(n-q)$, and*

$$\left[\int_{B_R} |v - v_R|^p dx \right]^{1/p} \leqslant C \left[\int_{B_R} |Dv|^q dx \right]^{1/q},$$

where $v_R = \fint_{B_R} v dx$, and C depends on n, q.

Proof. For the case $R = 1$, the Sobolev embedding theorem implies that

$$\|v - v_{B_1}\|_{L^p(B_1)} \leqslant C(n) \|v - v_{B_1}\|_{W^{1,q}(B_1)},$$

where $v_{B_1} = \fint_{B_1} v dx$. By Poincaré's inequality,

$$\|v - v_{B_1}\|_{L^q(B_1)} \leqslant C(n, q) \|Dv\|_{L^q(B_1)}.$$

Combining these two inequalities, we obtain the desired result. For general R, the result can be obtained by rescaling. □

We set $p = 2nr/(n+2)$. Then $p > 2$, and

$$\left[\fint_{B_{R/2}} |Du|^p dx\right]^{1/p} \leqslant C\left[\fint_{B_R} |Du|^2 dx\right]^{1/2},$$

where $C = C(n, N, \lambda, \Lambda)$. The proof is complete. □

It should be pointed out the exponent p in Theorem 4.2 cannot be arbitrarily large. Here is a counterexample.

Example (N. G. Meyers, 1963). Let $N = 1$, $n = 2$, $\Omega = B_1(0)$. Consider the equation

$$(au_x + bu_y)_x + (bu_x + cu_y)_y = 0, \quad (x, y) \in B_1(0), \tag{4.5}$$

where

$$a = 1 - (1-\mu^2)\frac{y^2}{x^2+y^2},$$
$$b = (1-\mu^2)\frac{xy}{x^2+y^2},$$
$$c = 1 - (1-\mu^2)\frac{x^2}{x^2+y^2},$$

where $\mu \in (0,1)$ is a fixed constant. It is easy to verify that the matrix $(A_{ij}) = \begin{pmatrix} a & b \\ c & d \end{pmatrix}$ has eigenvalues μ^2 and 1. It can be shown that the function

$$u(x, y) = x(x^2+y^2)^{(\mu-1)/2}$$

is a weak solution of (4.5), and

$$|Du| \in L^p_{loc}(B_1(0)), \quad 2 \leqslant p < \frac{2}{1-\mu}.$$

Since $\int_{B_1(0)} |Du|^{2/(1-\mu)} dxdy = +\infty$, $|Du| \in L^p_{loc}(B_1(0))$ only if $p < 2/(1-\mu)$. It is clear that $p \to 2$ as $\mu \to 0$.

Now we consider the nonlinear elliptic system

$$-D_\alpha A_i^\alpha(x, u, Du) + B_i(x, u, Du) = 0 \quad (i = 1, \cdots, N). \tag{4.6}$$

First we study the case where A_i^α, B_i satisfy the controllable structure conditions.

Theorem 4.3. *Suppose that A_i^α, B_i satisfy the controllable structure conditions:*

$$A_i^\alpha(x, u, p)p_\alpha^i \geqslant \lambda|p|^2 - \Lambda|u|^r - f^2(x), \tag{4.7}$$

$$\begin{cases} |A_i^\alpha(x, u, p)| \leqslant \Lambda_1(|p| + |u|^{r/2} + f_i^\alpha(x)), \\ |B_i(x, u, p)| \leqslant \Lambda(|p|^{2(1-1/r)} + |u|^{r-1} + f_i(x)), \end{cases} \tag{4.8}$$

where $\lambda, \Lambda, \Lambda_1$ are positive constants, $r = 2^*$ is the Sobolev conjugate of 2 (i.e., $r = 2n/(n-2)$ when $n > 2$, $2 \leqslant r < +\infty$ when $n = 2$), $f_i^\alpha, f_i \geqslant 0$, $f, f_i^\alpha \in L^\sigma(\Omega)$, $\sigma > 2$, and $f_i \in L^s(\Omega)$, $s > r/(r-1)$. If $u \in H^1(\Omega, \mathbb{R}^N)$ is a weak solution of (4.6), then there exists $p > 2$ such that $u \in W^{1,p}_{loc}(\Omega, \mathbb{R}^N)$; furthermore, there exists *a small $R_0 > 0$ such that for any concentric balls $B_{R/2} \subset B_R \subset \Omega$ with $R < R_0$, we have*

$$\left\{ \fint_{B_{R/2}} \left(|u|^r + |Du|^2\right)^{p/2} dx \right\}^{1/p}$$

$$(4.9) \qquad \leqslant C\left\{ \left[\fint_{B_R} \left(|u|^r + |Du|^2\right) dx \right]^{1/2} + \left[\fint_{B_R} \left(f^2 + \left|\widetilde{\widetilde{f}}\right|^2\right)^{p/2} dx \right]^{1/p} + R\left[\fint_{B_R} |\widetilde{f}|^{pr/[2(r-1)]} dx \right]^{2(r-1)/(pr)} \right\},$$

where $\widetilde{\widetilde{f}} = (f_i^\alpha)$, $\widetilde{f} = (f_i)$; C depends on $n, N, \lambda, \Lambda, \Lambda_1$; p depends on $n, N, \lambda, \Lambda, \Lambda_1$, σ, s; and R_0 depends on u.

Proof. For all $B \subset \Omega$, all $x^0 \in B$, and all R, $0 < R < \operatorname{dist}(x^0, \partial B)$, we choose in the integral identity

$$\int_\Omega \left[A_i^\alpha(x, u, Du) D_\alpha \varphi^i + B_i(x, u, Du)\varphi^i\right] dx = 0, \quad \forall \varphi \in H_0^1(\Omega, \mathbb{R}^N)$$

the test function $\varphi = \eta^2(u - u_R)$, where $\eta \in C_0^\infty(B_R(x^0))$ is a cutoff function:

$$(4.10) \qquad 0 \leqslant \eta \leqslant 1 \quad \text{in } B_R(x^0), \quad \eta \equiv 1 \quad \text{on } B_{R/2}(x^0), \quad |D\eta| \leqslant \frac{C}{R}.$$

Then

$$\begin{aligned} &\int_{B_R} \eta^2 A_i^\alpha(x, u, Du) D_\alpha u^i dx \\ &\quad = -\int_{B_R} 2\eta D_\alpha \eta \cdot (u^i - u_R^i) A_i^\alpha(x, u, Du) dx \\ &\qquad - \int_{B_R} \eta^2 (u^i - u_R^i) B_i(x, u, Du) dx. \end{aligned}$$

By using the structure conditions (4.7) and (4.8), we get

$$\begin{aligned} &\lambda \int_{B_R} \eta^2 |Du|^2 dx \\ &\leqslant \Lambda \int_{B_R} \eta^2 |u|^r dx + \int_{B_R} \eta^2 f^2 dx + \Lambda_1 \int_{B_R} 2\eta |D\eta| \cdot |u - u_R| \cdot |Du| dx \\ (4.11) \quad &+ \Lambda_1 \int_{B_R} 2\eta |D\eta| \cdot |u - u_R| \cdot |u|^{r/2} dx + \Lambda_1 \int_{B_R} 2\eta |D\eta| \cdot |u - u_R| \cdot \left|\widetilde{\widetilde{f}}\right| dx \\ &+ \Lambda \int_{B_R} \eta^2 |u - u_R| \cdot |Du|^{2(1-1/r)} dx + \Lambda \int_{B_R} \eta^2 |u - u_R| \cdot |u|^{r-1} dx \\ &+ \Lambda \int_{B_R} \eta^2 |u - u_R| \cdot |\widetilde{f}| dx. \end{aligned}$$

For definiteness, we consider the case $n > 2$. In this case $r = 2n/(n-2)$, and therefore $2(1 - 1/r) = (n+2)/n$, $r - 1 = (n+2)/(n-2)$. Using the Hölder inequality and the Sobolev-Poincaré inequality, we obtain

$$\begin{aligned}
&\int_{B_R} \eta^2 |u-u_R| \cdot |\widetilde{f}| dx \\
\leqslant\ & \left[\int_{B_R} |u-u_R|^{2n/(n-2)} dx\right]^{(n-2)/(2n)} \left[\int_{B_R} |\widetilde{f}|^{2n/(n+2)} dx\right]^{(n+2)/(2n)} \\
\leqslant\ & C(n)\left[\int_{B_R} |Du|^2 dx\right]^{1/2} \left[\int_{B_R} |\widetilde{f}|^{2n/(n+2)} dx\right]^{(n+2)/(2n)} \\
\leqslant\ & \varepsilon \int_{B_R} |Du|^2 dx + \frac{C(n)}{\varepsilon}\left[\int_{B_R} |\widetilde{f}|^{2n/(n+2)} dx\right]^{(n+2)/n},
\end{aligned}$$

$$\begin{aligned}
&\int_{B_R} \eta^2 |u-u_R| \cdot |u|^{r-1} dx \\
=\ & \int_{B_R} \eta^2 |u-u_R| \cdot |u|^{(n+2)/(n-2)} dx \\
\leqslant\ & \left[\int_{B_R} |u-u_R|^{2n/(n-2)} dx\right]^{(n-2)/(2n)} \left[\int_{B_R} |u|^{2n/(n-2)} dx\right]^{(n+2)/(2n)} \\
\leqslant\ & \frac{n-2}{2n}\int_{B_R} |u-u_R|^{2n/(n-2)} dx + \frac{n+2}{2n}\int_{B_R} |u|^{2n/(n-2)} dx \\
\leqslant\ & C\left[\int_{B_R} |Du|^2 dx\right]^{n/(n-2)} + C\int_{B_R} |u|^{2n/(n-2)} dx,
\end{aligned}$$

$$\begin{aligned}
&\int_{B_R} \eta^2 |u-u_R| \cdot |Du|^{2(1-1/r)} dx \\
=\ & \int_{B_R} \eta^2 |u-u_R| \cdot |Du|^{(n+2)/n} dx \\
\leqslant\ & \left[\int_{B_R} |u-u_R|^{2n/(n-2)} dx\right]^{(n-2)/(2n)} \left[\int_{B_R} |Du|^2 dx\right]^{(n+2)/(2n)} \\
\leqslant\ & C\left[\int_{B_R} |Du|^2 dx\right]^{1/2+(n+2)/(2n)},
\end{aligned}$$

$$\begin{aligned}
&\int_{B_R} \eta^2 |u|^r dx \leqslant \int_{B_R} |u|^{2n/(n-2)} dx \\
\leqslant\ & C\int_{B_R} |u-u_R|^{2n/(n-2)} dx + C\int_{B_R} |u_R|^{2n/(n-2)} dx \\
\leqslant\ & C\left[\int_{B_R} |Du|^2 dx\right]^{n/(n-2)} + C|B_R| \cdot \left|\fint_{B_R} u dx\right|^{2n/(n-2)} \\
\leqslant\ & C\left[\int_{B_R} |Du|^2 dx\right]^{n/(n-2)} + CR^n\left[\fint_{B_R} |u|^{[2n/(n-2)]\cdot[n/(n+2)]} dx\right]^{(n+2)/n},
\end{aligned}$$

$$\begin{aligned}
&\int_{B_R} 2\eta|D\eta|\cdot|u-u_R|\cdot|Du|dx \\
&\leqslant \varepsilon\int_{B_R}\eta^2|Du|^2dx+\frac{C}{\varepsilon}\frac{1}{R^2}\int_{B_R}|u-u_R|^2dx \\
&\leqslant \varepsilon\int_{B_R}\eta^2|Du|^2dx+\frac{C}{\varepsilon}\frac{1}{R^2}\left[\int_{B_R}|Du|^{2n/(n+2)}dx\right]^{(n+2)/n},
\end{aligned}$$

$$\begin{aligned}
&\int_{B_R} 2\eta|D\eta|\cdot|u-u_R|\cdot|u|^{r/2}dx \\
&= \int_{B_R} 2\eta|D\eta|\cdot|u-u_R|\cdot|u|^{n/(n-2)}dx \\
&\leqslant C\int_{B_R}|u|^{2n/(n-2)}dx+\frac{C}{R^2}\int_{B_R}|u-u_R|^2dx \\
&\leqslant C\left[\int_{B_R}|Du|^2dx\right]^{n/(n-2)}+CR^n\left[\fint_{B_R}|u|^{[2n/(n-2)]\cdot[n/(n+2)]}dx\right]^{(n+2)/n} \\
&\quad +\frac{C}{R^2}\left[\int_{B_R}|Du|^{2n/(n+2)}dx\right]^{(n+2)/n},
\end{aligned}$$

$$\begin{aligned}
&\int_{B_R} 2\eta|D\eta|\cdot|u-u_R|\cdot\left|\widetilde{\widetilde{f}}\right|dx \\
&\leqslant \int_{B_R}\left|\widetilde{\widetilde{f}}\right|^2dx+\frac{C}{R^2}\int_{B_R}|u-u_R|^2dx \\
&\leqslant \int_{B_R}\left|\widetilde{\widetilde{f}}\right|^2dx+\frac{C}{R^2}\left[\int_{B_R}|Du|^{2n/(n+2)}dx\right]^{(n+2)/n}.
\end{aligned}$$

Substituting all these inequalities into (4.11) and then adding $\int_{B_R}\eta^2|u|^{2n/(n-2)}dx$ on both sides, we find that, for any $\varepsilon>0$,

$$\begin{aligned}
&\int_{B_{R/2}}\left(|Du|^2+|u|^{2n/(n-2)}\right)dx \\
&\leqslant \frac{C}{R^2}\left[\int_{B_R}|Du|^{2n/(n+2)}dx\right]^{(n+2)/n} \\
&\quad +CR^n\left[\fint_{B_R}|u|^{[2n/(n-2)]\cdot[n/(n+2)]}dx\right]^{(n+2)/n} \\
&\quad +C\left\{\varepsilon+\left[\int_{B_R}|Du|^2dx\right]^{2/(n-2)}+\left[\int_{B_R}|Du|^2dx\right]^{1/n}\right\}\cdot\int_{B_R}|Du|^2dx \\
&\quad +C\int_{B_R}\left(|f|^2+\left|\widetilde{\widetilde{f}}\right|^2+|\widetilde{F}|^2\right)dx,
\end{aligned}$$

where $\widetilde{F} = \left[\int_{B_R} |\widetilde{f}|^{2n/(n+2)} dx\right]^{1/n} \widetilde{f}^{n/(n+2)}$, and C depends on $n, N, \lambda, \Lambda, \Lambda_1$ and ε.

By the absolute continuity of Lebesgue integrals,

$$\int_{B_R} |Du|^2 dx \to 0 \quad \text{as } R \to 0.$$

Thus if $\varepsilon > 0$ and $R_0 > 0$ (R_0 depends on u) are sufficiently small, then for $R < R_0$, we have

$$\begin{aligned}
&\int_{B_{R/2}} \left(|Du|^2 + |u|^{2n/(n-2)}\right) dx \\
\leqslant\ & \frac{C}{R^2}\left[\int_{B_R} |Du|^{2n/(n+2)} dx\right]^{(n+2)/n} \\
&+ CR^n \left[⨍_{B_R} |u|^{[2n/(n-2)]\cdot[n/(n+2)]} dx\right]^{(n+2)/n} \\
&+ \frac{1}{2}\int_{B_R} |Du|^2 dx + C\int_{B_R} \left(|f|^2 + \left|\widetilde{\widetilde{f}}\right|^2 + |\widetilde{F}|^2\right) dx.
\end{aligned}$$

Dividing both sides by R^n, we obtain

$$\begin{aligned}
&⨍_{B_{R/2}} \left(|Du|^2 + |u|^{2n/(n-2)}\right) dx \\
\leqslant\ & C\left[⨍_{B_R} |Du|^{2n/(n+2)} dx\right]^{(n+2)/n} \\
&+ C\left[⨍_{B_R} |u|^{[2n/(n-2)]\cdot[n/(n+2)]} dx\right]^{(n+2)/n} \\
&+ \frac{1}{2}⨍_{B_R} |Du|^2 dx + C⨍_{B_R} \left(|f|^2 + \left|\widetilde{\widetilde{f}}\right|^2 + |\widetilde{F}|^2\right) dx \\
\leqslant\ & C\left[⨍_{B_R} \left(|Du|^{2n/(n+2)} + |u|^{[2n/(n-2)]\cdot[n/(n+2)]}\right) dx\right]^{(n+2)/n} \\
&+ C⨍_{B_R} \left(|f|^2 + \left|\widetilde{\widetilde{f}}\right|^2 + |\widetilde{F}|^2\right) dx + \frac{1}{2}⨍_{B_R} |Du|^2 dx.
\end{aligned}$$

Now in Theorem 4.1 we choose $g = |Du|^{2n/(n+2)} + |u|^{[2n/(n-2)]\cdot[n/(n+2)]}$, $q = (n+2)/n$, $\theta = 1/2$, and choose f in Theorem 4.1 as $\left(|f|^2 + \left|\widetilde{\widetilde{f}}\right|^2 + |\widetilde{F}|^2\right)^{n/(n+2)}$. Then there exists $\widehat{p} > (n+2)/n$ such that

$$\begin{aligned}
&\left\{⨍_{B_{R/2}} \left(|Du|^{2\widehat{p}n/(n+2)} + |u|^{[2\widehat{p}n/(n+2)]\cdot[n/(n-2)]}\right) dx\right\}^{1/p} \\
\leqslant\ & C\left\{\left[⨍_{B_R} \left(|Du|^2 + |u|^{2n/(n-2)}\right) dx\right]^{n/(n+2)}\right. \\
&\left. + \left[⨍_{B_R} \left(|f|^2 + \left|\widetilde{\widetilde{f}}\right|^2 + |\widetilde{F}|^2\right)^{\widehat{p}n/(n+2)} dx\right]^{1/\widehat{p}}\right\}.
\end{aligned}$$

Set $p = 2\widehat{p}n/(n+2)$. Since $\widehat{p} > (n+2)/n$, we have $p > 2$, and

$$
\begin{aligned}
&\left\{ \fint_{B_{R/2}} \left(|Du|^p + |u|^{pn/(n-2)} \right) dx \right\}^{[1/p]\cdot[2n/(n+2)]} \\
(4.12) \qquad &\leqslant C \left\{ \left[\fint_{B_R} \left(|Du|^2 + |u|^{2n/(n-2)} \right) dx \right]^{n/(n+2)} \right. \\
&\quad \left. + \left[\fint_{B_R} \left(|f|^p + \left| \widetilde{\widetilde{f}} \right|^p + |\widetilde{F}|^p \right) dx \right]^{[1/p]\cdot[2n/(n+2)]} \right\}.
\end{aligned}
$$

Now we estimate $\fint_{B_R} |\widetilde{F}|^p dx$ in terms of $\widetilde{f}$:

$$
\begin{aligned}
\fint_{B_R} |\widetilde{F}|^p dx &= R^p \left[\fint_{B_R} |\widetilde{f}|^{2n/(n+2)} dx \right]^{p/n} \fint_{B_R} |\widetilde{f}|^{pn/(n+2)} dx \\
&\leqslant CR^p \left[\fint_{B_R} |\widetilde{f}|^{pn/(n+2)} dx \right]^{(n+2)/n}.
\end{aligned}
$$

Taking the p-th root on both sides, we obtain

$$
\left[\fint_{B_R} |\widetilde{F}|^p dx \right]^{1/p} \leqslant CR \left[\fint_{B_R} |\widetilde{f}|^{pn/(n+2)} dx \right]^{(n+2)/(pn)}.
$$

It follows that, for $R < R_0$,

$$
\begin{aligned}
&\fint_{B_{R/2}} \left(|Du|^p + |u|^{pn/(n-2)} \right) dx \Big\}^{1/p} \\
&\leqslant C \left\{ \left[\fint_{B_R} \left(|Du|^2 + |u|^{2n/(n-2)} \right) dx \right]^{1/2} \right. \\
&\quad \left. + \left[\fint_{B_R} \left(|f|^p + \left| \widetilde{\widetilde{f}} \right|^p \right) dx \right]^{1/p} + R \left[\fint_{B_R} |\widetilde{f}|^{pn/(n+2)} dx \right]^{(n+2)/(pn)} \right\},
\end{aligned}
$$

where C depends on $n, N, \lambda, \Lambda, \Lambda_1$; R_0 depends on u; and p depends on σ, s in addition to the dependence on $n, N, \lambda, \Lambda, \Lambda_1$. (In order for the integrals $\int_{B_R} \left| \widetilde{\widetilde{f}} \right|^p dx$ and $\int_{B_R} |\widetilde{f}|^{pn/(n+2)} dx$ to make sense, we also require $p \leqslant \sigma$ and $pn/(n+2) \leqslant s$; i.e., $p \leqslant \min[\sigma, s(n+2)/n]$.)

The proof for the case $n > 2$ is now complete. The proof for the case $n = 2$ is left as an exercise for the reader, or see [GQ1]. □

Using a similar method, we can also prove

Theorem 4.3′. *Suppose that A_i^α, B_i satisfy*

$$
\begin{aligned}
&A_i^\alpha(x,u,p)p_\alpha^i \geqslant \lambda |p|^2 - f^2(x), \\
&|A_i^\alpha(x,u,p)| \leqslant \Lambda_1(|p| + f_i^\alpha(x)), \\
&|B_i(x,u,p)| \leqslant \Lambda(|p|^{2(1-1/r)} + f_i(x)),
\end{aligned}
$$

where $\lambda, \Lambda, \Lambda_1$ are positive constants, $r = 2^*$ is the Sobolev conjugate of 2 (i.e., $r = 2n/(n-2)$ when $n > 2$, $2 \leqslant r < +\infty$ when $n = 2$), $f_i^\alpha, f_i \geqslant 0$, $f, f_i^\alpha \in L^\sigma(\Omega)$,

$\sigma > 2$, and $f_i \in L^s(\Omega)$, $s > r/(r-1)$. If $u \in H^1(\Omega, \mathbb{R}^N)$ is a weak solution of (4.6), then there exists $p > 2$ such that $u \in W^{1,p}_{loc}(\Omega, \mathbb{R}^N)$; furthermore, there exists a small $R_0 > 0$ such that for any concentric balls $B_{R/2} \subset B_R \subset \Omega$ with $R < R_0$, we have

$$\left\{ \fint_{B_{R/2}} |Du|^p dx \right\}^{1/p} \leqslant C\left\{ \left[\fint_{B_R} |Du|^2 dx \right]^{1/2} + \left[\fint_{B_R} \left(f^2 + \left| \tilde{\tilde{f}} \right|^2 \right)^{p/2} dx \right]^{1/p} + R \left[\fint_{B_R} |\tilde{f}|^{pr/[2(r-1)]} dx \right]^{2(r-1)/(pr)} \right\},$$

where $\tilde{\tilde{f}} = (f_i^\alpha)$, $\tilde{f} = (f_i)$; C depends on $n, N, \lambda, \Lambda, \Lambda_1$; p depends on $n, N, \lambda, \Lambda, \Lambda_1$, σ, s; and R_0 depends on u.

Now we study the L^p $(p > 2)$ estimate for the gradient of a weak solution of the elliptic system (4.6) under the natural structure conditions.

Theorem 4.4. *Assume that*

1° *for $|u| \leqslant M$,*

$$A_i^\alpha(x, u, p) p_\alpha^i \geqslant \lambda |p|^2 - f^2(x), \tag{4.13}$$

$$\begin{cases} |A_i^\alpha(x, u, p)| \leqslant \Lambda_1 |p| + f_i^\alpha(x), \\ |B_i(x, u, p)| \leqslant \Lambda |p|^2 + f_i(x), \end{cases} \tag{4.14}$$

where $\lambda, \Lambda, \Lambda_1$ are positive constants depending on M, $f_i, f_i^\alpha \geqslant 0$, $f, f_i^\alpha \in L^\sigma(\Omega)$, $\sigma > 2$, and $f_i \in L^s(\Omega)$, $s > 1$;

2° *$u \in H^1 \cap L^\infty(\Omega, \mathbb{R}^N)$ satisfies*

$$\int_\Omega \left[A_i^\alpha(x, u, Du) D_\alpha \varphi^i + B(x, u, Du) \varphi^i \right] dx = 0, \quad \forall \varphi \in H_0^1 \cap L^\infty(\Omega, \mathbb{R}^N); \tag{4.15}$$

3° *$2\Lambda M < \lambda$.*

Under these assumptions, there exists $p > 2$ such that $u \in W^{1,p}_{loc}(\Omega, \mathbb{R}^N)$; furthermore, for any concentric balls $B_{R/2} \subset B_R \subset \Omega$, we have

$$\left\{ \fint_{B_{R/2}} |Du|^p dx \right\}^{1/p} \leqslant C\left\{ \left[\fint_{B_R} |Du|^2 dx \right]^{1/2} + \left[\fint_{B_R} \left(|f|^2 + |\tilde{f}| + \left| \tilde{\tilde{f}} \right|^2 \right)^{p/2} dx \right]^{1/p} \right\}, \tag{4.16}$$

where $\tilde{f} = (f_i)$, $\tilde{\tilde{f}} = (f_i^\alpha)$; C depends on $n, N, \lambda, \Lambda, \Lambda_1, M$; and p depends on σ, s in addition to the dependence on $n, N, \lambda, \Lambda, \Lambda_1, M$.

Proof. For all $B \subset \Omega$, all $x^0 \in B$, and all R, $0 < R < \operatorname{dist}(x^0, \partial B)$, we choose in the integral identity (4.15) $\varphi = \eta^2(u - u_R)$, where $\eta \in C_0^\infty(B_R(x^0))$ is a cutoff function satisfying (4.10). Then

$$\int_{B_R} \eta^2 A_i^\alpha(x,u,Du) D_\alpha u^i dx = -\int_{B_R} 2\eta D_\alpha \eta \cdot (u^i - u_R^i) A_i^\alpha(x,u,Du) dx - \int_{B_R} \eta^2 (u^i - u_R^i) B_i(x,u,Du) dx.$$

Using the structure conditions (4.13), (4.14), we get

$$\begin{aligned} \lambda \int_{B_R} \eta^2 |Du|^2 dx & \\ \leqslant & \int_{B_R} \eta^2 f^2 dx + \Lambda_1 \int_{B_R} 2\eta |D\eta| \cdot |u - u_R| \cdot |Du| dx \\ & + \int_{B_R} 2\eta |D\eta| \cdot |u - u_R| \cdot \left| \widetilde{\widetilde{f}} \right| dx + \int_{B_R} \eta^2 |u - u_R| \cdot \left| \widetilde{f} \right| dx \\ & + \Lambda \int_{B_R} \eta^2 |u - u_R| \cdot |Du|^2 dx. \end{aligned} \tag{4.17}$$

Using Cauchy's inequality and the assumptions of the theorem, we easily find that

$$\begin{aligned} & \int_{B_R} 2\eta |D\eta| \cdot |u - u_R| \cdot |Du| dx \\ & \qquad \leqslant \varepsilon \int_{B_R} \eta^2 |Du|^2 dx + \frac{C}{\varepsilon} \frac{1}{R^2} \int_{B_R} |u - u_R|^2 dx, \\ & \int_{B_R} 2\eta |D\eta| \cdot |u - u_R| \cdot \left| \widetilde{\widetilde{f}} \right| dx \leqslant \int_{B_R} \eta^2 \left| \widetilde{\widetilde{f}} \right|^2 dx + \frac{C}{R^2} \int_{B_R} |u - u_R|^2 dx, \\ & \int_{B_R} \eta^2 |u - u_R| \cdot \left| \widetilde{f} \right| dx \leqslant 2M \int_{B_R} \left| \widetilde{f} \right| dx, \\ & \int_{B_R} \eta^2 |u - u_R| \cdot |Du|^2 dx \leqslant 2M \int_{B_R} \eta^2 |Du|^2 dx. \end{aligned}$$

Substituting these estimates into (4.17) and using the Sobolev-Poincaré inequality, we obtain

$$\begin{aligned} \lambda \int_{B_R} \eta^2 |Du|^2 dx \leqslant & (2\Lambda M + \varepsilon \Lambda_1) \int_{B_R} \eta^2 |Du|^2 dx \\ & + \frac{C}{\varepsilon} \frac{\Lambda_1}{R^2} \left[\int_{B_R} |Du|^{2n/(n+2)} dx \right]^{(n+2)/(2n)} \\ & + \int_{B_R} \left(|f|^2 + \left| \widetilde{\widetilde{f}} \right|^2 \right) dx + 2M \int_{B_R} \left| \widetilde{f} \right| dx. \end{aligned}$$

By condition 3°, $\lambda - 2\Lambda M > 0$; and therefore if we choose $\varepsilon = (\lambda - 2\Lambda M)/(2\Lambda_1)$, then

$$\begin{aligned} \int_{B_R} \eta^2 |Du|^2 dx \leqslant & C \left\{ \frac{1}{R^2} \left[\int_{B_R} |Du|^{2n/(n+2)} dx \right]^{(n+2)/(2n)} \right. \\ & \left. + \int_{B_R} \left(|f|^2 + \left| \widetilde{\widetilde{f}} \right|^2 + \left| \widetilde{f} \right| \right) dx \right\}, \end{aligned}$$

where $C = C(n, N, \lambda, \Lambda, \Lambda_1, M)$. Dividing both sides by R^n, we conclude that

$$\begin{aligned}\fint_{B_{R/2}(x_0)} |Du|^2 dx \leqslant C\Big\{\Big[\fint_{B_R(x_0)} |Du|^{2n/(n+2)} dx\Big]^{(n+2)/(2n)} \\ + \fint_{B_R(x_0)} \Big(|f|^2 + \Big|\widetilde{\widetilde{f}}\Big|^2 + |\widetilde{f}|\Big) dx\Big\}.\end{aligned}$$

Now we can proceed as in Theorem 4.3. Applying the reverse Hölder inequality (Theorem 4.1), we conclude the proof. □

Remark 4.1. Condition 3° ($2\Lambda M < \lambda$) in Theorem 4.4 cannot be dropped. There is an example (see [HB], [GQ1]) showing that Theorem 4.4 is not valid if $\Lambda M < \lambda$ is not satisfied. However, whether this condition can be relaxed to $\Lambda M < \lambda$ is still an open question.

Remark 4.2. If the growth condition (4.14) in Theorem 4.4 is replaced by the stronger condition

$$(4.14)' \qquad \begin{cases} |A_i^\alpha(x, u, p)| \leqslant \Lambda_1(M)|p| + f_i^\alpha(x), \\ |B_i(x, u, p)| \leqslant \Lambda(M)|p|^{2-\delta} + f_i(x), \end{cases}$$

where δ is a small positive constant, then condition 3° can be dropped (we still discuss a weak solution in $H^1 \cap L^\infty(\Omega, \mathbb{R}^N)$).

In fact, (4.17) now becomes

$$(4.17)' \qquad \begin{aligned} &\lambda \int_{B_R} \eta^2 |Du|^2 dx \\ &\leqslant \int_{B_R} \eta^2 f^2 dx + \Lambda_1 \int_{B_R} 2\eta|D\eta| \cdot |u - u_R| \cdot |Du| dx \\ &\quad + \int_{B_R} 2\eta|D\eta| \cdot |u - u_R| \cdot \Big|\widetilde{\widetilde{f}}\Big| dx + \int_{B_R} \eta^2 |u - u_R| \cdot |\widetilde{f}| dx \\ &\quad + \Lambda \int_{B_R} \eta^2 |u - u_R| \cdot |Du|^{2-\delta} dx. \end{aligned}$$

The last term on the right-hand side of the above inequality can be estimated as follows:

$$\begin{aligned} &\Lambda \int_{B_R} \eta^2 |u - u_R| \cdot |Du|^{2-\delta} dx \\ &\leqslant C\Big[\int_{B_R} |u - u_R|^{2/\delta} dx\Big]^{\delta/2} \Big[\int_{B_R} |Du|^2 dx\Big]^{(2-\delta)/2} \\ &\leqslant C(M)\Big[\int_{B_R} |u - u_R|^2 dx\Big]^{\delta/2} \Big[\int_{B_R} |Du|^2 dx\Big]^{(2-\delta)/2} \\ &\leqslant C(M)\Big[R^2 \int_{B_R} |Du|^2 dx\Big]^{\delta/2} \Big[\int_{B_R} |Du|^2 dx\Big]^{(2-\delta)/2} \\ &= C(M) R^\delta \int_{B_R} |Du|^2 dx. \end{aligned}$$

From (4.17)′, we deduce that, for $R < R_0$,

$$\begin{aligned}\fint_{B_{R/2}(x^0)} |Du|^2 dx \leqslant\ & C\Big\{\Big[\fint_{B_R(x^0)} |Du|^{2n/(n+2)} dx\Big]^{(n+2)/(2n)} \\ &+ \fint_{B_R(x^0)} \Big(|f|^2 + \big|\tilde{\tilde{f}}\big|^2 + |\tilde{f}|\Big) dx\Big\} \\ &+ \frac{1}{2} \fint_{B_R(x^0)} |Du|^2 dx,\end{aligned}$$

where R_0 depends on M, δ. We can then apply Theorem 4.1 to obtain $|Du| \in L^p(B_{R/2})$ for some $p > 2$ and all $B_R \subset \Omega$, where $B_{R/2}$ and B_R are concentric balls.

5. Direct methods for studying regularity

In this section, we use the following quasilinear elliptic system as an example to introduce the direct method for studying partial regularity. Consider

$$(5.1) \qquad -D_\alpha(A_{ij}^{\alpha\beta}(x,u) D_\beta u^j) = -D_\alpha f_i^\alpha(x) + f_i(x) \quad (i = 1, \cdots, N).$$

We assume throughout this section that

$$(5.2) \qquad A_{ij}^{\alpha\beta}(x,u)\xi_\alpha^i \xi_\beta^j \geqslant \lambda |\xi|^2, \quad \lambda > 0,$$

$$(5.3) \qquad |A_{ij}^{\alpha\beta}(x,u)| \leqslant \Lambda \quad (\alpha, \beta = 1, \cdots, n;\ i, j = 1, \cdots, N),$$

where λ, Λ are constants.

Theorem 5.1. *Suppose that the $A_{ij}^{\alpha\beta}(x,u)$ satisfy* (5.2), (5.3) *and are uniformly continuous on $\overline{\Omega} \times \mathbb{R}^N$, $f_i^\alpha \in L^p(\Omega)$, $p > n$, and $f_i \in L^q(\Omega)$, $q > n/2$. If $u \in H^1_{loc}(\Omega, \mathbb{R}^N)$ is a weak solution of* (5.1), *then there exists an open set $\Omega_0 \subset \Omega$ such that $u \in C^{0,\delta}_{loc}(\Omega_0, \mathbb{R}^N)$, where $\delta = \min(1-n/p, 2-n/q)$, and* $\operatorname{meas}(\Omega \setminus \Omega_0) = 0$.

Proof. The proof is divided into two steps.

Step 1. Show that $\forall \tilde{\Omega} \subset\subset \Omega$, $\forall x^0 \in \tilde{\Omega}$, $\forall \rho, R: 0 < \rho < R < \operatorname{dist}(\tilde{\Omega}, \partial\Omega) \wedge R_0$, we have

$$(5.4) \qquad \int_{B_\rho(x^0)} |Du|^2 dx \leqslant C\Big[\Big(\frac{\rho}{R}\Big)^n + \chi(x^0, R)\Big] \int_{B_R(x^0)} |Du|^2 dx + CR^{n-2+2\delta},$$

where $\delta = \min(1-n/p, 2-n/q)$, $\chi(x^0, R) = \omega^{(\sigma-2)/\sigma}\Big(R^2 + R^{2-n}\displaystyle\int_{B_R(x^0)} |Du|^2 dx\Big)$, $\sigma > 2$, $\omega(t)$ is a continuous, bounded, nonnegative, nondecreasing and concave function, $\omega(0) = 0$; C depends on $n, N, \lambda, \Lambda, p, q$, $\operatorname{dist}(\tilde{\Omega}, \partial\Omega)$, $\|\tilde{\tilde{f}}\|_{L^p}$, $\|\tilde{f}\|_{L^q}$ $\Big($here $\tilde{\tilde{f}} = (f_i^\alpha)$, $\tilde{f} = (f_i)\Big)$, and $R_0 < 1$ depends on u.

For $x^0 \in \tilde{\Omega}$, for each R, $0 < R < \operatorname{dist}(\tilde{\Omega}, \partial\Omega) \wedge 1$, we choose v to be the solution of the following Dirichlet problem:

$$\begin{cases} \displaystyle\int_{B_{R/2}(x^0)} A_{ij}^{\alpha\beta}(x^0, u_R) D_\beta v^j D_\alpha \varphi^i dx = 0, & \forall \varphi \in H^1_0(B_{R/2}(x^0), \mathbb{R}^N), \\ v - u \in H^1_0(B_{R/2}(x^0), \mathbb{R}^N). \end{cases}$$

Since v satisfies a homogeneous elliptic system of constant coefficients, so does Dv. Therefore, by the estimate (2.4) in Lemma 2.2 in Chapter 9, we have

$$\int_{B_\rho(x^0)} |Dv|^2 dx \leqslant C\Big(\frac{\rho}{R}\Big)^n \int_{B_{R/2}(x^0)} |Dv|^2 dx \quad \text{for } \rho < \frac{R}{2}. \tag{5.5}$$

Let $w = u - v$. Then

$$\int_{B_\rho(x^0)} |Du|^2 dx \leqslant C\Big(\frac{\rho}{R}\Big)^n \int_{B_{R/2}(x^0)} |Du|^2 dx + C\int_{B_{R/2}(x^0)} |Dw|^2 dx, \tag{5.6}$$

where $C = C(n, N, \lambda, \Lambda, \operatorname{dist}(\widetilde{\Omega}, \partial\Omega))$. Now we estimate $\int_{B_{R/2}(x^0)} |Dw|^2 dx$. It is clear that $w \in H_0^1(B_{R/2}(x^0), \mathbb{R}^N)$ satisfies

$$\begin{aligned}
&\int_{B_{R/2}(x^0)} A_{ij}^{\alpha\beta}(x^0, u_R) D_\beta w^j D_\alpha \varphi^i dx \\
&\quad = \int_{B_{R/2}(x^0)} \Big[A_{ij}^{\alpha\beta}(x^0, u_R) - A_{ij}^{\alpha\beta}(x, u)\Big] D_\beta u^j D_\alpha \varphi^i dx \\
&\qquad + \int_{B_{R/2}(x^0)} \Big[f_i^\alpha D_\alpha \varphi^i + f_i \varphi^i\Big] dx, \quad \forall \varphi \in H_0^1(B_{R/2}(x^0), \mathbb{R}^N).
\end{aligned}$$

Choosing $\varphi = w$ in the above equality, we obtain

$$\begin{aligned}
&\int_{B_{R/2}(x^0)} |Dw|^2 dx \\
&\quad \leqslant C\int_{B_{R/2}(x^0)} \sum |A_{ij}^{\alpha\beta}(x^0, u_R) - A_{ij}^{\alpha\beta}(x, u)|^2 |Du|^2 dx \\
&\qquad + C\int_{B_{R/2}(x^0)} \Big|\widetilde{\widetilde{f}}\Big|^2 dx \\
&\qquad + \Big[\int_{B_{R/2}(x^0)} |w|^r dx\Big]^{1/r} \cdot \Big[\int_{B_{R/2}(x^0)} |\widetilde{f}|^{r/(r-1)} dx\Big]^{1-1/r}.
\end{aligned} \tag{5.7}$$

We take $r = 2^*$ (i.e., $r = 2n/(n-2)$ when $n > 2$, $2 \leqslant r < +\infty$ when $n = 2$) in the above inequality. For definiteness we consider only the case $n > 2$; the case $n = 2$ is left as an exercise for the reader. Applying the Sobolev embedding theorem, we obtain

$$\begin{aligned}
&\Big[\int_{B_{R/2}(x^0)} |w|^r dx\Big]^{1/r} \cdot \Big[\int_{B_{R/2}(x^0)} |\widetilde{f}|^{r/(r-1)} dx\Big]^{1-1/r} \\
&\quad = \Big[\int_{B_{R/2}(x^0)} |w|^{2n/(n-2)} dx\Big]^{(n-2)/(2n)} \Big[\int_{B_{R/2}(x^0)} |\widetilde{f}|^{2n/(n+2)} dx\Big]^{(n+2)/(2n)} \\
&\quad \leqslant C\Big[\int_{B_{R/2}(x^0)} |Dw|^2 dx\Big]^{1/2} \Big[\int_{B_{R/2}(x^0)} |\widetilde{f}|^{2n/(n+2)} dx\Big]^{(n+2)/(2n)} \\
&\quad \leqslant \varepsilon\int_{B_{R/2}(x^0)} |Dw|^2 dx + \frac{C}{\varepsilon}\Big[\int_{B_{R/2}(x^0)} |\widetilde{f}|^{2n/(n+2)} dx\Big]^{(n+2)/n}.
\end{aligned}$$

Substituting this inequality into (5.7) and choosing $\varepsilon = \lambda/2$, we then obtain

$$
\begin{aligned}
&\int_{B_{R/2}(x^0)} |Dw|^2 dx \\
(5.8) \qquad &\leqslant C\int_{B_{R/2}(x^0)} \sum |A_{ij}^{\alpha\beta}(x^0, u_R) - A_{ij}^{\alpha\beta}(x,u)|^2 |Du|^2 dx \\
&+C\int_{B_{R/2}(x^0)} \left|\tilde{\tilde{f}}\right|^2 dx + C\left[\int_{B_{R/2}(x^0)} |\tilde{f}|^{2n/(n+2)} dx\right]^{(n+2)/n},
\end{aligned}
$$

where $C = C(n, N, \lambda)$.

Using Hölder's inequality, we easily get

$$
\int_{B_{R/2}(x^0)} \left|\tilde{\tilde{f}}\right|^2 dx \leqslant \left\|\tilde{\tilde{f}}\right\|^2_{L^p(\Omega, \mathbb{R}^{nN})} R^{n-2+2(1-n/p)},
$$

$$
\left[\int_{B_{R/2}(x^0)} |\tilde{f}|^{2n/(n+2)} dx\right]^{(n+2)/n} \leqslant \|\tilde{f}\|^2_{L^q(\Omega, \mathbb{R}^N)} R^{n-2+2(2-n/q)}.
$$

Now we estimate the first term on the right-hand side of (5.8). Since the $A_{ij}^{\alpha\beta}(x,u)$ are uniformly bounded and uniformly continuous on $\overline{\Omega} \times \mathbb{R}^N$, there exists a continuous, nonnegative, nondecreasing function $\omega(t)$ such that

$$
|A_{ij}^{\alpha\beta}(x,u) - A_{ij}^{\alpha\beta}(y,v)| \leqslant \omega(|x-y|^2 + |u-v|^2), \quad \forall x, y \in \Omega, \ \forall u, v \in \mathbb{R}^N,
$$

with $\omega(0) = 0$. We shall assume without loss of generality that $\omega(t)$ is a concave function. In fact, if $\omega(t)$ is not concave, we can replace it by the function

$$
\inf\left\{\lambda(t) \;\middle|\; \lambda(t) \text{ is a continuous concave function, } \lambda(t) \geqslant \omega(t)\right\};
$$

it can be verified that this function is a concave function possessing all the remaining properties. Thus, we obtain

$$
\begin{aligned}
&\int_{B_{R/2}(x^0)} \sum |A_{ij}^{\alpha\beta}(x^0, u_R) - A_{ij}^{\alpha\beta}(x,u)|^2 |Du|^2 dx \\
&\qquad \leqslant C\int_{B_{R/2}(x^0)} \omega^2(R^2 + |u-u_R|^2)|Du|^2 dx.
\end{aligned}
$$

Substituting this inequality into (5.8), we get

(5.9)

$$
\int_{B_{R/2}(x^0)} |Dw|^2 dx \leqslant C\int_{B_{R/2}(x^0)} \omega^2(R^2 + |u-u_R|^2)|Du|^2 dx + CR^{n-2+2\delta},
$$

where $\delta = \min(1 - n/p, 2 - n/q)$, and C depends on $n, N, \lambda, \left\|\tilde{\tilde{f}}\right\|_{L^p}, \|\tilde{f}\|_{L^q}$.

Now we estimate the first term on the right-hand side of (5.9). First, Theorem 4.3′ implies that there exists $\sigma > 2$ such that $|Du| \in L^\sigma_{loc}(\Omega)$ and for any $B_{R/2} \subset B_R \subset \Omega$, there exists R_0 such that for $R < R_0$, we have the estimate

$$
\begin{aligned}
\left[\fint_{B_{R/2}} |Du|^\sigma dx\right]^{1/\sigma} \leqslant C\Bigg\{&\left[\fint_{B_R} |Du|^2 dx\right]^{1/2} + \left[\fint_{B_R} \left|\tilde{\tilde{f}}\right|^\sigma dx\right]^{1/\sigma} \\
&+ R\left[\fint_{B_R} |\tilde{f}|^{\sigma r/[2(r-1)]} dx\right]^{2(r-1)/(\sigma r)}\Bigg\},
\end{aligned}
$$

where $r = 2^*$, $C = C(n, N, \lambda, \Lambda)$, $\sigma = \sigma(n, N, \lambda, \Lambda, p, q)$, and R_0 depends on u.

Next, applying Hölder's inequality and the above estimate, we obtain

$$\begin{aligned}
&\int_{B_{R/2}(x^0)} \omega^2(R^2 + |u - u_R|^2)|Du|^2 dx \\
&\leqslant \left\{\int_{B_{R/2}(x^0)} |Du|^\sigma dx\right\}^{2/\sigma} \left\{\int_{B_{R/2}(x^0)} \omega^{2\sigma/(\sigma-2)} dx\right\}^{(\sigma-2)/\sigma} \\
&\leqslant CR^{2n/\sigma}\left\{ \fint_{B_R(x^0)} |Du|^2 dx + \left[\fint_{B_R(x^0)} \left|\tilde{\tilde{f}}\right|^\sigma dx\right]^{2/\sigma} \right. \\
&\quad \left. +R^2\left[\fint_{B_R(x^0)} |\tilde{f}|^{\sigma r/[2(r-1)]} dx\right]^{4(r-1)/(\sigma r)}\right\} \cdot \left\{\int_{B_R(x^0)} \omega^{2\sigma/(\sigma-2)} dx\right\}^{(\sigma-2)/\sigma}.
\end{aligned}$$

Therefore

$$\begin{aligned}
&\int_{B_{R/2}(x^0)} \omega^2(R^2 + |u - u_R|^2)|Du|^2 dx \\
&\leqslant CR^{-n(\sigma-2)/\sigma} \int_{B_R(x^0)} |Du|^2 dx \left\{\int_{B_R(x^0)} \omega^{2\sigma/(\sigma-2)} dx\right\}^{(\sigma-2)/\sigma} \\
&\quad +C\left\{\int_{B_R(x^0)} \left|\tilde{\tilde{f}}\right|^\sigma dx\right\}^{2/\sigma} \left\{\int_{B_R(x^0)} \omega^{2\sigma/(\sigma-2)} dx\right\}^{(\sigma-2)/\sigma} \qquad (5.10) \\
&\quad +CR^{2+2n/\sigma}\left\{\fint_{B_R} |\tilde{f}|^{\sigma r/[2(r-1)]} dx\right\}^{4(r-1)/(r\sigma)} \\
&\quad \times\left\{\int_{B_R(x^0)} \omega^{2\sigma/(\sigma-2)} dx\right\}^{(\sigma-2)/\sigma} \\
&\triangleq I + II + III.
\end{aligned}$$

By Jensen's inequality[3], we get

$$\fint_{B_R(x^0)} \omega(R^2 + |u - u_R|^2) dx \leqslant \omega\left(R^2 + \fint_{B_R(x^0)} |u - u_R|^2 dx\right).$$

---

[3]**Jensen's inequality.** *Let $\omega : \mathbb{R} \to \mathbb{R}$ be a concave function ($\omega'' \leqslant 0$ in the distribution sense), and let $f \in L^1(\Omega)$. Then*

$$\frac{1}{|\Omega|}\int_\Omega \omega(f(z))dz \leqslant \omega\left(\frac{1}{|\Omega|}\int_\Omega f(z)dz\right).$$

Proof. Let $\tilde{f} = \frac{1}{|\Omega|}\int_\Omega f(z)dz$. Since ω is concave, there is a real number m (depending on $\tilde{f}$) such that

$$\omega(x) \leqslant m \cdot (x - \tilde{f}) + \omega(\tilde{f}), \quad x \in \mathbb{R}.$$

It follows that

$$\omega(f(z)) \leqslant m \cdot (f(z) - \tilde{f}) + \omega(\tilde{f}), \quad \text{a.e. } \in \Omega.$$

Integrating over $z \in \Omega$ on both sides, we get

$$\int_\Omega \omega(f(z))dz \leqslant m\int_\Omega (f(z) - \tilde{f})dz + \omega(\tilde{f})|\Omega| = \omega(\tilde{f})|\Omega|.$$

Dividing by $|\Omega|$ on both sides, we obtain the desired result. □

The function ω is bounded. Therefore, by Jensen's inequality and Poincaré's inequality,

$$\begin{aligned} I &\leqslant C\int_{B_R(x^0)}|Du|^2dx\Big\{ ⨍_{B_R(x^0)} \omega\cdot\omega^{(\sigma+2)/(\sigma-2)}dx\Big\}^{(\sigma-2)/\sigma} \\ &\leqslant C\int_{B_R(x^0)}|Du|^2dx\Big\{ ⨍_{B_R(x^0)} \omega(R^2+|u-u_R|^2)dx\Big\}^{(\sigma-2)/\sigma} \\ &\leqslant C\omega^{(\sigma-2)/\sigma}\Big(R^2+⨍_{B_R(x^0)}|u-u_R|^2dx\Big)\int_{B_R(x^0)}|Du|^2dx \\ &\leqslant C\omega^{(\sigma-2)/\sigma}\Big(R^2+R^{2-n}\int_{B_R(x^0)}|Du|^2dx\Big)\int_{B_R(x^0)}|Du|^2dx. \end{aligned}$$

As for II and III, using Hölder's inequality and the boundedness of ω, we easily find that

$$\begin{aligned} II &= C\Big\{\int_{B_R(x^0)}\Big|\widetilde{\widetilde{f}}\Big|^\sigma dx\Big\}^{2/\sigma}\Big\{\int_{B_R(x^0)}\omega^{2\sigma/(\sigma-2)}dx\Big\}^{(\sigma-2)/\sigma} \\ &\leqslant C\Big\{\Big[\int_{B_R(x^0)}\Big|\widetilde{\widetilde{f}}\Big|^p dx\Big]^{\sigma/p}R^{n(1-\sigma/p)}\Big\}^{2/\sigma}R^{n(\sigma-2)/\sigma} \\ &\leqslant C\Big\|\widetilde{\widetilde{f}}\Big\|^2_{L^p(\Omega,\mathbb{R}^{nN})}R^{n-2+2(1-n/p)}, \\ III &= CR^{2+2n/\sigma}\Big\{R^{-n}\int_{B_R(x_0)}|\widetilde{f}|^{\sigma r/[2(r-1)]}dx\Big\}^{4(r-1)/(\sigma r)} \\ &\qquad\times\Big\{\int_{B_R(x^0)}\omega^{2\sigma/(\sigma-2)}dx\Big\}^{(\sigma-2)/\sigma} \\ &\leqslant CR^{2-2n/q+2n/\sigma}\Big\{\int_{B_R(x_0)}|\widetilde{f}|^q dx\Big\}^{2/q}\Big\{\int_{B_R(x^0)}\omega^{2\sigma/(\sigma-2)}dx\Big\}^{(\sigma-2)/\sigma} \\ &\leqslant C\|\widetilde{f}\|^2_{L^q(\Omega,\mathbb{R}^N)}R^{n-2+2(2-n/q)}. \end{aligned}$$

Now we substitute I, II, III into (5.10) to obtain

$$\begin{aligned} &\int_{B_{R/2}(x^0)}\omega^2(R^2+|u-u_R|^2)|Du|^2dx \\ &\quad\leqslant C\omega^{(\sigma-2)/2}\Big(R^2+R^{2-n}\int_{B_R(x^0)}|Du|^2dx\Big)\int_{B_R(x^0)}|Du|^2dx \\ &\qquad+CR^{n-2+2\delta}. \end{aligned}\tag{5.11}$$

Substituting (5.11) into (5.9), we get

$$\begin{aligned} &\int_{B_{R/2}(x^0)}|Dw|^2dx \\ &\quad\leqslant C\omega^{(\sigma-2)/2}\Big(R^2+R^{2-n}\int_{B_R(x^0)}|Du|^2dx\Big)\int_{B_R(x^0)}|Du|^2dx \\ &\qquad+CR^{n-2+2\delta}, \end{aligned}\tag{5.12}$$

where $\delta = \min(1 - n/p, 2 - n/q)$, and C depends on $n, N, \lambda, \Lambda, p, q$, $\left\|\widetilde{\widetilde{f}}\right\|_{L^p}$, $\|\widetilde{f}\|_{L^q}$.

Finally, substituting (5.12) into (5.6), we conclude that (5.4) is valid for $\rho < R/2$. It is obvious that (5.4) is also valid for $\rho \geqslant R/2$.

Step 2. Show that $u \in C^{0,\delta}_{loc}(\Omega_0, \mathbb{R}^N)$ for some open set $\Omega_0 \subset \Omega$ such that $\text{meas}(\Omega \setminus \Omega_0) = 0$.

Let $\widetilde{\Omega}$ be the same as before, $x^0 \in \widetilde{\Omega}$, and $0 < R < \text{dist}(\widetilde{\Omega}, \partial\Omega) \wedge R_0$. Set

$$\Phi(x^0, R) = R^{2-n} \int_{B_R(x^0)} |Du|^2 dx.$$

From the result in Step 1, we deduce that, for any $\tau \in (0, 1)$,

$$\Phi(x^0, \tau R) \leqslant C_1 \Big[1 + \chi(x^0, R)\tau^{-n}\Big]\tau^2 \Phi(x^0, R) + C_2 \tau^{2-n} R^{2\delta},$$

where $\delta = \min(1 - n/p, 2 - n/q)$, and C_1, C_2 depend on $n, N, \lambda, \Lambda, p, q$, $\left\|\widetilde{\widetilde{f}}\right\|_{L^p}$, and $\|\widetilde{f}\|_{L^q}$. We assume without loss of generality that $C_1 > 1$.

First, we choose γ such that $\delta < \gamma < 1$. Then, we choose τ such that $2C_1\tau^2 = \tau^{2\gamma}$. For this τ, there exist $\varepsilon_1 = \varepsilon_1(\tau)$ and $R_1 = R_1(\tau) \leqslant \text{dist}(\widetilde{\Omega}, \partial\Omega) \wedge R_0$ such that if $R < R_1$ and $\Phi(x^0, R) < \varepsilon_1^2$, then

$$\chi(x^0, R) = \omega^{(\sigma-2)/2}\Big(R^2 + \Phi(x^0, R)\Big) < \tau^n,$$

and therefore

$$\Phi(x^0, \tau R) \leqslant \tau^{2\gamma}\Phi(x^0, R) + H_0 R^{2\delta},$$

where $H_0 = C_2 \tau^{2-n}$.

Using iteration, we obtain

$$\begin{aligned} \Phi(x^0, \tau^k R) &\leqslant \tau^{2k\gamma}\Phi(x^0, R) + H_0(\tau^{k-1}R)^{2\delta} \sum_{j=0}^{\infty} [\tau^{2(\gamma-\delta)}]^j \\ &\leqslant \left[\Phi(x^0, R) + H_0 \frac{R^{2\delta}}{\tau^{2\delta} - \tau^{2\gamma}}\right]\tau^{2k\delta}. \end{aligned}$$

Now we choose R_2, $0 < R_2 < R_1(\tau)$, such that

$$H_0 \frac{R^{2\delta}}{\tau^{2\delta} - \tau^{2\gamma}} \leqslant \varepsilon_1^2 \quad \text{for } R < R_2.$$

Thus for this τ, we have $\varepsilon_1(\tau)$ and $R_2(\tau) > 0$ such that if $R < R_2$ and $\Phi(x^0, R) < \varepsilon_1^2$, then

$$\Phi(x^0, \tau^k R) \leqslant 2\varepsilon_1^2 \tau^{2k\delta}.$$

For any ρ, $0 < \rho < R$, we choose the positive integer k such that $\tau^{k+1}R < \rho \leqslant \tau^k R$. Then

$$\begin{aligned} \Phi(x^0, \rho) &= \rho^{2-n} \int_{B_\rho(x^0)} |Du|^2 dx \\ &\leqslant (\tau^{k+1}R)^{2-n} \int_{B_{\tau^k R}(x^0)} |Du|^2 dx \\ &\leqslant C\Phi(x^0, \tau^k R) \leqslant C\tau^{2k\delta} \leqslant C\Big(\frac{\rho}{R}\Big)^{2\delta}. \end{aligned}$$

Since $\Phi(x^0, R)$ is a continuous function of x^0, for any $x^0 \in \widetilde{\Omega}$ there exists some neighborhood $B_r(x^0)$ such that $\Phi(x, R) < \varepsilon_1^2$ for $x \in B_r(x^0)$. Hence,

$$\Phi(x,\rho) \leqslant C\Big(\frac{\rho}{R}\Big)^{2\delta}, \quad \forall x \in B_r(x^0),$$

i.e.,

$$\int_{B_\rho(x)} |Du|^2 dz \leqslant C\rho^{n-2+2\delta},$$

where C depends on $n, N, \lambda, \Lambda, p, q, \left\|\widetilde{\widetilde{f}}\right\|_{L^p}, \|\widetilde{f}\|_{L^q}$ and $\operatorname{dist}(\widetilde{\Omega}, \partial\Omega)$.

Applying Morrey's theorem on Hölder continuity (Theorem 1.1 in Chapter 9), we conclude that

$$u \in C^{0,\delta}_{loc}(B_r(x^0), \mathbb{R}^N).$$

Set

$$\Omega_0 = \{x^0 \in \Omega \mid \Phi(x^0, R) < \varepsilon_1^2 \text{ for some } R < R_2\},$$

where ε_1 and R_2 are constants determined as above. We can proceed as in §3 to prove that Ω_0 is an open subset of Ω, $u \in C^{0,\delta}_{loc}(\Omega_0, \mathbb{R}^N)$, and

$$\Omega_0 = \left\{x \in \Omega \,\middle|\, \liminf_{\rho\to 0} \rho^{2-n} \int_{B_\rho(x)} |Du|^2 dz = 0\right\},$$

and therefore $\operatorname{meas}(\Omega \setminus \Omega_0) = 0$. □

6. The singular set

In §§3 and 5 we proved that a weak solution of the elliptic system (3.1) or (5.1) with uniformly continuous coefficients is locally Hölder continuous in an open set Ω_0 such that $\operatorname{meas}(\Omega \setminus \Omega_0) = 0$. Now we give a further estimate on the Hausdorff dimension of the singular set $\Omega \setminus \Omega_0$.

6.1. Hausdorff measure.

Definition 6.1. Let $E \subset \mathbb{R}^n$, $0 \leqslant k < +\infty$, $0 < \delta \leqslant +\infty$, and let $\{F_j\}$ be a family of open sets in $\mathbb{R}^n$. Set

$$\mathscr{H}_k^\delta(E) = \omega_k 2^{-k} \inf\left\{\sum_j (\operatorname{diam} F_j)^k \,\middle|\, \bigcup_j F_j \supset E,\ \operatorname{diam} F_j < \delta\right\}, \tag{6.1}$$

and

$$\mathscr{H}_k(E) = \lim_{\delta\to 0} \mathscr{H}_k^\delta(E) = \sup_{\delta>0} \mathscr{H}_k^\delta(E), \tag{6.2}$$

where $\omega_k = \Gamma^k(1/2)/\Gamma(1+k/2)$. $\mathscr{H}_k(E)$ is said to be the k-dimensional *Hausdorff measure* of E.

Notice that if $0 < \delta_1 < \delta_2$, then $\mathscr{H}_k^{\delta_1}(E) \geqslant \mathscr{H}_k^{\delta_2}(E)$; therefore the limit in (6.2) exists (it may be equal to $+\infty$).

Example 6.1. Let E be a Lebesgue measurable set in $\mathbb{R}^n$. Then $\mathscr{H}_n(E) = \mathscr{L}_n(E)$, where $\mathscr{L}_n(E)$ denotes the Lebesgue measure of E.

Example 6.2. Let E be a set consisting of finitely many points P_i $(i = 1, \cdots, m)$. Then $\mathscr{H}_0(E) = m$, while $\mathscr{H}_1(E) = 0$.

Example 6.3. Let E be a rectifiable curve l in $\mathbb{R}^n$ $(n \geqslant 2)$. Then $\mathscr{H}_1(E) =$ the length of l, while $\mathscr{H}_2(E) = 0$.

Example 6.4. Let E be a rectifiable surface S in $\mathbb{R}^n$ $(n \geqslant 3)$. Then $\mathscr{H}_2(E) =$ the area of S, while $\mathscr{H}_3(E) = 0$.

From Definition 6.1, we easily deduce

Theorem 6.1. *If $\mathscr{H}_k(E) < +\infty$, then for any $\varepsilon > 0$,*

$$\mathscr{H}_{k+\varepsilon}(E) = 0.$$

We now introduce the following definition:

Definition 6.2. The real number

$$\inf\{k \in \mathbb{R}^+ \mid \mathscr{H}_k(E) = 0\}$$

is said to be the *Hausdorff dimension* of the set E and is denoted by $\dim_{\mathscr{H}} E$.

In the above examples, we have $\dim_{\mathscr{H}}\left(\bigcup_{i=1}^m P_i\right) = 0$, $\dim_{\mathscr{H}}(l) = 1$, $\dim_{\mathscr{H}} S = 2$.

6.2. Estimate for the Hausdorff dimension of the singular set.

Now we estimate the Hausdorff dimension of the singular set of a weak solution of the elliptic system (3.1) or (5.1). We have shown in Theorem 3.1 (or Theorem 5.1, respectively) that a weak solution of the elliptic system (3.1) (or (5.1)) is locally Hölder continuous in an open subset Ω_0 of Ω such that

$$\Omega_0 = \left\{x \in \Omega \;\middle|\; \liminf_{\rho \to 0} \rho^{2-n} \int_{B_\rho(x)} |Du|^2 dz = 0\right\}, \tag{6.3}$$

i.e., $\Omega \setminus \Omega_0 \subset \Sigma$, where

$$\Sigma = \left\{x \in \Omega \;\middle|\; \liminf_{\rho \to 0} \rho^{2-n} \int_{B_\rho(x)} |Du|^2 dz > 0\right\}. \tag{6.4}$$

Theorem 4.3′ implies that $|Du| \in L^p_{loc}(\Omega)$ for some $p > 2$. By Hölder's inequality,

$$\left\{\rho^{2-n} \int_{B_\rho(x)} |Du(z)|^2 dz\right\}^{1/2} \leqslant \left\{\rho^{p-n} \int_{B_\rho(x)} |Du(z)|^p dx\right\}^{1/p}.$$

Therefore $\Sigma \subset E_{n-p}$, where

$$E_{n-p} = \left\{x \in \Omega \;\middle|\; \limsup_{\rho \to 0} \rho^{p-n} \int_{B_\rho(x)} |Du(z)|^p dx > 0\right\}. \tag{6.5}$$

In order to estimate the Hausdorff dimension of $\Omega \setminus \Omega_0$, it suffices to estimate the Hausdorff dimension of E_{n-p}. We first prove the following lemma:

Lemma 6.2 (Covering lemma). *Let G be a bounded set in $\mathbb{R}^n$. If $r : x \mapsto r(x)$ is a function defined on G such that $0 < r(x) < 1$, then there exists a sequence of points $\{x_i\}$, $x_i \in G$ $(i = 1, 2, \cdots)$, such that*

$$B(x_i, r(x_i)) \cap B(x_j, r(x_j)) = \emptyset, \quad i \neq j, \tag{6.6}$$

$$\bigcup_i B(x_i, 3r(x_i)) \supset G. \tag{6.7}$$

Proof. Consider the following family of balls:

$$B_{2^{-k},2^{-k-1}} = \left\{B(x, r(x))\Big|\, x \in G,\ 2^{-k-1} \leqslant r(x) < 2^{-k}\right\}, \qquad k = 0, 1, 2, \cdots.$$

Since G is bounded, we can choose finitely many (say n_1) balls which comprise a subfamily of $B_{1,1/2}$, given by

$$\widetilde{B}_{1,1/2} = \left\{B(x_i, r(x_i))\Big|\, x_i \in G,\ \frac{1}{2} \leqslant r(x_i) < 1,\ i = 1, \cdots, n_1\right\},$$

such that

(1) $B(x_i, r(x_i)) \cap B(x_j, r(x_j)) = \emptyset$, $i \neq j$, $i, j = 1, \cdots, n_1$;

(2) every ball in $B_{1,1/2}$ must intersect with at least one ball in $\widetilde{B}_{1,1/2}$.

Next, we choose a subfamily of $B_{1/2,1/4}$, given by

$$\widetilde{B}_{1/2,1/4} = \left\{B(x_i, r(x_i))\Big|\, x_i \in G,\ \frac{1}{4} \leqslant r(x_i) < \frac{1}{2},\ i = n_1 + 1, \cdots, n_2\right\}, \quad n_2 \geqslant n_1,$$

such that

(1) $B(x_i, r(x_i)) \cap B(x_j, r(x_j)) = \emptyset$, $i \neq j$, $i, j = 1, \cdots, n_2$;

(2) every ball in $B_{1/2,1/4}$ must intersect with at least one ball in

$$\widetilde{B}_{1,1/2} \cup \widetilde{B}_{1/2,1/4} = \left\{B(x_i, r(x_i))\Big|\, x_i \in G, \quad i = 1, \cdots, n_2\right\}.$$

(It could happen that $n_2 = n_1$; in this case $\widetilde{B}_{1/2,1/4} = \emptyset$.)

Continuing this process, if $x_1, \cdots, x_{n_k}$ are chosen, then we choose a subfamily of $B_{2^{-k},2^{-k-1}}$, given by

$$\begin{aligned}\widetilde{B}_{2^{-k},2^{-k-1}} \quad = \quad &\Big\{B(x_i, r(x_i))\Big|\, x_i \in G,\ 2^{-k-1} \leqslant r(x_i) < 2^{-k}, \\ &\qquad i = n_k + 1, \cdots, n_{k+1}\Big\}, \qquad n_{k+1} \geqslant n_k,\end{aligned}$$

with the following properties:

(1) $B(x_i, r(x_i)) \cap B(x_j, r(x_j)) = \emptyset$, $i \neq j$, $i, j = 1, \cdots, n_{k+1}$;

(2) every ball in $B_{2^{-k},2^{-k-1}}$ must intersect with at least one ball in

$$\bigcup_{j=0}^{k} \widetilde{B}_{2^{-j},2^{-j-1}} = \left\{B(x_i, r(x_i))\Big|\, x_i \in G, \quad i = 1, \cdots, n_{k+1}\right\}.$$

(It could happen that $n_{k+1} = n_k$; in this case $\widetilde{B}_{2^{-k},2^{-k-1}} = \emptyset$.)

By our selection, it is obvious that (6.6) is satisfied. We next prove (6.7). In fact, for $x \in G$, we can find some x_i such that

$$B(x, r(x)) \cap B(x_i, r(x_i)) \neq \emptyset,$$

and $2r(x_i) \geqslant r(x)$. Thus

$$|x - x_i| \leqslant r(x) + r(x_i) \leqslant 3r(x_i),$$

i.e., $x \in B(x_i, 3r(x_i))$. The proof is complete. □

Lemma 6.3. *Let Ω be an open set in $\mathbb{R}^n$, $v \in L^1_{loc}(\Omega)$, and $0 \leqslant \alpha < n$. For the set*

$$E_\alpha = \left\{ x \in \Omega \;\middle|\; \limsup_{\rho \to 0} \rho^{-\alpha} \int_{B_\rho(x)} |v(z)| dz > 0 \right\},$$

we have $\mathscr{H}_\alpha(E_\alpha) = 0$.

Proof. It suffices to prove that for any compact $K \subset \Omega$,

$$\mathscr{H}_\alpha(E_\alpha \cap K) = 0.$$

Setting $F = E_\alpha \cap K$ and

$$F^{(s)} = \left\{ x \in F \;\middle|\; \limsup_{\rho \to 0} \rho^{-\alpha} \int_{B_\rho(x)} |v(z)| dz > \frac{1}{s} \right\} \quad (s = 1, 2, \cdots);$$

we then have $F^{(1)} \subset F^{(2)} \subset \cdots$ and

$$F = \bigcup_{s=1}^{\infty} F^{(s)}.$$

Therefore it suffices to show that $\mathscr{H}_\alpha(F^{(s)}) = 0$ for each positive integer s. Let Q be a bounded open set satisfying $K \subset Q \subset \overline{Q} \subset \Omega$. For any $x \in F^{(s)}$ and any δ, $0 < \delta < \operatorname{dist}(K, \partial Q) \wedge 1$, we deduce from the definition of $F^{(s)}$ that there exists $r(x)$, $0 < r(x) < \delta$, such that

$$r^{-\alpha}(x) \int_{B(x, r(x))} |v(z)| dz \geqslant \frac{1}{2s}. \tag{6.8}$$

We choose in the covering lemma $G = F^{(s)}$ and $r = r(x)$ as in (6.8). Then there exists $\{x_j\} \subset G$ such that $B(x_i, r(x_i)) \cap B(x_j, r(x_j)) = \emptyset$ $(i \neq j)$, and $\{B(x_j, 3r(x_j))\}$ covers $F^{(s)}$. It is also clear that $\bigcup B_j \subset Q$, where $B_j = B(x_j, r(x_j))$.

Now we estimate

$$\sum_j \omega_\alpha 2^{-\alpha} (\operatorname{diam} B(x_j, 3r_j))^\alpha = \omega_\alpha 2^{-\alpha} \sum_j (6r_j)^\alpha, \tag{6.9}$$

where $r_j = r(x_j)$. It suffices to estimate $\sum_j r_j^\alpha$. Choosing $x = x_j$ $(j = 1, 2 \cdots)$ in (6.8), we obtain

$$r_j^\alpha \leqslant 2s \int_{B_j} |v(z)| dz.$$

Summing over all j's, we get

$$\sum_j r_j^\alpha \leqslant 2s \sum_j \int_{B_j} |v(z)| dz. \tag{6.10}$$

As the B_j are disjoint, we have

$$\sum_j \int_{B_j} |v(z)|dz = \int_{\bigcup_j B_j} |v(z)|dz. \tag{6.11}$$

We can show that for $0 < 6r_j < \delta$,

$$\text{meas}\Big(\bigcup_j B_j\Big) \to 0, \quad \text{as } \delta \to 0. \tag{6.12}$$

In fact,

$$\begin{aligned}
\text{meas}\Big(\bigcup_j B_j\Big) &= \text{meas}\Big(\bigcup_j B(x_j, r_j)\Big) \\
&= \omega_n \sum_j r_j^n \leqslant \omega_n \Big(\frac{\delta}{6}\Big)^{n-\alpha} \sum_j r_j^\alpha \\
&\leqslant C(n,\alpha)\delta^{n-\alpha} \cdot 2s \int_Q |v(z)|dz \to 0, \quad \text{as } \delta \to 0.
\end{aligned}$$

Using (6.12) and the absolute continuity of Lebesgue integrals, we obtain

$$\int_{\bigcup_j B_j} |v(z)|dz \to 0, \quad \text{as } \delta \to 0. \tag{6.13}$$

Combining (6.10), (6.11), (6.13), we conclude that

$$\sum_j r_j^\alpha \to 0, \quad \text{as } \delta \to 0. \tag{6.14}$$

By the definition 6.1, (6.9) and (6.14),

$$\mathscr{H}_\alpha(F^{(s)}) = 0.$$

The proof is complete. □

Now we prove

Theorem 6.4. *Suppose that Ω is a bounded domain in $\mathbb{R}^n$, that u is a weak solution of the elliptic system* (3.1) (*or* (5.1), *respectively*), *and that the assumptions of Theorem 3.1* (*or Theorem 5.1, respectively*) *hold. Then there exists an open set $\Omega_0 \subset \Omega$ such that $u \in C^{0,\delta}_{loc}(\Omega_0, \mathbb{R}^N)$, $\delta > 0$, and for some $p > 2$ we have*

$$\mathscr{H}_{n-p}(\Omega \setminus \Omega_0) = 0.$$

Proof. By Theorem 4.3′, we deduce that $|Du| \in L^p_{loc}$ for some $p > 2$. If $p > n$, then the embedding theorem implies that $u \in C^{0,\delta}_{loc}(\Omega, \mathbb{R}^N)$, $\delta > 0$, and hence $\Omega_0 = \Omega$. If $2 < p \leqslant n$, then in Lemma 6.3 we choose $v = |Du|^p$, $\alpha = n-p$ to conclude that $\mathscr{H}_{n-p}(E_{n-p}) = 0$, where E_{n-p} is defined in (6.5). Since $\Omega \setminus \Omega_0 \subset E_{n-p}$, we have $\mathscr{H}_{n-p}(\Omega \setminus \Omega_0) = 0$. □

APPENDIX 1

Sobolev Spaces

Sobolev spaces are fundamental in modern partial differential equation theory. Other than Poincaré's inequality, we shall list only the results used in this book and refer readers to [AD], [MJ] for the proofs.

1. Weak derivatives and the Sobolev space $W^{k,p}(\Omega)$

We first introduce the concept of weak derivatives. We use the following notation:

1° Multi-index. We say $\alpha = (\alpha_1, \cdots, \alpha_n)$ is a multi-index, where α_i $(i = 1, \cdots, n)$ are nonnegative integers. We use the notation $|\alpha| = \alpha_1 + \cdots + \alpha_n$ and $\alpha! = \alpha_1! \cdots \alpha_n!$; $\alpha \leqslant \beta$ means $\alpha_i \leqslant \beta_i$ $(i = 1, \cdots, n)$, and $\binom{\beta}{\alpha} = \frac{\beta!}{\alpha!(\beta-\alpha)!}$ $(\alpha \leqslant \beta)$.

2° Notation for higher order derivatives. Let Ω be a bounded domain in $\mathbb{R}^n$, $x = (x_1, \cdots, x_n)$ a point in Ω, and $u : \Omega \to \mathbb{R}$ a function. We write $D^\alpha u(x) = D_1^{\alpha_1} D_2^{\alpha_2} \cdots D_n^{\alpha_n} u(x) = \frac{\partial^{|\alpha|} u}{\partial x_1^{\alpha_1} \cdots \partial x_n^{\alpha_n}}$ to denote the α-th derivative of u (if it exists).

Definition 1.1. Let $u \in L^1_{loc}(\Omega)$, and let α be a multi-index. If $v \in L^1_{loc}(\Omega)$ satisfies

$$\int_\Omega u D^\alpha \varphi dx = (-1)^{|\alpha|} \int_\Omega v \varphi dx, \quad \forall \varphi \in C_0^\infty(\Omega),$$

then v is said to be the *α-th weak derivative* of u, and we write $v = D^\alpha u$.

It can be shown that a weak derivative $D^\alpha u$ is uniquely determined up to a zero measure set.

If all first order weak derivatives exist for a function, this function is said to be weakly differentiable. If all its weak derivatives up to (and including) k-th order exist, then it is said to be k times weakly differentiable. We denote by $W^k(\Omega)$ the space of all k times weakly differentiable functions.

Definition 1.2. Let k be a nonnegative integer, $p \geqslant 1$ a real number, and Ω a bounded domain in $\mathbb{R}^n$. The set

$$\left\{ u \in W^k(\Omega) \,\middle|\, D^\alpha u \in L^p(\Omega),\ \forall |\alpha| \leqslant k \right\},$$

with the norm

$$\|u\|_{W^{k,p}(\Omega)} = \left\{ \int_\Omega \sum_{|\alpha|\leqslant k} |D^\alpha u|^p dx \right\}^{1/p}, \tag{1.1}$$

forms a normed linear space and is called the Sobolev space $W^{k,p}(\Omega)$.

It can be shown that $W^{k,p}(\Omega)$ under the norm (1.1) is a Banach space. If $p=2$, we simply denote $W^{k,2}(\Omega)$ by $H^k(\Omega)$.

Definition 1.3. $W_0^{k,p}(\Omega)$ is the closure of $C_0^\infty(\Omega)$ in $W^{k,p}(\Omega)$.

Proposition 1.1. *$W_0^{k,p}(\mathbb{R}^n) = W^{k,p}(\mathbb{R}^n)$, and $W_0^{0,p}(\Omega) = W^{0,p}(\Omega) = L^p(\Omega)$. In general, however, $W_0^{k,p}(\Omega)$ is a nontrivial subspace of $W^{k,p}(\Omega)$.*

Next, we state two propositions on approximations in $W^{k,p}(\Omega)$.

Proposition 1.2. *For $1 \leqslant p < \infty$, the subspace $C^\infty(\Omega) \cap W^{k,p}(\Omega)$ is dense in $W^{k,p}(\Omega)$.*

In general, however, the subspace $C^k(\overline{\Omega})$ is not dense in $W^{k,p}(\Omega)$.

Definition 1.4. A domain Ω is said to have the *line segment property*, if for any $x \in \partial\Omega$, there exist an open set U_x and a nonzero vector y_x such that $x \in U_x$, and $z \in \overline{\Omega} \cap U_x$ imply $z + ty_x \in \Omega$ $(\forall\, 0 < t < 1)$.

Any domain Ω satisfying the line segment property has its $n-1$ dimensional boundary, and Ω cannot lie on both sides of the boundary at the same time.

Proposition 1.3. *If Ω satisfies the line segment property, then for $1 \leqslant p < \infty$, the subspace $C^k(\overline{\Omega})$ is dense in $W^{k,p}(\Omega)$.*

Using approximation, we can generalize some rules in classical calculus to weak derivatives. For example, we have

Proposition 1.4. *Let $u \in W^{k,p}(\Omega)$, $\varphi \in C_0^k(\Omega)$. Then $\varphi u \in W^{k,p}(\Omega)$, and*

$$D^\alpha(\varphi u) = \sum_{\beta \leqslant \alpha} \binom{\alpha}{\beta} D^\beta u D^{\alpha-\beta}\varphi, \quad \forall |\alpha| \leqslant k.$$

Proposition 1.5. *Suppose that f is continuous on $\mathbb{R}$ and piecewise continuously differentiable, $f' \in L^\infty(\mathbb{R})$. If $u \in W^1(\Omega)$, then $f \circ u \in W^1(\Omega)$, and*

$$D(f \circ u) = \begin{cases} f'(u)Du & \text{if } u \notin L, \\ 0 & \text{if } u \in L, \end{cases}$$

where L is the set of all corner points of f (i.e., those points where f is not differentiable).

Next, we state the embedding theorem and the compact embedding theorem. First we introduce the following definition.

Definition 1.5. Let X and Y be Banach spaces, $X \subset Y$. If the identity map $I : X \to Y$ is bounded, i.e.,

$$\|Ix\|_Y \leqslant C\|x\|_X, \quad \forall x \in X,$$

for some constant C independent of x, then we say that X can be *embedded* in Y, and we write $X \hookrightarrow Y$; the constant C is said to be the *embedding constant.* If furthermore $I : X \to Y$ is compact, then we say that X is *compactly embedded* in Y.

Definition 1.6. Let Ω be a bounded domain in $\mathbb{R}^n$. Suppose that for any $x_0 \in \partial\Omega$, there exist $r > 0$, $i \in \{1, \cdots, n\}$ and a Lipschitz function $f : \mathbb{R}^{n-1} \to \mathbb{R}$, such that

$$\Omega \cap B(x_0, r) = \{x \in B(x_0, r),\ x_i > f(x_1, \cdots, x_{i-1}, x_{i+1}, \cdots, x_n)\}$$

and

$$(\mathbb{R}^n \setminus \overline{\Omega}) \cap B(x_0, r) = \{x \in B(x_0, r),\ x_i < f(x_1, \cdots, x_{i-1}, x_{i+1}, \cdots, x_n)\}.$$

Then Ω is said to be *locally Lipschitz.*

Proposition 1.6 (Embedding theorem [AD, Theorem 5.4]).

Let Ω be a bounded domain, k a nonnegative integer, and $1 \leqslant p \leqslant +\infty$.

1° *If Ω satisfies an interior cone condition (see Definition 1.3 in Chapter 2), then*

(1) for $kp < n$, $W^{k,p}(\Omega) \hookrightarrow L^q(\Omega)$, $\forall q \in [p, p^*]$, where $p^* = np/(n - kp)$;

(2) for $kp = n$, $W^{k,p}(\Omega) \hookrightarrow L^q(\Omega)$, $\forall q \in [p, +\infty)$.

2° *If Ω is locally Lipschitz, then for $kp > n$,*

$$W^{k,p}(\Omega) \hookrightarrow C^{k-1-[n/p],\alpha}(\overline{\Omega}), \qquad \forall \alpha \in [0, \alpha_0],$$

where $\alpha_0 = [n/p] + 1 - (n/p)$ if n/p is not an integer, and α_0 can be chosen to be any positive number less than 1 if n/p is an integer.

In all the above cases, the embedding constants depend on n, k, p, Ω.

Proposition 1.7 (Compact embedding theorem [AD, Theorem 6.2]).

Let Ω be a bounded domain, k a nonnegative integer, and $1 \leqslant p \leqslant +\infty$.

1° *If Ω satisfies an interior cone condition (see Definition 1.3 in Chapter 2), then the following embeddings are compact:*

(1) for $kp < n$, $W^{k,p}(\Omega) \hookrightarrow L^q(\Omega)$, $\forall q \in [p, p^*)$, where $p^* = np/(n - kp)$;

(2) for $kp = n$, $W^{k,p}(\Omega) \hookrightarrow L^q(\Omega)$, $\forall q \in [p, +\infty)$.

2° *If Ω is locally Lipschitz, then for $kp > n$, the following embedding is compact:*

$$W^{k,p}(\Omega) \hookrightarrow C^{k-1-[n/p],\alpha}(\overline{\Omega}), \qquad \forall \alpha \in [0, \alpha_0),$$

where α_0 is the same as in Proposition 1.6.

Remark 1.1. For the space $W_0^{k,p}(\Omega)$, Propositions 1.6 and 1.7 for embedding and compact embedding are also valid; furthermore, they are valid for arbitrary bounded domains Ω, and the embedding constants are independent of Ω.

Next, we introduce propositions about difference quotients. These propositions enable us to study the weak differentiability through difference quotients.

Let e_s be the unit vector in the x_s direction, $s = 1, \cdots, n$. We denote by

$$\Delta_{h,s}u(x) = \frac{1}{h}[u(x + he_s) - u(x)]$$

the difference quotient of u in the x_s direction.

Proposition 1.8 (cf. [GT] or [GQ2]).

1° *Let $u \in W^{1,p}(\Omega)$, $1 \leqslant p < +\infty$, $\Omega' \subset\subset \Omega$. Then there exists a constant $C = C(\Omega', \Omega)$, such that $\Delta_{h,s}u \in L^p(\Omega')$ for sufficiently small $|h|$; furthermore,*

$$\|\Delta_{h,s}u\|_{L^p(\Omega')} \leqslant C\|D_s u\|_{L^p(\Omega)}.$$

2° *Let $u \in L^p(\Omega)$, $1 < p < +\infty$. Suppose that there exists a constant K such that for any $\Omega' \subset\subset \Omega$,*

$$\|\Delta_{h,s}u\|_{L^p(\Omega')} \leqslant K,$$

for sufficiently small $|h|$, where K is independent of h. Then $u \in W^{1,p}(\Omega')$, and

$$\|D_s u\|_{L^p(\Omega')} \leqslant K.$$

2. Real exponent Sobolev spaces $H^s(\mathbb{R}^n)$

Definition 2.1. Let $s \geqslant 0$ be a real number. The set

$$\left\{u \in L^2(\mathbb{R}^n) \,\middle|\, (1 + |\xi|^2)^{s/2}\widehat{u}(\xi) \in L^2(\mathbb{R}^n)\right\},$$

with the inner product

$$(u, v)_s = \int_{\mathbb{R}^n} \widehat{u}(\xi)\overline{\widehat{v}}(\xi)(1 + |\xi|^2)^s d\xi$$

forms a Hilbert space and is called the Sobolev space $H^s(\mathbb{R}^n)$ with real exponent s.

Proposition 2.1 (Trace theorem). *Let k be a positive integer. Then there exists a linear operator*

$$\begin{aligned} \gamma : H^k(\mathbb{R}^n_+) \quad &\rightarrow \quad \prod_{i=0}^{k-1} H^{k-i-1/2}(\mathbb{R}^{n-1}), \\ u \quad &\mapsto \quad \gamma u = (\gamma_0 u, \gamma_1 u, \cdots, \gamma_{k-1}u), \end{aligned}$$

such that

(*i*) *for $u \in C^\infty(\overline{\mathbb{R}^n_+})$,*

$$\gamma u = \left(u\Big|_{x_n=0}, \frac{\partial u}{\partial x_n}\Big|_{x_n=0}, \cdots, \frac{\partial^{k-1}u}{\partial x_n^{k-1}}\Big|_{x_n=0}\right),$$

(*ii*) *for all $u \in H^k(\mathbb{R}^n_+)$,*

$$\sum_{j=0}^{k-1} \|\gamma_j u\|^2_{H^{k-j-1/2}(\mathbb{R}^{n-1})} \leqslant C\|u\|^2_{H^k(\mathbb{R}^n_+)}.$$

Proposition 2.2 (Inverse trace theorem). *There exists a bounded linear operator*

$$\gamma^{-1} : \prod_{i=0}^{k-1} H^{k-i-1/2}(\mathbb{R}^{n-1}) \to H^k(\mathbb{R}^n_+),$$

such that $\gamma\gamma^{-1} = I$, *where* I *is the identity operator.*

3. Poincaré's inequality

Theorem 3.1. *Let* Ω *be a bounded domain in* $\mathbb{R}^n$.

1° *If* $u \in W_0^{1,p}(\Omega)$, $1 \leqslant p < +\infty$, *then*

$$\int_\Omega |u|^p dx \leqslant C(n, p, \Omega) \int_\Omega |Du|^p dx. \tag{1.8}$$

2° *If* Ω *is a bounded domain with* $\partial\Omega$ *Lipschitz continuous, then for* $u \in W^{1,p}(\Omega)$, $1 \leqslant p < +\infty$,

$$\int_\Omega |u - u_\Omega|^p dx \leqslant C(n, p, \Omega) \int_\Omega |Du|^p dx, \tag{1.9}$$

where $u_\Omega = \fint_\Omega u dx = \dfrac{1}{|\Omega|} \int_\Omega u dx$.

Proof.

1° We first consider the case $u \in C_0^1(\Omega)$. Assume without loss of generality that $\Omega \subset\subset Q = \{x \in \mathbb{R}^n \mid |x_i| < a,\ i = 1, \cdots, n\}$. Set

$$\tilde{u}(x) = \begin{cases} u(x), & x \in \Omega, \\ 0, & x \in Q \setminus \Omega. \end{cases}$$

For all $x \in Q$,

$$\begin{aligned} |\tilde{u}(x)|^p &= \left| \int_{-a}^{x_1} D_1 \tilde{u}(t, x_2, \cdots, x_n) dt \right|^p \\ &\leqslant (2a)^{p-1} \int_{-a}^{a} |D_1 \tilde{u}|^p dx_1 \leqslant (2a)^{p-1} \int_{-a}^{a} |D\tilde{u}|^p dx_1. \end{aligned}$$

It follows that

$$\begin{aligned} \int_\Omega |u(x)|^p dx &= \int_Q |\tilde{u}(x)|^p dx \leqslant (2a)^{p-1} \int_Q dx \int_{-a}^{a} |D\tilde{u}|^p dx_1 \\ &\leqslant (2a)^p \int_Q |D\tilde{u}|^p dx = (2a)^p \int_\Omega |Du|^p dx. \end{aligned}$$

Letting $C(n, p, \Omega) = (2a)^p$, we obtain (3.1).

For $u \in W_0^{1,p}(\Omega)$, we can simply use the fact that $C_0^1(\Omega)$ is dense in $W_0^{1,p}(\Omega)$ to obtain (1.8).

2° Here we only show the case $p > 1$. For the general case, see [MJ].

Since (1.9) is invariant if we add a constant to u, we can assume without loss of generality that $u_\Omega = 0$. If (1.9) is not valid, then for any positive integer k, there exists $u_k \in W^{1,p}(\Omega)$ satisfying $\fint_\Omega u_k dx = 0$, and

$$\int_\Omega |u_k|^p dx > k \fint_\Omega |Du_k|^p dx.$$

Set

$$w_k = \frac{u_k}{\|u_k\|_{L^p(\Omega)}} \qquad (k = 1, 2, \cdots).$$

Then $w_k \in W^{1,p}(\Omega)$ satisfies

(i) $\fint_\Omega w_k dx = 0$,

(ii) $\|w_k\|_{L^p(\Omega)} = 1$,

(iii) $\int_\Omega |Dw_k|^p dx \leqslant \frac{1}{k}$.

From (ii) and (iii), we deduce that $\|w_k\|_{W^{1,p}(\Omega)}$ is bounded. By using the compact embedding theorem, we can find a subsequence $w_{k_j} \in W^{1,p}(\Omega)$ and $w \in W^{1,p}(\Omega)$ such that

$$w_{k_j} \to w \quad \text{strongly in } L^p(\Omega), \tag{1.10}$$

$$Dw_{k_j} \rightharpoonup Dw \quad \text{weakly in } L^p(\Omega, \mathbb{R}^n). \tag{1.11}$$

By (iii) and (1.11), $Dw(x) = 0$ (a.e. $x \in \Omega$). It follows that

$$w(x) \equiv \text{constant}, \qquad \text{a.e. } x \in \Omega.$$

By (i) and (1.10), $\fint_\Omega w dx = 0$; therefore,

$$w(x) \equiv 0, \qquad \text{a.e. } x \in \Omega. \tag{1.12}$$

However, (ii) and (1.10) imply that

$$\|w\|_{L^p(\Omega)} = 1,$$

which contradicts (1.12). The proof is complete. □

Corollary 3.1. *Let B_R be a ball in $\mathbb{R}^n$ with radius R.*

1° *If $u \in W_0^{1,p}(B_R)$, $1 \leqslant p < +\infty$, then*

$$\int_{B_R} |u|^p dx \leqslant C(n,p) R^p \int_{B_R} |Du|^p dx.$$

2° *If $u \in W^{1,p}(B_R)$, $1 \leqslant p < +\infty$, then*

$$\int_{B_R} |u - u_R|^p dx \leqslant C(n,p) R^p \int_{B_R} |Du|^p dx,$$

where $u_R = \fint_{B_R} u dx = \frac{1}{|B_R|} \int_{B_R} u dx.$

Proof. If $R = 1$, Theorem 3.1 implies the conclusion. If $R \neq 1$, the conclusion can be established through a rescaling. □

APPENDIX 2

Sard's Theorem

Sard's theorem. *Let Ω be an open set in $\mathbb{R}^n$. Suppose that $f : \Omega \to \mathbb{R}^n$ has continuous first derivatives. Then $|f(E)| = 0$, where $E = \{x \in \Omega \mid \det Df(x) = 0\}$.*

Proof. Take a countable number of closed cubes $Q_i \subset \Omega$ such that

$$\bigcup_{i=1}^{\infty} Q_i = \Omega.$$

Then

$$f(E) = \bigcup_{i=1}^{\infty} f(E \cap Q_i).$$

Now it suffices to prove that $|f(E \cap Q)| = 0$ for all cubes $Q \subset \Omega$.

Denote the side length of Q by l. Divide Q into N^n congruent smaller cubes with diameter $\delta = \sqrt{n}l/N$. We denote by K a typical smaller cube.

Since $Df(x)$ is uniformly continuous on Q, we have $M \triangleq \sup_Q |Df(x)| < \infty$. Furthermore, for any $\varepsilon > 0$,

$$|f(x) - f(x_0) - Df(x_0)(x - x_0)| < \varepsilon|x - x_0| \leqslant \delta\varepsilon \quad \text{for } x, x_0 \in K, \tag{2.1}$$

provided N is large enough. If $x_0 \in E$, then the linear transform $Df(x_0)$ is degenerate. It follows that $Df(x_0)K$ is contained in an $n-1$ dimensional parallel-polyhedron of side length no more than $M\delta$. The inequality (2.1) implies that the distance between the sets $f(K) - f(x_0) + x_0 Df(x_0)$ and $Df(x_0)K$ is no more than $\varepsilon\delta$. It follows that $f(K) - f(x_0) + x_0 Df(x_0)$ is contained in a cylinder with height $2\varepsilon\delta$; the base of this cylinder is an $n-1$ dimensional parallel-polyhedron of side length no more than $M\delta + 2\varepsilon\delta$. Since the Lebesgue measure is translation invariant,

$$|f(K)| \leqslant (M\delta + 2\varepsilon\delta)^{n-1}(2\varepsilon\delta) \leqslant \frac{A\varepsilon}{N^n}, \tag{2.2}$$

where A depends only on n and l. The above estimate is valid for any smaller cubes K containing at least one critical point (points in E are called critical points). It follows that

$$\text{meas}\{f(E \cap Q)\} \leqslant N^n |f(K)| \leqslant A\varepsilon.$$

Since ε is arbitrary, $|f(E \cap Q)| = 0$. $\square$

APPENDIX 3

Proof of the John-Nirenberg Theorem

The John-Nirenberg theorem is stated in §1 of Chapter 10. We give a proof here (cf. [JN]).

Proof of the John-Nirenberg theorem. We assume without loss of generality that $Q = Q_0$. Since the inequality is homogeneous, we can assume that $|u|_{*,Q_0} = 1$.

For $\alpha > 1 \geqslant ⨍_{Q_0} |u - u_{Q_0}|dx$, we apply the Calderón-Zygmund decomposition (Lemma 2.1 in Chapter 3) to the function $|u(x) - u_{Q_0}|$. There exist nonoverlapping cubes $\{Q_j^{(1)}\}$ such that

$$\alpha < ⨍_{Q_j^{(1)}} |u - u_{Q_0}|dx \leqslant 2^n \alpha, \tag{3.1}$$

$$|u(x) - u_{Q_0}| \leqslant \alpha, \quad a.e.\ x \in Q_0 \setminus \bigcup_j Q_j^{(1)}. \tag{3.2}$$

It follows that

$$\sum_j |Q_j^{(1)}| \leqslant \frac{1}{\alpha} \int_{Q_0} |u - u_{Q_0}|dx \leqslant \frac{1}{\alpha}|Q_0|, \tag{3.3}$$

$$|u_{Q_j^{(1)}} - u_{Q_0}| \leqslant ⨍_{Q_j^{(1)}} |u - u_{Q_0}|dx \leqslant 2^n \alpha. \tag{3.4}$$

Since $|u|_{*,Q_0} = 1$, we still have $\alpha > 1 \geqslant ⨍_{Q_j^{(1)}} |u - u_{Q_j^{(1)}}|dx$ for all cubes $Q_j^{(1)}$. Applying the Calderón-Zygmund decomposition again to the function $|u(x) - u_{Q_j^{(1)}}|$, we obtain a sequence of nonoverlapping cubes $\{Q_j^{(2)}\}$ (collecting all cubes obtained for each $Q_j^{(1)}$) such that

$$\sum_j |Q_j^{(2)}| \leqslant \frac{1}{\alpha} \sum_j \int_{Q_j^{(1)}} |u - u_{Q_j^{(1)}}|dx, \tag{3.5}$$

$$|u(x) - u_{Q_j^{(1)}}| \leqslant \alpha, \quad \text{a.e. } x \in Q_j^{(1)} \setminus \bigcup_i Q_i^{(2)}. \tag{3.6}$$

Notice that we have $|u|_{*,Q_0} = 1$. (3.3) and (3.5) imply that

$$\sum_j |Q_j^{(2)}| \leqslant \frac{1}{\alpha} \sum_j |Q_j^{(1)}| \leqslant \frac{1}{\alpha^2}|Q_0|. \tag{3.7}$$

We shall show that

$$|u(x) - u_{Q_0}| \leqslant 2 \cdot 2^n \alpha, \quad \text{a.e. } x \in Q_0 \setminus \bigcup_i Q_i^{(2)}. \tag{3.8}$$

In fact, if $x \in Q_0 \setminus \bigcup_j Q_j^{(1)}$, then (3.2) implies (3.8). Now suppose that $x \in \bigcup_j Q_j^{(1)} \setminus \bigcup_k Q_k^{(2)}$. Then $x \in Q_j^{(1)}$ for some j, and therefore, by (3.6) and (3.4),

$$|u(x) - u_{Q_0}| \leqslant |u(x) - u_{Q_j^{(1)}}| + |u_{Q_j^{(1)}} - u_{Q_0}| \leqslant 2 \cdot 2^n \alpha.$$

We inductively repeat the above decomposition. For any integer $k \geqslant 1$, there exist nonoverlapping cubes $Q_j^{(k)}$ such that

$$\sum_j |Q_j^{(k)}| \leqslant \frac{1}{\alpha^k} |Q_0|,$$

$$|u(x) - u_{Q_0}| \leqslant k \cdot 2^n \alpha, \quad \text{a.e. } x \in Q_0 \setminus \bigcup_j Q_j^{(k)}.$$

It follows that

$$\begin{aligned} &\text{meas}\left\{x \in Q_0 \,\middle|\, |u(x) - u_{Q_0}| > 2^n k\alpha\right\} \\ &\leqslant \sum_j |Q_j^{(k)}| \leqslant \frac{1}{\alpha^k} |Q_0| \qquad (k = 1, 2, \cdots). \end{aligned}$$

The above inequality is obviously valid for $k = 0$. For any $t \in (0, +\infty)$, we choose $k \geqslant 0$ such that

$$2^n \alpha k < t \leqslant 2^n \alpha (k+1).$$

Then

$$\begin{aligned} &\text{meas}\left\{x \in Q_0 \,\middle|\, |u(x) - u_{Q_0}| > t\right\} \\ &\leqslant \text{meas}\left\{x \in Q_0 \,\middle|\, |u(x) - u_{Q_0}| > 2^n k\alpha\right\} \\ &\leqslant \frac{1}{\alpha^k} |Q_0| \leqslant \alpha e^{-At} |Q_0|, \end{aligned}$$

where $A = (\log \alpha)/(2^n \alpha)$. The proof is complete. $\square$

APPENDIX 4

Proof of the Stampacchia Interpolation Theorem

In this appendix, we shall use $Q(x,r)$ to denote the cube in $\mathbb{R}^n$ centered at x, with side length $2r$ and sides parallel to the coordinate axes. We often drop the x and r and simply denote the cube by Q, $Q_0, \cdots$, if there is no confusion.

Stampacchia interpolation theorem. *Let Q_0 be a cube in $\mathbb{R}^n$. Suppose that for some $p \geqslant 1$, T is both a bounded linear operator: $L^p(Q_0) \to L^p(Q_0)$ and a bounded linear operator: $L^\infty(Q_0) \to \mathrm{BMO}(Q_0)$, i.e.,*

$$\|Tu\|_{L^p(Q_0)} \leqslant B_p\|u\|_{L^p(Q_0)}, \quad \forall u \in L^p(Q_0), \tag{4.1}$$

$$|Tu|_{*,Q_0} \leqslant B_\infty\|u\|_{L^\infty(Q_0)}, \quad \forall u \in L^\infty(Q_0). \tag{4.2}$$

Then for any $q \in (p,\infty)$, T is a bounded linear operator: $L^q(Q_0) \to L^q(Q_0)$, and

$$\|Tu\|_{L^q(Q_0)} \leqslant C\|u\|_{L^q(Q_0)}, \quad \forall u \in L^q(Q_0), \tag{4.3}$$

where C depends only on n, p, q, B_p, B_∞.

In order to prove this theorem, we first introduce the Hardy-Littlewood maximal function and the maximal mean oscillation function (sharp function).

For $f \in L(Q_0)$, the function

$$M_0 f(x) = \sup_{r>0} \fint_{Q(x,r)\cap Q_0} |f(y)|dy, \quad x \in Q_0, \tag{4.4}$$

is called the *centered Hardy-Littlewood maximal function.*

Sometimes it is more convenient to define

$$Mf(x) = \sup_{x\in Q} \fint_{Q\cap Q_0} |f(y)|dy, \quad x \in Q_0, \tag{4.5}$$

where Q is a cube with center inside Q_0. Mf is called the *Hardy-Littlewood maximal function* on Q_0. It is not difficult to verify that

$$M_0 f(x) \leqslant Mf(x) \leqslant 2^n M_0 f(x), \quad x \in Q_0. \tag{4.6}$$

A maximal function has the following elementary properties:

(1) Mf is a measurable function on Q_0;

(2) M is sublinear, i.e.,

$$|M(f+g)| \leqslant Mf + Mg;$$

(3) if $f \in L^\infty(Q_0)$, then $Mf \in L^\infty(Q_0)$, and

$$\|Mf\|_{L^\infty(Q_0)} \leqslant \|f\|_{L^\infty(Q_0)}.$$

We next give a deeper property of the maximal function:

Hardy-Littlewood maximal theorem. *The operator M is of weak type $(1,1)$, and*

$$\|Mf\|_{L^1_w(Q_0)} \leqslant C(n)\|f\|_{L^1(Q_0)}, \quad \forall f \in L^1(Q_0), \tag{4.7}$$

where $\|\cdot\|_{L^1_w(Q_0)}$ is defined in §1 of Chapter 3.

Proof. Let $x \in \{x \in Q_0 \mid M_0 f(x) > s\}$. From the definition of M_0, we deduce that there exists $Q(x, r(x))$ such that

$$⨍_{Q(x,r(x))\cap Q_0} |f(y)|dy > s,$$

i.e.,

$$|Q(x,r(x)) \cap Q_0| < \frac{1}{s}\int_{Q(x,r(x))\cap Q_0} |f(y)|dy. \tag{4.8}$$

By the covering lemma in §6 of Chapter 12 (although here we have cubes instead of balls, the lemma is still valid), there exist countably many cubes $Q(x_i, r(x_i))$ $(i = 1, 2\cdots)$ such that

(1) $Q(x_i, r(x_i)) \cap Q(x_j, r(x_j)) = \emptyset$ for $i \neq j$,

(2) $\{x \in Q_0 \mid M_0 f(x) > s\} \subset \bigcup_{i=1}^{\infty} Q(x_i, 3r(x_i)) \cap Q_0$.

It easily follows that

$$\begin{aligned}\text{meas}\{x \in Q_0 \mid M_0 f(x) > s\} &\leqslant \sum_{i=1}^{\infty} |Q(x_i, 3r(x_i)) \cap Q_0| \\ &\leqslant 3^n \sum_{i=1}^{\infty} |Q(x_i, r(x_i)) \cap Q_0|.\end{aligned}$$

By (4.8),

$$\text{meas}\{x \in Q_0 \mid M_0 f(x) > s\} \leqslant \frac{3^n}{s}\sum_{i=1}^{\infty}\int_{Q(x_i,r(x_i))\cap Q_0} |f(y)|dy \leqslant \frac{3^n}{s}\|f\|_{L^1(Q_0)}.$$

This proves that M_0 is of weak type $(1,1)$. Using (4.6), we see that M is also of weak type $(1,1)$. □

Corollary. *For $1 < p < \infty$, the maximal operator M is of strong type (p,p).*

This is a direct consequence of the Hardy-Littlewood maximal theorem, the Marcinkiewicz interpolation theorem, and the elementary property (3).

Now we introduce the maximal mean oscillation function (sharp function). For $f \in L^1(Q_0)$, the function

$$f^{\#}(x) = \sup_{x\in Q} ⨍_{Q\cap Q_0} |f - f_{Q\cap Q_0}|dy, \quad x \in Q_0, \tag{4.9}$$

is called the *maximal mean oscillation function* of f on Q_0, where Q has center inside Q_0, and

$$f_{Q\cap Q_0} = ⨍_{Q\cap Q_0} f(y)dy.$$

The maximal mean oscillation function has the following elementary properties:

(1) $f \in \mathrm{BMO}(Q_0)$ if and only if $f^{\#} \in L^\infty(Q_0)$;

(2) if $1 < p < \infty$, then

$$\|f^{\#}\|_{L^p(Q_0)} \leqslant C(n,p)\|f\|_{L^p(Q_0)} \quad \text{for } f \in L^p(Q_0). \tag{4.10}$$

Proof. Obviously, $f^{\#} \leqslant 2Mf$. By the corollary of the Hardy-Littlewood maximal theorem, M is of strong type (p,p). It follows that the sharp operator $(\cdot)^{\#}$ must also be of strong type (p,p). □

(3) If $f \in L^1(Q_0)$, then $f^{\#} \in L^1_w(Q_0)$.

The most important property of the maximal mean oscillation function was discovered by Fefferman and Stein (cf. [FS]):

Fefferman-Stein theorem. *Let $f \in L^1(Q_0)$. If $f^{\#} \in L^p(Q_0)$ for some $1 < p < \infty$, then $f \in L^p(Q_0)$, and*

$$\|f\|_{L^p(Q_0)} \leqslant C(n,p)\Big(\|f^{\#}\|_{L^p(Q_0)} + |f|_{Q_0}|Q_0|^{1/p}\Big), \tag{4.11}$$

where

$$|f|_{Q_0} = ⨍_{Q_0} |f|.$$

Proof. For any fixed α such that

$$\alpha > ⨍_{Q_0} |f|dx, \tag{4.12}$$

we apply the Calderón-Zygmund decomposition (§2 of Chapter 3) to the function $|f|$. There exists a sequence of nonoverlapping cubes $\{Q_j^\alpha\}$ such that

$$\alpha < ⨍_{Q_j^\alpha} |f|dx \leqslant 2^n\alpha \qquad (j = 1,2,\cdots), \tag{4.13}$$

$$|f(x)| \leqslant \alpha, \quad \text{a.e. } x \in Q_0 \setminus \bigcup_j Q_j^\alpha. \tag{4.14}$$

If we do the decomposition simultaneously for all values of α satisfying (4.12), then $\{Q_j^{\alpha_1}\}$ are subcubes of $\{Q_j^{\alpha_2}\}$ when $\alpha_1 > \alpha_2$. Set

$$\mu(\alpha) = \sum_j |Q_j^\alpha| \quad \text{for } \alpha > ⨍_{Q_0} |f|dx.$$

Then $\mu(\alpha)$ is a nonincreasing function. From (4.14), we easily deduce that

$$\lambda(\alpha) \triangleq \operatorname{meas}\{x \in Q_0 \mid |f| > \alpha\} \leqslant \mu(\alpha). \tag{4.15}$$

We claim that if

$$\frac{\alpha}{2^{n+1}} > ⨍_{Q_0} |f|dx = |f|_{Q_0}, \tag{4.16}$$

then

$$\mu(\alpha) \leqslant \operatorname{meas}\Big\{x \;\Big|\; f^{\#}(x) > \frac{\alpha}{A}\Big\} + \frac{2}{A}\mu\Big(\frac{\alpha}{2^{n+1}}\Big), \tag{4.17}$$

where A is an arbitrary positive number.

Denote any cube $Q^{\beta}_{j_0}$ ($\beta = \alpha/2^{n+1}$) by $\widetilde{Q}_0$. For any cubes Q^{α}_j contained in $\widetilde{Q}_0$, there are two cases:

Case 1: $\widetilde{Q}_0 \subset \left\{x \in Q_0 \,\middle|\, f^{\#}(x) > \alpha/A\right\}$. Then

$$\sum_{Q^{\alpha}_j \subset \widetilde{Q}_0} |Q^{\alpha}_j| \leqslant \operatorname{meas}\left\{x \in Q_0 \,\middle|\, f^{\#}(x) > \frac{\alpha}{A}\right\} \cap \widetilde{Q}_0. \tag{4.18}$$

Case 2: $\widetilde{Q}_0 \not\subset \left\{x \in Q_0 \,\middle|\, f^{\#}(x) > \alpha/A\right\}$. Then there exists $x_0 \in \widetilde{Q}_0$ such that $f^{\#}(x_0) \leqslant \alpha/A$. By the definition of $f^{\#}$, we have

$$⨍_{\widetilde{Q}_0} |f(x) - f_{\widetilde{Q}_0}| dx \leqslant \frac{\alpha}{A}. \tag{4.19}$$

On the other hand, (4.13) implies that

$$|f|_{\widetilde{Q}_0} \leqslant 2^n \left(\frac{\alpha}{2^{n+1}}\right) \leqslant \frac{\alpha}{2}, \qquad |f|_{Q^{\alpha}_j} > \alpha.$$

It follows that

$$\int_{Q^{\alpha}_j} |f(x) - f_{\widetilde{Q}_0}| dx \geqslant (|f|_{Q^{\alpha}_j} - |f|_{\widetilde{Q}_0})|Q^{\alpha}_j| > \frac{\alpha}{2}|Q^{\alpha}_j|,$$

where Q^{α}_j is an arbitrary element in $\{Q^{\alpha}_j\}$ which is contained in $\widetilde{Q}_0$. We sum over all such cubes. Then

$$\begin{aligned} \sum_{Q^{\alpha}_j \subset \widetilde{Q}_0} |Q^{\alpha}_j| &< \frac{2}{\alpha} \sum_{Q^{\alpha}_j \subset \widetilde{Q}_0} \int_{Q^{\alpha}_j} |f(x) - f_{\widetilde{Q}_0}| dx \\ &\leqslant \frac{2}{\alpha} \int_{\widetilde{Q}_0} |f(x) - f_{\widetilde{Q}_0}| dx \leqslant \frac{2}{A} |\widetilde{Q}_0|. \end{aligned} \tag{4.20}$$

In deriving the last inequality, we also made use of (4.19).

Combining the two cases, we deduce from (4.18) and (4.20) that

$$\sum_{Q^{\alpha}_j \subset \widetilde{Q}_0} |Q^{\alpha}_j| \leqslant \operatorname{meas}\left\{x \in Q_0 \,\middle|\, f^{\#}(x) > \frac{\alpha}{A}\right\} \cap \widetilde{Q}_0 + \frac{2}{A}|\widetilde{Q}_0|,$$

where $\widetilde{Q}_0$ is an arbitrary cube from the set $\{Q^{\beta}_j \mid \beta = \alpha/2^{n+1}\}$. By summing over all such cubes, we immediately obtain (4.17).

For any $s > 2^{n+1}|f|_{Q_0}$, we set

$$I_s = p \int_0^s \alpha^{p-1} \mu(\alpha) d\alpha.$$

Since $\mu(\alpha) \leqslant |Q_0|$, the above integral is well defined. Using (4.17), we get

$$\begin{aligned} I_s &= p \int_0^{2^{n+1}|f|_{Q_0}} \alpha^{p-1}\mu(\alpha) d\alpha + p \int_{2^{n+1}|f|_{Q_0}}^s \alpha^{p-1}\mu(\alpha) d\alpha \\ &\leqslant p \int_0^{\infty} \operatorname{meas}\left\{x \in Q_0 \,\middle|\, f^{\#}(x) > \frac{\alpha}{A}\right\} \alpha^{p-1} d\alpha \\ &\quad + \frac{2p}{A} \int_0^s \alpha^{p-1} \mu\left(\frac{\alpha}{2^{n+1}}\right) d\alpha + |Q_0|(2^{n+1}|f|_{Q_0})^p \\ &\leqslant A^p \|f^{\#}\|^p_{L^p(Q_0)} + \frac{2}{A} \cdot 2^{(n+1)p} I_s + 2^{(n+1)p} |Q_0| \, |f|^p_{Q_0}. \end{aligned}$$

We choose $A = 4 \cdot 2^{(n+1)p}$ in the above inequality. Then

$$I_s \leqslant 2A^p \|f^\#\|^p_{L^p(Q_0)} + 2^{(n+1)p+1}|Q_0|\, |f|^p_{Q_0}.$$

By (4.15),

$$p\int_0^s \lambda(\alpha)\alpha^{p-1}d\alpha \leqslant 2A^p \|f^\#\|^p_{L^p(Q_0)} + 2^{(n+1)p+1}|Q_0|\, |f|^p_{Q_0}.$$

Letting $s \to \infty$, we conclude (4.11). The proof is complete. □

Proof of the Stampacchia interpolation theorem. Define

$$T^\# u = (Tu)^\#.$$

From property (2) of $f^\#$ and the assumptions, we deduce that

$$\|T^\# u\|_{L^p(Q_0)} \leqslant C(n,p)\|Tu\|_{L^p(Q_0)} \leqslant C(n,p)B_p\|u\|_{L^p(Q_0)}.$$

By the definition of $f^\#$,

$$\|T^\# u\|_{L^\infty(Q_0)} = |Tu|_{*,Q_0} \leqslant B_\infty \|u\|_{L^\infty(Q_0)}.$$

Thus the Marcinkiewicz interpolation theorem implies that $T^\#$ is of strong type (q,q) for $q \in (p,\infty)$, and

$$\|T^\# u\|_{L^q(Q_0)} \leqslant C B_p^{p/q} B_\infty^{(q-p)/q} \|u\|_{L^q(Q_0)}, \quad \forall u \in L^q(Q_0),$$

where C depends only on n, p, q. Now we use the Fefferman-Stein theorem to conclude that

$$\begin{aligned}
\|Tu\|_{L^q(Q_0)} &\leqslant C\Big(\|T^\# u\|_{L^q(Q_0)} + |Q_0|^{1/q}|Tu|_{Q_0}\Big)\\
&\leqslant C\Big(\|u\|_{L^q(Q_0)} + |Q_0|^{1/q-1/p}\|Tu\|_{L^p(Q_0)}\Big)\\
&\leqslant C\Big(\|u\|_{L^q(Q_0)} + |Q_0|^{1/q-1/p}B_p\|u\|_{L^p(Q_0)}\Big)\\
&\leqslant C\|u\|_{L^q(Q_0)},
\end{aligned}$$

where C depends only on n, p, q, B_p, B_∞. □

APPENDIX 5

Proof of the Reverse Hölder Inequality

In order to prove the reverse Hölder inequality, we first introduce some lemmas.

Lemma 5.1 (Covering lemma). *Suppose that E is a measurable set covered by a family of balls $\{B_j\}$ with their radii bounded above. Then there exists a subsequence of disjoint balls $B_{j_1}, B_{j_2}, \cdots$ (finitely many or infinitely many) such that*

$$|E| \leqslant 5^n \sum_k |B_{j_k}|, \tag{5.1}$$

where $|E|$ denotes the measure of E.

Proof. Choose $B_{j_1} \in \{B_j\}$ such that

$$\operatorname{diam} B_{j_1} \geqslant \frac{1}{2} \sup_j \operatorname{diam} B_j.$$

We use induction. Suppose that $B_{j_1}, \cdots, B_{j_k}$ have been chosen. We want to choose $B_{j_{k+1}}$ such that $B_{j_{k+1}}$ is disjoint from $B_{j_1}, \cdots, B_{j_k}$, and

$$\operatorname{diam} B_{j_{k+1}} \geqslant \frac{1}{2} \sup \left\{ \operatorname{diam} B_j \;\middle|\; B_j \cap B_{j_s} = \emptyset,\ s = 1, 2, \cdots, k \right\}. \tag{5.2}$$

We stop the process if no such ball exists.

If $\sum_k |B_{j_k}| = \infty$, there is nothing to prove. Thus we assume $\sum_k |B_{j_k}| < \infty$. Denote by $B^*_{j_k}$ the ball with the same center as B_{j_k}, but 5 times the radius of B_{j_k}. We claim that

$$E \subset \bigcup_k B^*_{j_k}. \tag{5.3}$$

Since $\{B_j\}$ is a cover for E, it suffices to prove that any B_j is a subset of $\bigcup_k B^*_{j_k}$. If B_j is an element of $\{B_{j_k}\}$, then the above conclusion is obvious. Now suppose that B_j is not an element of $\{B_{j_k}\}$. Since $\sum_k |B_{j_k}| < \infty$, we must have $\lim_{k\to\infty} \operatorname{diam} B_{j_k} = 0$. It follows that there exists a k such that

$$\operatorname{diam} B_{j_{k+1}} < \frac{1}{2} \operatorname{diam} B_j; \tag{5.4}$$

we choose the smallest k for which (5.4) is valid. (5.4) implies that B_j must intersect with at least one of $B_{j_1}, \cdots, B_{j_k}$, for otherwise it would contradict (5.2). Since k is the smallest for which (5.4) is valid,

$$\operatorname{diam} B_{j_s} \geqslant \frac{1}{2} \operatorname{diam} B_j \qquad (s = 1, 2, \cdots, k).$$

It follows that

$$B_j \subset \bigcup_{s=1}^{k} B_j^*,$$

which implies (5.3). The proof is similar when $\{B_{j_k}\}$ is a finite set. Using (5.3), we conclude that

$$|E| \leqslant \sum_k |B_{j_k}^*| \leqslant 5^n \sum_k |B_{j_k}|.$$

The proof is complete. □

Let $h(t)$ be a function defined on $(0,\infty)$ of bounded variation. For $p \in (0,\infty)$, we set

$$\Phi_p(t;h) = -\int_t^\infty s^p dh(s); \tag{5.5}$$

this integral is well defined.

Lemma 5.2. *Let $q \in (0,\infty)$, $a \in (1,\infty)$. Suppose that $h(t)$ and $H(t)$ are nonnegative nonincreasing functions defined on $[t_0,\infty)$ such that*

(1) $\lim_{t\to\infty} h(t) = \lim_{t\to\infty} H(t) = 0$,

(2) $\Phi_q(t;h) \leqslant a[t^q h(t) + t^q H(t)]$.

Then for $p \in [q, aq/(a-1))$, we have

$$\Phi_p(t_0;h) \leqslant \frac{t_0^{p-q}}{aq-(a-1)p}\Big[q\Phi_q(t_0;h) + a(p-q)\Phi_p(t_0;H)\Big]. \tag{5.6}$$

Proof. Rescaling if necessary, we may assume without loss of generality that $t_0 = 1$. For simplicity, we set

$$I(r) = \Phi_r(1;h), \qquad J(r) = \Phi_r(1;H).$$

First we assume that there exists $j \in (1,+\infty)$ such that

$$h(t) = 0, \qquad t \in [j,+\infty).$$

For $p \in (0,\infty)$, we use integration by parts to obtain

$$\begin{aligned} I(p) &= -\int_1^j t^p dh(t) = \int_1^j t^{p-q} d\Big(\int_t^j s^q dh(s)\Big) \\ &= I(q) + (p-q)\int_1^j t^{p-q-1}\Phi_q(t;h)dt. \end{aligned} \tag{5.7}$$

By the assumption (2),

$$\begin{aligned} \int_1^j t^{p-q-1}\Phi_q(t;h)dt &\leqslant a\Big[\int_1^j t^{p-1}h(t)dt + \int_1^j t^{p-1}H(t)dt\Big] \\ &\leqslant -\frac{a}{p}h(1) + \frac{a}{p}I(p) - \frac{a}{p}H(1) + \frac{a}{p}J(p). \end{aligned} \tag{5.8}$$

In deriving the last inequality, we also made use of integration by parts and the assumption $\lim_{t\to\infty} H(t) = 0$. The assumption (2) also implies that

$$a[h(1) + H(1)] \geqslant I(q). \tag{5.9}$$

Substituting (5.9) into (5.8), we then obtain

$$\int_1^j t^{p-q-1}\Phi_q(t;h)dt \leqslant -\frac{1}{p}I(q) + \frac{a}{p}I(p) + \frac{a}{p}J(p).$$

Substituting this inequality into (5.7), we get

$$I(p) \leqslant I(q) - \frac{p-q}{p}I(q) + \frac{a}{p}(p-q)I(p) + \frac{a}{p}(p-q)J(p).$$

Rearranging the terms in the above inequality, we obtain (5.6).

In the general case, we set

$$h_j(t) = \begin{cases} h(t), & t \in [1, j), \\ 0, & t \in [j, +\infty). \end{cases}$$

Then

$$-\int_t^\infty s^q dh_j(s) = -\int_t^j s^q dh(s) + j^q h(j), \qquad t < j. \tag{5.10}$$

Since $h(t)$ is nonincreasing, for $\sigma > j$ we have

$$j^q h(j) - j^q h(\sigma) \leqslant -\int_j^\sigma t^q dh(t).$$

Letting $\sigma \to +\infty$, and using assumption (1), we obtain

$$j^q h(j) \leqslant -\int_j^\infty t^q dh(t).$$

Substituting into (5.10), we get

$$\Phi_q(t;h_j) \leqslant \Phi_q(t;h) \leqslant a[t^q h(t) + t^q H(t)].$$

We can now follow the previous proof to derive

$$\begin{aligned} -\int_1^j t^q dh(t) &\leqslant -\int_1^j t^p dh_j(t) \\ &\leqslant \frac{q}{aq-(a-1)p}\Phi_q(1;h) + \frac{a(p-q)}{aq-(a-1)p}\Phi_p(1;H). \end{aligned}$$

We now obtain the lemma by letting $j \to \infty$. □

From now on we shall use Q to denote a cube with sides parallel to the coordinate axes. For $f \in L^1_{loc}(\mathbb{R}^n)$, Mf denotes the Hardy-Littlewood maximal function (cf. Appendix 4). For any cube Q and any $q \geqslant 1$, we have

$$⨍_Q (Mf)^{1/q} dx \geqslant \left(⨍_Q f dx \right)^{1/q}. \tag{5.11}$$

In fact, for any $x \in Q$,

$$⨍_Q f(y)dy \leqslant Mf(x).$$

Taking the q-th root on both sides and then integrating over $x \in Q$, we derive (5.11). This fact will be used later on.

Proof of the reverse Hölder inequality (Theorem 4.1 in Chapter 12; cf. [GQ1]). For simplicity, the balls in the theorem will be replaced by cubes. Assume without loss of generality that Q is a cube centered at the origin, with side length 2. We introduce the following decomposition:

$$
\begin{aligned}
Q_1 &= \left\{x \in \mathbb{R}^n \,\middle|\, |x_i| < \frac{3}{2}, \quad i = 1, 2, \cdots, n\right\}, \\
C_0 &= \left\{x \in \mathbb{R}^n \,\middle|\, |x_i| < \frac{1}{2}, \quad i = 1, 2, \cdots, n\right\}, \\
C_k &= \left\{x \in Q_1 \,\middle|\, 2^{-k} \leqslant \operatorname{dist}\{x, \partial Q_1\} \leqslant 2^{-k+1}\right\}, \quad k \geqslant 1.
\end{aligned} \tag{5.12}
$$

Obviously,

$$Q_1 = C_0 \cup \Big(\bigcup_{k \geqslant 1} C_k\Big).$$

For C_k, we can decompose it into nonoverlapping congruent cubes $\{P_{k,j}\}$ of side length 2^{-k} such that $\bigcup_j P_{k,j} = C_k$.

We define the function

$$\varphi(x) = [\operatorname{dist}(x, \partial Q_1)]^n, \quad x \in Q_1. \tag{5.13}$$

Obviously, there exists a constant $\sigma > 0$ such that for any $x \in C_{k-1} \cup C_k \cup C_{k+1}$,

$$\sigma^{-1}|P_{k,j}| \leqslant \varphi(x) \leqslant \sigma |P_{k,j}|. \tag{5.14}$$

We choose

$$\lambda \geqslant \gamma_0 \|g\|_{L^q(Q_1)}, \tag{5.15}$$

where $\gamma_0 \geqslant 1$ is to be determined. Notice that

$$⨍_{P_{k,j}} (g\varphi)^q dx \leqslant \sigma^q |P_{k,j}|^q ⨍_{P_{k,j}} g^q dx \leqslant \sigma^q \|g\|^q_{L^q(Q_1)};$$

therefore, if γ_0 is chosen so that $\gamma_0 > \sigma$, then

$$\lambda^q > ⨍_{P_{k,j}} (g\varphi)^q dx.$$

Now we apply the Calderón-Zygmund decomposition (Lemma 2.1 in Chapter 3) to each $P_{k,j}$, and we obtain nonoverlapping cubes $\{Q^l_{k,j}\}$ with the following properties:

$$
\begin{aligned}
& Q^l_{k,j} \subset P_{k,j}, \quad \forall l, k, j, \\
& \lambda^q < ⨍_{Q^l_{k,j}} (g\varphi)^q dx \leqslant 2^n \lambda^q, \quad \forall l, k, j,
\end{aligned} \tag{5.16}
$$

$$g\varphi \leqslant \lambda, \quad \text{a.e. } x \in P_{k,j} \setminus \bigcup_l Q^l_{k,j}. \tag{5.17}$$

From (5.17), we deduce that

$$g\varphi \leqslant \lambda, \quad \text{a.e. } x \in Q_1 \setminus \bigcup_l Q^l_{k,j},$$

which implies that

$$A_\lambda(g\varphi) \triangleq \{x \in Q_1 \mid g\varphi > \lambda\} \subset \bigcup_{k,j,l} Q^l_{k,j}.$$

It follows that

$$(5.18)\qquad \int_{A_\lambda(g\varphi)} (g\varphi)^q dx \leqslant \sum_{k,j,l} |Q^l_{k,j}| \fint_{Q^l_{k,j}} (g\varphi)^q dx \leqslant 2^n \lambda^q \sum_{k,j,l} |Q^l_{k,j}|.$$

In deriving the last inequality, we made use of (5.16). Now we estimate the right-hand side of (5.18). For any $x \in Q^l_{k,j}$, (5.16) implies that

$$(5.19)\qquad \lambda^q \leqslant \fint_{Q^l_{k,j}} (g\varphi)^q dy \leqslant \sigma^q |P_{k,j}|^q \fint_{Q^l_{k,j}} g^q dy \leqslant \sigma^q |P_{k,j}|^q \overline{M}(g^q)(x),$$

where $\overline{M}(g^q)$ is the maximal function restricted on Q_1, i.e.,

$$(5.20)\qquad \overline{M}(g^q)(x) = \sup_{x\in Q\subset Q_1} \fint_Q g^q(y)dy.$$

Since $\theta < 1$ (see the statement of the theorem), there exists a cube $Q_0 \subset Q_1$ such that

$$(5.21)\qquad \sqrt{\theta}\,\overline{M}(g^q)(x) \leqslant \fint_{Q_0} g^q(y)dy, \quad x \in Q_0.$$

Utilizing (5.19), we get

$$\frac{\lambda^q\sqrt{\theta}}{\sigma^q|P_{k,j}|^q} \leqslant \sqrt{\theta}\,\overline{M}(g^q)(x) \leqslant \fint_{Q_0} g^q(y)dy.$$

Notice that λ satisfies (5.15); the above inequality implies

$$|Q_0| \leqslant \frac{\sigma^q|P_{k,j}|^q}{\lambda^q\sqrt{\theta}} \int_{Q_0} g^q(y)dy \leqslant \frac{\sigma^q|P_{k,j}|^q}{\gamma_0^q\sqrt{\theta}},$$

where we have assumed $\theta > 0$ (we replace θ by a positive θ if it is zero, noticing that the conditions of the theorem are still valid). It follows that

$$\text{the side length of } Q_0 \leqslant \Big[\frac{\sigma^q}{\gamma_0^q\sqrt{\theta}}\Big]^{1/n} 2^{-kq}.$$

We choose γ_0 such that

$$\Big(\frac{\sigma^q}{\gamma_0^q\sqrt{\theta}}\Big)^{1/n} \leqslant \frac{1}{8}.$$

Denote by $\widetilde{Q}_0$ the cube with the same center as Q_0 and twice the diameter of Q_0. It is obvious (since $x \in Q^l_{k,j} \cap Q_0$) that

$$\widetilde{Q}_0 \subset C_{k-1} \cup C_k \cup C_{k+1}.$$

By (5.21) and assumption (2) of the theorem,

$$\begin{aligned} \overline{M}(g^q)(x) &\leqslant \frac{1}{\sqrt{\theta}} \fint_{Q_0} g^q(y)dy \\ &\leqslant \frac{b}{\sqrt{\theta}}\Big(\fint_{\widetilde{Q}_0} g(y)dy\Big)^q + \frac{1}{\sqrt{\theta}} \fint_{\widetilde{Q}_0} f^q(y)dy + \sqrt{\theta}\,\overline{M}(g^q)(x). \end{aligned}$$

We can rewrite this inequality as

$$\overline{M}(g^q)(x) \leqslant \frac{1}{\sqrt{\theta}(1-\sqrt{\theta})}\Big[b\Big(\fint_{\widetilde{Q}_0} g(y)dy\Big)^q + \fint_{\widetilde{Q}_0} f^q(y)dy\Big].$$

Substituting into (5.19), we get

$$\lambda \leqslant \sigma|P_{k,j}| \cdot C\left[\fint_{\widetilde{Q}_0} g(y)dy + \left(\fint_{\widetilde{Q}_0} f^q(y)dy \right)^{1/q} \right],$$

where C depends on θ, b, n, q. Using (5.13), we obtain

$$\lambda \leqslant C\left[\fint_{\widetilde{Q}_0} g\varphi dx + \left(\fint_{\widetilde{Q}_0} (f\varphi)^q dx \right)^{1/q} \right]. \tag{5.22}$$

We extend $f\varphi$ to be zero on $\mathbb{R}^n \setminus Q_1$. Set

$$F = [M(f\varphi)^q]^{1/q},$$

where M is the Hardy-Littlewood maximal function on $\mathbb{R}^n$. By (5.11),

$$\left(\fint_{\widetilde{Q}_0} (f\varphi)^q dx \right)^{1/q} \leqslant \fint_{\widetilde{Q}_0} F dx. \tag{5.23}$$

Combining (5.22) and (5.23), we derive

$$\begin{aligned} \lambda|\widetilde{Q}_0| &\leqslant C\left[\int_{\widetilde{Q}_0} g\varphi dx + \int_{\widetilde{Q}_0} F dx \right] \\ &\leqslant C\left[\int_{\widetilde{Q}_0 \cap \{g\varphi > \beta\lambda\}} g\varphi dx + \int_{\widetilde{Q}_0 \cap \{F > \beta\lambda\}} F dx + 2\beta\lambda|\widetilde{Q}_0| \right]. \end{aligned}$$

We choose β so that $\beta C = 1/4$. Then

$$\lambda|\widetilde{Q}_0| \leqslant C\left[\int_{\widetilde{Q}_0 \cap \{g\varphi > \beta\lambda\}} g\varphi dx + \int_{\widetilde{Q}_0 \cap \{F > \beta\lambda\}} F dx \right]. \tag{5.24}$$

For any $Q_{k,l}^l$ and any $x \in Q_{k,l}^l$, we can find the corresponding $\widetilde{Q}_0 \ni x$ such that the above inequality is valid. Collecting all such cubes, we obtain a cover $\{\widetilde{Q}_0\}$ for $\bigcup_{k,j,l} Q_{k,j}^l$. By the covering lemma (Lemma 5.1), there exists a disjoint subsequence $\widetilde{Q}_{0,1}, \widetilde{Q}_{0,2}, \cdots$ such that

$$\sum_{k,j,l} |Q_{k,j}^l| \leqslant 5^n \sum_m |\widetilde{Q}_{0,m}|. \tag{5.25}$$

For each $\widetilde{Q}_{0,m}$, (5.24) holds. From (5.18), (5.24) and (5.25), we deduce that

$$\int_{Q_1 \cap \{g\varphi > \lambda\}} (g\varphi)^q dx \leqslant C\lambda^{q-1}\left[\int_{Q_1 \cap \{g\varphi > \beta\lambda\}} g\varphi dx + \int_{Q_1 \cap \{F > \beta\lambda\}} F dx \right].$$

In addition, it is obvious that

$$\int_{Q_1 \cap \{\beta\lambda < g\varphi \leqslant \lambda\}} (g\varphi)^q dx \leqslant C\lambda^{q-1} \int_{Q_1 \cap \{g\varphi > \beta\lambda\}} g\varphi dx.$$

Summing the above inequalities, we see that, for $t = \beta\lambda$,

$$\int_{Q_1 \cap \{g\varphi > t\}} (g\varphi)^q dx \leqslant Ct^{q-1}\left[\int_{Q_1 \cap \{g\varphi > t\}} g\varphi dx + \int_{Q_1 \cap \{F > t\}} F dx \right]. \tag{5.26}$$

According to (5.15), the above inequality is valid for all $t \geqslant t_0 \triangleq (\gamma_0/\beta)\|g\|_{L^q(Q_1)}$. Set

$$h(t) = \int_{Q_1 \cap \{g\varphi > t\}} g\varphi dx, \qquad H(t) = \int_{Q_1 \cap \{F > t\}} F dx.$$

Just as in the proof of Lemma 1.1 in Chapter 3, we have

$$\int_{Q_1 \cap \{g\varphi > t\}} (g\varphi)^q dx = -\int_t^\infty s^{q-1} dh(s) = \Phi_{q-1}(t; h), \tag{5.27}$$

where $\Phi_p(t; h)$ is defined in (5.5). Thus we can rewrite (5.26) as

$$\Phi_{q-1}(t; h) \leqslant Ct^{q-1}[h(t) + H(t)], \quad t \geqslant t_0.$$

Applying Lemma 5.2, we find that, for $\varepsilon = (q-1)/(C-1)$, $p \in [q, q+\varepsilon)$,

$$\Phi_{p-1}(t_0; h) \leqslant Ct_0^{p-q}[\Phi_{q-1}(t_0; h) + \Phi_{p-1}(t_0; H)]. \tag{5.28}$$

By (5.27),

$$\begin{aligned} \int_{Q_1} (g\varphi)^p dx &\leqslant t_0^p |Q_1| + \Phi_{p-1}(t_0; h) \\ &\leqslant t_0^p |Q_1| + Ct_0^{p-q} \Big[\|g\varphi\|_{L^q(Q_1)}^q + \|F\|_{L^p(\mathbb{R}^n)}^p \Big]. \end{aligned}$$

Using Young's inequality and the definition of t_0 and F, we obtain

$$\|g\varphi\|_{L^p(Q_1)} \leqslant C\Big[\|g\|_{L^q(Q_1)} + \|M(f\varphi)^q\|_{L^{p/q}(\mathbb{R}^n)}^{1/q} \Big].$$

We assume that $p/q > 1$ (the conclusion of the theorem is obvious if $p = q$). The operator M is of strong type $(p/q, p/q)$ (cf. Appendix 4, the corollary of the Hardy-Littlewood maximal theorem). It follows that

$$\|g\varphi\|_{L^p(Q_1)} \leqslant C\Big[\|g\|_{L^q(Q_1)} + \|f\|_{L^p(Q_1)} \Big],$$

which implies the conclusion of the theorem. □

Bibliographic Notes

Part I

Detailed bibliographic notes for the theory of second order elliptic equations can be found in [GT]. Here we give brief comments on the materials that we choose to present in our textbook.

Chapter 1. L^2 Theory

Since the work of K. O. Friedrichs [FR], there have been major developments in the theory of variational methods for elliptic equations. The Fredholm alternative theory for equations of divergence form is due primarily to Ladyženskaja and Ural′ceva [LU] and G. Stampacchia [ST1]. Here we use their assumptions on the coefficients and their proofs. The maximum principle for weak solutions is due to M. Chicco [CH] and N. S. Trudinger [TR1]; the proof given here is a combination of their methods. We have also benefited from the lecture notes of L. C. Evans [EV1].

Chapter 2. Schauder Theory

The Schauder theory for the Dirichlet problem of second order elliptic equations was established as early as the 1930s by J. Schauder [SC1], [SC2]. Around the same time, E. Hopf [HO] and R. Caccioppoli [CA] established some similar results. The method here is due to N. S. Trudinger [TR4], who lectured in 1985 at the Institute of Mathematics at Nankai University, Tianjin, China. He established the equivalent Hölder norm of a solution using its mollified function; his proof avoids the tedious computations of potentials. In fact, this equivalent norm can be simplified further; another equivalent Hölder norm was introduced in [CY2] without using the derivatives with respect to the parameter of the mollified function.

Chapter 3. L^p Theory

Based on the Calderón-Zygmund decomposition lemma [CZ] and the singular integral operator theory, the L^p theory for the Dirichlet problem of second order elliptic equations was first established by D. Greco [GR] and A. I. Koselev [KO]. The results were generalized later to higher order elliptic equations as well as more general boundary conditions by S. Agmon, A. Douglis and L. Nirenberg [ADN]. Here, we use the special technique in [GT] on potential operators, which avoids the Fourier transform.

Chapter 4. De Giorgi-Nash-Moser Estimates

The Hölder estimates for solutions of elliptic equations with bounded measurable coefficients are fundamental in studying quasilinear elliptic equations. For divergence equations, such estimates were obtained by E. De Giorgi [DG1]. J. Nash independently obtained similar estimates for elliptic and parabolic equations. A new proof was given by Moser [MJ1], [MJ2], who also established Harnack's inequality for nonnegative solutions. We use Moser's proof. For the proof of the weak Harnack inequality, we have also consulted lecture notes of R. Schoen.

Chapter 5. Quasilinear Equations of Divergence Form

After the establishment of the De Giorgi-Nash-Moser estimates, there were rapid developments for the theory of quasilinear equations in divergence form. The major works are: Ladyženskaja and Ural′ceva [LU], Serrin [SE], Trudinger [TR1], etc. Ladyženskaja and Ural′ceva also reduced general quasilinear equations to divergence form so that the De Giorgi-Nash-Moser estimates are applicable. This chapter is based on [GT]. For gradient estimates of the solutions, we employ a technique from [CY1], which simplifies the proof in [GT].

Chapter 6. Krylov-Safonov Estimates

The maximum principle for strong solutions was obtained by A.D. Alexandroff [AL1], [AL2], I. Ya. Bakelman [BA], and C. Pucci [PU] in different forms. Applying this maximum principle and L^p-estimates, the existence of a $W^{2,p}$-strong solution can be established. Based on Pucci's paper and [GT], we give a clear, rigorous and detailed proof for the maximum principle.

The Ladyženskaja-Ural′ceva techniques for solving general quasilinear elliptic equations cannot be generalized to fully nonlinear elliptic equations, as these equations cannot be reduced to equations of divergence form. To study fully nonlinear elliptic equations, the estimates of De Giorgi-Nash-Moser's type for equations in nondivergence form are indispensable. Based on Alexandroff-Bakelman-Pucci's maximum principle and techniques similar to the Calderón-Zygmund decomposition, N. V. Krylov and M. V. Safonov [KS] established Harnack's inequality for nonnegative solutions of uniformly elliptic equations in nondivergence form with bounded measurable coefficients (in fact, they gave a proof for more complicated parabolic equations). Using their ideas, Trudinger[TR2] gave a different proof; the proof here is his ([GT]).

Chapter 7. Fully Nonlinear Elliptic Equations

Since this is a textbook, we discuss only fully nonlinear uniformly elliptic equations under the natural structure conditions and the concavity condition. A typical example is the Bellman equation from control theory. The Monge-Ampère equation, which is a fully nonlinear nonuniformly elliptic equation, is not discussed in this book. For the Bellman equation, partial differential equation techniques were first used by H. Brezis and L. C. Evans [BE], L. C. Evans and A. Friedman [EF],

P. L. Lions [LP], and L. C. Evans and P. L. Lions [EL]. In the last two papers, $C^{1,1}$-strong solutions were established. Afterwards, L. C. Evans established the interior Hölder estimates for second order derivatives of the solution and obtained the classical solution. These results were generalized by N. S. Trudinger [TR3] to fully nonlinear uniformly elliptic equations under the natural structure conditions. In his lectures at Nankai University in 1985, Trudinger [TR4] uses the mollified function and freezing coefficients techniques to derive the Schauder estimate; for the Bellman equation, an a priori estimate for the lower order derivatives is no longer necessary. This chapter is primarily based on Trudinger [TR3]. For the gradient estimate, our technique simplifies the Bernstein method. The Hölder estimate for the gradient near the boundary was first obtained by N. V. Krylov; the simplified proof given here is due to L. A. Caffarelli. The results for the estimate of Schauder type for fully nonlinear elliptic equations are chosen from [TR4].

Part II

Detailed bibliographic notes for the theory of second order elliptic systems can be found in [GQ1]. Here we give a brief commentary on the materials that we chose for our textbook.

Chapter 8. L^2 Theory for Linear Elliptic Systems of Divergence Form

The Hilbert space method for linear elliptic systems dates back to 1900 (D. Hilbert [HI]) and 1907 (H. Lebesgue [LE]). A brief introduction to the developments thereafter can be found in the bibliographical notes for Chapter 1. In this chapter, we study the existence and differentiability of weak solutions of linear divergence elliptic systems; only those systems without lower order terms are studied.

Chapter 9. Schauder Theory for Linear Elliptic Systems of Divergence Form

In the 1930s, the Schauder theory was established using delicate estimates of the $C^{2,\alpha}$ norm of the potential integrals. In Chapter 2, we introduced the method from [TR4] to establish an equivalent C^α norm using the mollified function, which is then used in establishing the Schauder theory without tedious computations of the potential integrals. In this chapter, we use the methods developed in the 1950s and 1960s by C. B. Morrey [MR3] and S. Campanato [CP] to establish the Schauder theory for elliptic systems in divergence form. They introduced an equivalent C^α norm by integral representations, which is very convenient in studying regularity of weak solutions; this gives another method for establishing the Schauder theory without tedious computations of the potential integrals. M. Giaquinta systematically introduced this method in his books [GQ1], [GQ2]. We introduce this method in a way similar to that in [GQ1].

Chapter 10. L^p Theory for Linear Elliptic Systems of Divergence Form

The L^p theory can be established by using the Calderón-Zygmund decomposition lemma and singular integral operator theory. In this chapter, however, we use the method given by S. Campanato and G. Stampacchia [CS] to establish the L^p theory; this method uses the Stampacchia interpolation theorem [ST2] rather than potential theory. We introduce this method in a way similar to that in [GQ2].

Chapter 11. Existence of Weak Solutions of Nonlinear Elliptic Systems

In this chapter, we introduce the variational method, which has a long history. Using this method, B. Riemann obtained many interesting results on geometric function theory in the nineteenth century. However, he assumed the Dirichlet principle to be valid; i.e., there exists a unique $u \in K$ such that

$$D[u] = \min_{v \in K} D[v],$$

where $D[v] = \int_\Omega |Dv|^2 dx$, $K = \{v \in C^1(\Omega) \cap C^0(\overline{\Omega}),\ v|_{\partial\Omega} = \varphi\}$, φ is a given function, and u is harmonic in Ω. With the development of rigorous analysis, Riemann's work was criticized. K. Weierstrass pointed out that $D[v]$ being bounded below does not imply the existence of a minimum point; furthermore, a minimum point, if it exists, need not be harmonic. Probably this criticism is why the variational method was not taken seriously until the 1900s, when D. Hilbert [HI] and H. Lebesgue [LE] rigorously established the Dirichlet principle for some special cases; the more general case was later obtained by L. Tonelli. Around 1940, the Sobolev spaces were employed by C. B. Morrey [MR1], and the existence of a minimum point was established for a large class of integral functionals. Nowadays, the variational method has become an effective method for studying the existence of weak solutions for elliptic equations and systems. We adopted the treatment in [GQ2], assuming that $f(x, u, p)$ is convex in p. This convexity condition is necessary when $N = 1$ and should be replaced by the quasi-convexity condition when $N > 1$. For further discussion, we refer readers to [MR2], [MR4], [GQ1], [GQ2] and the references therein.

Chapter 12. Regularity for Weak Solutions of Nonlinear Elliptic Systems

In 1968, De Giorgi [DG2] gave an example illustrating that the De Giorgi-Nash-Moser theorem is in fact not valid for linear elliptic systems, therefore the regularity of a weak solution of a nonlinear elliptic system cannot be established in the same way as in the single equation case. But the question whether a weak solution of a nonlinear elliptic system can be regular was still open. Later on, E. Giusti and M. Miranda [GD1] and other mathematicians gave many examples (cf. [GQ1]) showing that, in general, a weak solution of a nonlinear elliptic system does not have regularity everywhere. After that, Morrey [MR5], Giusti and M. Miranda [GD2], Giusti [GI], and Pepe [PE] showed that weak solutions of some nonlinear elliptic systems have partial regularity (i.e., regularity in some open subset $\Omega_0 \subset \Omega$, and $\text{meas}(\Omega \setminus \Omega_0) = 0$); they used the indirect method. For more general nonlinear elliptic systems, partial regularity of weak solutions was established by M. Giaquinta and E. Giusti [GG], and by M. Giaquinta and G. Modica [GM1], [GM2]; they used the direct method based on the reverse Hölder inequality. Here, our main purpose is to introduce the methods, and we do not pursue the most general results. Therefore, we study only two special classes of quasilinear elliptic systems. We introduce the indirect and direct methods in a way similar to that in [GQ1]. For more general results, we refer readers to [GQ1] and the references therein, and [AF2], [EV2], [EG], [FU], [FH], [GM3], [HM], etc.

For nonlinear elliptic systems of some special structures, such as nonlinear elliptic systems of diagonal form, a weak solution has regularity everywhere. We do not present such results in this book, but refer interested readers to [HW1], [HW2], [WI1], [WI2], [HB], [GH], etc.

Bibliography

[AF1] Acerbi, E. & Fusco, N., *Semicontinuity problems in the calculus of variations*, Arch. Rational Mech. Anal., **86** (1984), 125–145.

[AF2] Acerbi, E. & Fusco, N., *A regularity theorem for minimizers of quasiconvex integrals*, Arch. Rational Mech. Anal., **99** (1987), 261–281.

[AD] Adams, R. A., *Sobolev Spaces*, Academic Press, New York, 1975.

[ADN] Agmon, S., Douglis, A. & Nirenberg, L., *Estimates near the boundary for solutions of elliptic partial differential equations satisfying general boundary conditions*, I, II, Comm. Pure Appl. Math., **12** (1959), 623–727, **17** (1964), 35–92.

[AL1] Alexandroff, A. D., *Certain estimates for the Dirichlet problem*, Dokl. Akad. Nauk. SSSR, **134** (1960), 1001–1004. English translation, Soviet Math. Dokl. **1** (1960), 1151–1154.

[AL2] Alexandroff, A. D., *Uniqueness conditions and estimates for the solution of the Dirichlet problem*, Vestnik Leningrad University, **18** (1963), 5–29. English translation, Amer. Math. Soc. Transl. **68** (1968), 89–119.

[BA] Bakelman, I. Ya., *Theory of quasilinear elliptic equations* (Russian), Sibirsk. Mat. Z., **2** (1961), 179–186.

[BS] Bers, L. & Schechter, M., *Elliptic equations*, in: Partial Differential Equations, Interscience, New York, 1964 (reprinted as Lectures in Appl. Math., vol. 3A, Amer. Math. Soc., Providence, RI, 1979), 131–299.

[BE] Brezis, H. & Evans, L. C., *A variational inequality approach to the Bellman-Dirichlet equation for two elliptic operators*, Arch. Rational Mech. Anal., **71** (1979), 1–13.

[CA] Caccioppoli, R., *Sulle equazioni ellittiche a derivate parziali con n variabili indipendenti*, Atti Accad. Naz. Lincei Rend. Cl. Sci. Fis. Mat. Natur. (6) **19** (1934), 83–89.

[CZ] Calderón, A. P. & Zygmund, A., *On the existence of certain singular integrals*, Acta Math., **88** (1952), 85–139.

[CP] Campanato, S., *Equazioni ellittiche del secondo ordine e spazi* $\mathscr{L}^{2,\lambda}$, Ann. Mat. Pura Appl., (4) **69** (1965), 321–381.

[CS] Campanato, S. & Stampacchia, G., *Sulle maggiorazioni in* L^p *nella teoria delle equazioni ellittiche*, Boll. UMI, (3) **20** (1965), 393–399.

[CY1] Chen, Y. Z. (陈亚浙), *Some methods of Krylov on a priori estimates for solutions of fully nonlinear equations* (Chinese), Adv. in Math. (Beijing), **15** (1986), 63–101.

[CY2] Chen, Y. Z. (陈亚浙), *Strong solution of the obstacle problem for fully nonlinear elliptic equations*, J. Partial Differential Equations, **4** (1991), no. 3, 15–34.

[CH] Chicco, M., *Principio di massimo forte per sottosoluzioni di equazioni ellittiche di tipo variazionale*, Boll. UMI, (3) **22** (1967), 368–372.

[DG1] De Giorgi, E., *Sulla differenziabilità e l'analiticità delle estremali degli integrali multipli regolari*, Mem. Accad. Sci. Torino Cl. Sci. Fis. Mat. Natur., (3) **3** (1957), 25–43.

[DG2] De Giorgi, E., *Un esempio di estremali discontinue per un problema variazionale di tipo ellittico*, Boll. UMI, (11) **1** (1968), 135–137.

[EV1] Evans, L. C., *Estimates for second order elliptic partial differential equations*, Manuscript.

[EV2] Evans, L. C., *Quasiconvexity and partial regularity in the calculus of variations*, Arch. Rational Mech. Anal., **95** (1986), 227–252.

[EF] Evans, L. C., & Friedman, A., *Optimal stochastic switching and the Dirichlet problem for the Bellman equation*, Trans. Amer. Math. Soc., **253** (1979), 365–389.

[EG] Evans, L. C. & Gariepy, R. F., *Blow-up, compactness and partial regularity in the calculus of variations*, Indiana Univ. Math. J., **36** (1987), 361–371.

[EL] Evans, L. C. & Lions, P. L., *Résolution des équations de Hamilton-Jacobi-Bellman pour des opérateurs uniformément elliptiques*, C. R. Acad. Sci. Paris Sér. A, **290** (1980), 1049–1052.

[FS] Fefferman, C & Stein, E. M., *H^p spaces of several variables*, Acta. Math., **129** (1972), 137–193.

[FR] Friedrichs, K. O., *The identity of weak and strong extensions of differential operators*, Trans. Amer. Math. Soc., **55** (1944), 132-151.

[FU] Fuchs, M., *Regularity theorems for nonlinear systems of partial differential equations under natural ellipticity conditions*, Analysis, **7** (1987), 83–93.

[FH] Fusco, N. & Hutchinson, J., *$C^{1,\alpha}$ partial regularity of functions minimizing quasiconvex integrals*, Manuscripta Math., **54** (1985), 121–143.

[GD] Gårding, L., *Dirichlet problem for linear elliptic partial differential equations*, Math. Scand., **1** (1953), 55–72.

[GE] Gehring, F. W., *The L^p-integrability of the partial derivatives of a quasiconformal mapping*, Acta Math. **130** (1973), 265-277.

[GQ1] Giaquinta, M., *Multiple Integrals in the Calculus of Variations and Nonlinear Elliptic Systems*, Princeton University Press, Princeton, NJ, 1983.

[GQ2] Giaquinta, M., *Introduction to the Regularity Theory for Nonlinear Elliptic Systems*, Birkhäuser Verlag, Basel, 1993.

[GG] Giaquinta, M. & Giusti, E., *Nonlinear elliptic systems with quadratic growth*, Manuscripta Math., **24** (1978), 323–349.

[GH] Giaquinta, M. & Hildebrandt, S., *Estimation a priori des solutions faibles de certains systèmes non lineaires elliptiques*, Sem. Goulaouic-Meyer-Schwartz 1980-81, exposé 17, École Polytech., Palaiseau, Mars 1981.

[GM1] Giaquinta, M. & Modica, G., *Regularity results for some classes of higher order nonlinear elliptic systems*, J. Reine Angew. Math., **311/312** (1979), 145–169.

[GM2] Giaquinta, M. & Modica, G., *Almost-everywhere regularity results for solutions of nonlinear elliptic systems*, Manuscripta Math., **28** (1979), 109–158.

[GM3] Giaquinta, M. & Modica, G., *Partial regularity of minimizers of quasiconvex integrals*, Ann. Inst. H Poincaré, Analyse Non Linéaire, **3** (1986), 185–208.

[GS] Giaquinta, M. & Souček, J., *Caccioppoli's inequality and Legendre-Hadamard condition*, Math. Ann., **270** (1985), 105–107.

[GT] Gilbarg, D. & Trudinger, N. S., *Elliptic Partial Differential Equations of Second Order*, 2nd ed., Springer-Verlag, New York, 1983.

[GI] Giusti, E., *Regolarità parziale delle soluzioni di sistemi ellittici quasi lineari di ordine arbitrario*, Ann. Scuola. Norm. Sup. Pisa, (3) **23** (1969), 115–141.

[GD1] Giusti, E. & Miranda, M., *Un esempio di soluzioni discontinue per un problema di minimo relativo ad un integrale regolare del calcolo delle variazioni*, Boll. UMI, (4) **1** (1968), 219–226.

[GD2] Giusti, E. & Miranda, M., *Sulla regolarità delle soluzioni deboli di una classe di sistemi ellittici quasilineari*, Arch. Rational Mech. Anal., **31** (1968), 173–184.

[GR] Greco, D., *Nuove formole integrali di maggiorazione per le soluzioni di un'equazione lineare di tipo ellittico ed applicazioni alla teoria del potenziale*, Ricerche Mat., **5** (1956), 126–149.

[HI] Hilbert, D., *Über das Dirichletsche prinzip*, Jber. Deutsch. Math.-Verein., **8** (1900), 184–188.

[HB] Hildebrandt, S., *Nonlinear elliptic systems and harmonic mappings*, Proc. Beijing Sympos. Differential Geometry and Differential Equations, Science Press, Beijing, and Gordon and Breach, New York, 1982, pp. 481–615.

[HW1] Hildebrandt, S. & Widman, K.-O., *Some regularity results for quasilinear elliptic systems of second order*, Math. Z., **142** (1975), 67–86.

[HW2] Hildebrandt, S. & Widman, K.-O., *On the Hölder continuity of weak solutions of quasilinear elliptic systems of second order*, Ann. Scuola Norm. Sup. Pisa Cl. Sci., (4) **4** (1977), 145–178.

[HM] Hong, M. C. (洪敏纯), *Existence and partial regularity in the calculus of variations*, Ann. Mat. Pura Appl., (4) **149** (1987), 311–328.

[HO] Hopf, E., *Über den Funktionalen, insbesondere den analytischen Charakter der Lösungen elliptischer Differentialgleichungen zweiter Ordnung*, Math. Z., **34** (1932), 194–233.

[JN] John, F. & Nirenberg, L., *On functions of bounded mean oscillation*, Comm. Pure Appl. Math., **14** (1961), 415–426.

[KO] Koševev, A. I., *On boundedness in L^p of derivatives of solutions of elliptic differential equations* (Russian), Mat. Sb. (N. S.), **38** (1956), 359–372.

[KL] Krylov, N. V., *Nonlinear Elliptic and Parabolic Equations of Second Order*, "Nauka", Moscow, 1985. English translation, Reidel, Dordrecht, 1987.

[KS] Krylov, N. V. & Safonov, M. V., *A certain property of solutions of parabolic equations with measurable coefficients*, Izv. Akad. Nauk. SSSR Ser. Mat., **44** (1980), 161–175. English translation, Math. USSR Izv. **16** (1981), 151–164.

[LU] Ladyženskaja, O. A. & Ural′ceva, N. N., *Linear and Quasilinear Elliptic Equations*, "Nauka", Moscow, 1964. English translation, Academic Press, New York, 1968.

[LE] Lebesgue, H., *Sur le problème de Dirichlet*, Rend. Circ. Mat. Palermo, **24** (1907), 371–402.

[LP] Lions, P. L., *Résolution analytic des problèmes de Bellman-Dirichlet*, Acta Math., **146** (1981), 151–166.

[MC] Marcellini, P., *Approximation of quasiconvex functionals and lower semicontinuity of multiple integrals*, Manuscripta Math., **51** (1985), 1–28.

[MJ] Maz′ja, V. G., *Sobolev Spaces*, Springer-Verlag, New York, 1985.

[MD] Miranda, C., *Partial Differential Equations of Elliptic Type*, 2nd ed., Springer-Verlag, Berlin, 1970.

[MR1] Morrey, C. B. Jr., *Existence and differentiability theorems for the solutions of variational problems for multiple integrals*, Bull. Amer. Math. Soc., **46** (1940), 439–458.

[MR2] Morrey, C. B. Jr., *Quasi-convexity and the lower semicontinuity of multiple integrals*, Pacific J. Math. **2** (1952), 25–53.

[MR3] Morrey, C. B. Jr., *Second order elliptic systems of differential equations*, in: Contributions to the Theory of Partial Differential Equations, Ann. of Math. Studies, No. 33, Princeton University Press, Princeton, NJ, 1954, pp. 101–159.

[MR4] Morrey, C. B. Jr., *Multiple Integrals in the Calculus of Variations*, Springer-Verlag, New York, 1966.

[MR5] Morrey, C. B. Jr., *Partial regularity results for nonlinear elliptic systems*, J. Math. and Mech., **17** (1968), 649–670.

[MJ1] Moser, J., *A new proof of De Giorgi's theorem concerning the regularity problem for elliptic differential equations*, Comm. Pure Appl. Math., **13** (1960), 457–468.

[MJ2] Moser, J., *On Harnack's theorem for elliptic differential equations*, Comm. Pure Appl. Math., **14** (1961), 577–591.

[NA] Nash, J., *Continuity of solutions of parabolic and elliptic equations*, Amer. J. Math., **80** (1958), 931–954.

[NC] Nečas, J., *Introduction to the Theory of Nonlinear Elliptic Equations*, Teubner, Leipzig, 1983.

[NS] Nečas, J. & Stará, J., *Principio di massimo per i sistemi ellittici quasilineari nondiagonali*, Boll. UMI, (4) **6** (1972), 1–10.

[NI] Nirenberg, L., *Remarks on strongly elliptic partial differential equations*, Comm. Pure Appl. Math., **8** (1955), 649–675.

[PE] Pepe, L., *Risultati di regolarità parziale per le soluzioni* $H^{1,p}(\Omega), 1 < p < 2$, *di sistemi ellittici quasilineari*, Ann. Univ. Ferrara Ser. VII (N.S.), **8** (1971), 129–148.

[PU] Pucci, C., *Limitazioni per soluzioni di equazioni ellittiche*, Ann. Mat. Pura Appl., (4) **74** (1966), 15–30.

[RG] Rogers, C. A., *Hausdorff Measure*, Cambridge University Press, Cambridge, 1970.

[RD] Royden, H. L., *Real Analysis*, Macmillan, New York, 1963.

[SC1] Schauder, J., *Über lineare elliptische Differentialgleichungen zweiter Ordnung*, Math. Z., **38** (1934), 257–282.

[SC2] Schauder, J., *Numerische Abschätzungen in elliptischen linearen Differentialgleichungen*, Studia Math., **5** (1935), 34–42.

[SW] Schwartz, J., *Nonlinear Functional Analysis*, Gordon and Breach, New York, 1969.

[SE] Serrin, J., *Local behavior of solutions of quasi-linear elliptic equations*, Acta Math., **111** (1964), 247–302.

[ST1] Stampacchia, G., *Équations Elliptiques du Second Ordre à Coefficients Discontinues*, Sém. Math. Sup., no. 16, Presses Univ. Montréal, Montréal, 1966.

[ST2] Stampacchia, G., *The spaces* $\mathscr{L}^{(p,\lambda)}, N^{(p,\lambda)}$ *and interpolation*, Ann. Scuola Norm. Sup. Pisa, (4) **19** (1965), 443–462.

[TR1] Trudinger, N. S., *On Harnack type inequalities and their application to quasilinear elliptic equations*, Comm. Pure Appl. Math., **20** (1967), 721–747.

[TR2] Trudinger, N. S., *Local estimates for subsolutions and supersolutions of general second order elliptic quasi-linear equations*, Invent. Math., **61** (1980), 67–79.

[TR3] Trudinger, N. S., *Fully nonlinear uniformly elliptic equations under natural structure conditions*, Trans. Amer. Math. Soc., **278** (1983), 751–769.

[TR4] Trudinger, N. S., *Lectures on nonlinear second order elliptic equations*, Nankai Institute of Math., Tianjin, China, 1985.

[WI1] Wiegner, M., *Ein optimaler Regularitätssatz für schwache Lösungen gewisser elliptischer Systeme*, Math. Z., **147** (1976), 21–28.

[WI2] Wiegner, M., *A-priori Schranken für Lösungen gewisser elliptischer Systeme*, Manuscripta Math., **18** (1976), 279–297.

[ZK] Zhang, K. W. (张克威), *On the Dirichlet problem of a class of elliptic systems*, in: Partial Differential Equations (Tianjin, 1986), Lecture Notes in Math., vol. 1306, Springer-Verlag, Berlin, 1988, pp. 262-277.

Index

Selected Titles in This Series

(Continued from the front of this publication)

139 **I. V. Skrypnik,** Methods for analysis of nonlinear elliptic boundary value problems, 1994
138 **Yu. P. Razmyslov,** Identities of algebras and their representations, 1994
137 **F. I. Karpelevich and A. Ya. Kreinin,** Heavy traffic limits for multiphase queues, 1994
136 **Masayoshi Miyanishi,** Algebraic geometry, 1994
135 **Masaru Takeuchi,** Modern spherical functions, 1994
134 **V. V. Prasolov,** Problems and theorems in linear algebra, 1994
133 **P. I. Naumkin and I. A. Shishmarev,** Nonlinear nonlocal equations in the theory of waves, 1994
132 **Hajime Urakawa,** Calculus of variations and harmonic maps, 1993
131 **V. V. Sharko,** Functions on manifolds: Algebraic and topological aspects, 1993
130 **V. V. Vershinin,** Cobordisms and spectral sequences, 1993
129 **Mitsuo Morimoto,** An introduction to Sato's hyperfunctions, 1993
128 **V. P. Orevkov,** Complexity of proofs and their transformations in axiomatic theories, 1993
127 **F. L. Zak,** Tangents and secants of algebraic varieties, 1993
126 **M. L. Agranovskiĭ,** Invariant function spaces on homogeneous manifolds of Lie groups and applications, 1993
125 **Masayoshi Nagata,** Theory of commutative fields, 1993
124 **Masahisa Adachi,** Embeddings and immersions, 1993
123 **M. A. Akivis and B. A. Rosenfeld,** Élie Cartan (1869–1951), 1993
122 **Zhang Guan-Hou,** Theory of entire and meromorphic functions: deficient and asymptotic values and singular directions, 1993
121 **I. B. Fesenko and S. V. Vostokov,** Local fields and their extensions: A constructive approach, 1993
120 **Takeyuki Hida and Masuyuki Hitsuda,** Gaussian processes, 1993
119 **M. V. Karasev and V. P. Maslov,** Nonlinear Poisson brackets. Geometry and quantization, 1993
118 **Kenkichi Iwasawa,** Algebraic functions, 1993
117 **Boris Zilber,** Uncountably categorical theories, 1993
116 **G. M. Fel′dman,** Arithmetic of probability distributions, and characterization problems on abelian groups, 1993
115 **Nikolai V. Ivanov,** Subgroups of Teichmüller modular groups, 1992
114 **Seizô Itô,** Diffusion equations, 1992
113 **Michail Zhitomirskiĭ,** Typical singularities of differential 1-forms and Pfaffian equations, 1992
112 **S. A. Lomov,** Introduction to the general theory of singular perturbations, 1992
111 **Simon Gindikin,** Tube domains and the Cauchy problem, 1992
110 **B. V. Shabat,** Introduction to complex analysis Part II. Functions of several variables, 1992
109 **Isao Miyadera,** Nonlinear semigroups, 1992
108 **Takeo Yokonuma,** Tensor spaces and exterior algebra, 1992
107 **B. M. Makarov, M. G. Goluzina, A. A. Lodkin, and A. N. Podkorytov,** Selected problems in real analysis, 1992
106 **G.-C. Wen,** Conformal mappings and boundary value problems, 1992
105 **D. R. Yafaev,** Mathematical scattering theory: General theory, 1992
104 **R. L. Dobrushin, R. Kotecký, and S. Shlosman,** Wulff construction: A global shape from local interaction, 1992
103 **A. K. Tsikh,** Multidimensional residues and their applications, 1992

(See the AMS catalog for earlier titles)